Stochastic Systems In Merging Phase Space

Stochastic Systems In Merging Phase Space

Vladimir S. Koroliuk
Institute of Mathematics, National Academy of Sciences, Ukraine

Nikolaos Limnios
Applied Mathematics Laboratory, University of Technology of Compiègne, France

NEW JERSEY · LONDON · SINGAPORE · BEIJING · SHANGHAI · HONG KONG · TAIPEI · CHENNAI

Published by

World Scientific Publishing Co. Pte. Ltd.
5 Toh Tuck Link, Singapore 596224
USA office: 27 Warren Street, Suite 401-402, Hackensack, NJ 07601
UK office: 57 Shelton Street, Covent Garden, London WC2H 9HE

British Library Cataloguing-in-Publication Data
A catalogue record for this book is available from the British Library.

STOCHASTIC SYSTEMS IN MERGING PHASE SPACE

Copyright © 2005 by World Scientific Publishing Co. Pte. Ltd.

All rights reserved. This book, or parts thereof, may not be reproduced in any form or by any means, electronic or mechanical, including photocopying, recording or any information storage and retrieval system now known or to be invented, without written permission from the Publisher.

For photocopying of material in this volume, please pay a copying fee through the Copyright Clearance Center, Inc., 222 Rosewood Drive, Danvers, MA 01923, USA. In this case permission to photocopy is not required from the publisher.

ISBN-13 978-981-256-591-4
ISBN-10 981-256-591-4

Printed in Singapore

Preface

"... the theory of systems should be built on the methods of simplification and is, essentially, the science of simplification".
Walter Ashby (1969)

The actual problem of systems theory is the development of mathematically justified methods of simplification of complicated systems whose mathematical analysis is difficult to perform even with help of modern computers. The main difficulties are caused by the complexity of the phase (state) space of the system, which leads to virtually boundedless mathematical models.

A simplified model for a system must satisfy the following conditions: (i) The local characteristics of the simplified model are determined by rather simple functions of the local characteristics of original model. (ii) The global characteristics describing the behavior of the stochastic system can be effectively calculated on large enough time intervals. (iii) The simplified model has an effective mathematical analysis, and the global characteristics of the simplified model are close enough to the corresponding characteristics of the original model for application.

Stochastic systems considered in the present book are *evolutionary systems* in *random medium*, that is, *dynamical systems* whose state space is subject to random variations. From a mathematical point of view, such systems are naturally described by operator-valued stochastic processes on Banach spaces and are nowadays known as *Random evolution*.

This book gives recent results on stochastic approximation of systems by weak convergence techniques. General and particular schemes of proofs for average, diffusion, diffusion with equilibrium, and Poisson approximations of stochastic systems are presented. The particular systems studied here are stochastic additive functionals, dynamical systems, stochastic integral functionals, increment processes, impulsive processes. As application we give the cases of absorption times, stationary phase merging, semi-Markov random walk, Lévy approximation, etc.

The main mathematical object of this book is a family of coupled stochastic processes $\xi^\varepsilon(t), x^\varepsilon(t), t \geq 0, \varepsilon > 0$ (where ε is the small series parameter) called the *switched* and *switching* processes. The switched process $\xi^\varepsilon(t), t \geq 0$, describes the system evolution, and, in general, is a stochastic functional of a third process. The switching process $x^\varepsilon(t), t \geq 0$, also called the *driving* or *modulation process*, is the perturbing process or the random medium, and can represent the environment, the technical structure, or any perturbation factor.

The two modes of switching considered here are Markovian and semi-Markovian. Of course, we could present only the semi-Markov case since the Markov is a special case. But we present both mainly for two reasons: the first is that proofs are simpler for the Markov case, and the second that most of the readers are mainly interested by the Markov case.

The switching processes are considered in *phase split and merging scheme*. The phase merging scheme is based on the split of the phase space into disjoint classes

$$E = \cup_{k \in V} E_k, \quad E_k \cap E_{k'} = \emptyset, \quad k \neq k' \qquad (0.1)$$

and further merging these classes E_k, $k \in V$, into distinct states $k \in V$. So the merged phase space of the simplified model of system is $\hat{E} = V$ (see Figure 4.1).[1]

The transitions (connections) between the states of the original system S are merged to yield the transitions between merged states of the merged system $\hat{S}$. The analysis of the merged system is thus significantly simplified.

It is important to note that the additional supporting system S^0 with the same phase space E but without connections between classes of states E_k is used.

Split the phase space (0.1) just means introducing a new supporting system consisting of isolated subsystems S_k, $k \in V$, defined on classes of states E_k, $k \in V$. The merged system $\hat{S}$ is constructed with respect to the ergodic property of the support system S^0.

It is worth noticing that the initial processes in the series scheme contain no diffusion part. Diffusion processes appear only as limit processes.

---

[1] Figures, theorems, lemmas, etc. are numbered by x.y, where x is the number of chapter, and y is the number of figure, theorem, etc. into the chapter.

The general scheme of proof of weak convergence for stochastic processes in series scheme is the following.

I. *Limit compensating operator:*
1. Construction of the compensating operator $\mathbb{L}^\varepsilon$ of the Markov additive process $\xi^\varepsilon(t), t \geq 0$.
2. Asymptotic form of $\mathbb{L}^\varepsilon$ acting on some kind of test functions φ^ε.
3. Singular perturbation problem: $\mathbb{L}^\varepsilon \varphi^\varepsilon = \Psi + \varepsilon \theta^\varepsilon$.

II. *Tightness:*
1. Compact containment condition

$$\lim_{M \to \infty} \sup_{0 < \varepsilon \leq \varepsilon_0} \mathbb{P}\left(\sup_{0 < t \leq T} |\xi^\varepsilon(t)| \geq M \right) = 0.$$

2. Submartingale condition. The process

$$\eta^\varepsilon(t) := \varphi(\xi^\varepsilon(t)) + C_\varphi t,$$

is a nonnegative submartingale with respect to some filtration.

In the case of semimartingales, used in proofs of Poisson approximation results, we follow the same scheme of weak convergence proof, but for their predictable characteristics presented under integral functional forms.

For Markov switching processes, step II is performed in a simpler and more adequate way by using predictable square characteristics of martingale characterization.

It is worth noticing that most of results presented in this book are new, or generalize results published previously by the authors.

The book is organized as follows.

In Chapter 1, we present shortly the basic families of stochastic processes used in this book. More specifically, we present the Markov, semi-Markov processes, and some of their subfamilies, and semimartingales.

In Chapter 2, we present the notions of switching and switched processes via additive functionals of processes with locally independent increments switched by Markov or semi-Markov processes.

In Chapter 3, we present stochastic systems in the series scheme. More specifically, we present the basic results of average and diffusion approximations; proofs are postponed until the next chapters. We start from the simplest case of integral functionals to end by the more complicated case of general additive functionals.

In Chapter 4, we present results concerning average and diffusion approximations, as in Chapter 3, but with the addition that the state space of the switching process is asymptotically merged (single and double merged). Ergodic and non-ergodic switching processes are considered.

In the next two chapters we present detailed proofs of the results stated in Chapters 3 and 4.

In Chapter 5, we present the algorithmic part of the proofs, realized by the singular perturbing reducible-invertible operator technique. This part corresponds to the finite-dimensional distribution convergence and is performed as in step I above.

In Chapter 6, we give the last step of the proofs concerning the tightness of the probability distribution of the processes. This is performed as in step II above.

In Chapter 7, we give Poisson approximation results for impulsive processes switched by Markov processes and for stochastic additive functionals switched by semi-Markov processes.

The next two chapters include applications of the theory presented in the previous chapters.

In Chapter 8, we present several applications: absorption time distributions, stationary phase merging, superposition of two independent renewal processes, semi-Markov random walks.

In Chapter 9, we present birth-and-death processes, repairable systems, and Lévy approximation of impulsive systems.

Finally, we give some problems to solve. These problems include results stated without proofs in previous chapters, additional results, and some extensions.

Three appendices are also included. The first gives some definitions about weak convergence of stochastic processes. The second gives some known basic theorems on convergence of semimartingales and composed processes needed in proofs. The third includes some additional results concerning intermediate proofs of theorems. Of course, even if these theorems are included in order that this book becomes as autonomous as possible, we encourage the readers to find the corresponding information directly to bibliography.

This book should serve as a textbook for graduate students, post-doctoral seminars or courses for applied scientists and engineers in stochastic approximation of complex systems: queuing, reliability, risk, finance, biology systems, etc.

Acknowledgment. Authors express their gratitude to DFG for a financial support of the project 436 UKR 113/70/2-5 and thank particularly *Sergio Albeverio* of the Institute of Applied Mathematics at University of Bonn, and *Yuri Kondratiev* of the BiBos Center of the University of Bielelfeld for their hospitality. They are also grateful to the University of Technology of Compiègne for some support and hospitality. They are also grateful to *Vladimir Anisimov* and *Anatoli Skorokhod* for several discussions about these problems.

Last, thanks are due also to Esther Tan Leng-Leng, Desk Editor in World Scientific, for her useful collaboration.

Kiev V.S. Koroliuk
Compiègne N. Limnios

July 2005

Contents

Preface   v

1. Markov and Semi-Markov Processes   1
   - 1.1 Preliminaries 1
   - 1.2 Markov Processes 2
     - 1.2.1 Markov Chains 2
     - 1.2.2 Continuous-Time Markov Processes 6
     - 1.2.3 Diffusion Processes 10
     - 1.2.4 Processes with Independent Increments 11
     - 1.2.5 Processes with Locally Independent Increments . . . 14
     - 1.2.6 Martingale Characterization of Markov Processes . . 15
   - 1.3 Semi-Markov Processes 19
     - 1.3.1 Markov Renewal Processes 19
     - 1.3.2 Markov Renewal Equation and Theorem 21
     - 1.3.3 Auxiliary Processes 23
     - 1.3.4 Compensating Operators 24
     - 1.3.5 Martingale Characterization of Markov Renewal Processes 25
   - 1.4 Semimartingales 25
   - 1.5 Counting Markov Renewal Processes 28
   - 1.6 Reducible-Invertible Operators 31

2. Stochastic Systems with Switching   35
   - 2.1 Introduction 35
   - 2.2 Stochastic Integral Functionals 36
   - 2.3 Increment Processes 40

- 2.4 Stochastic Evolutionary Systems 43
- 2.5 Markov Additive Processes 46
- 2.6 Stochastic Additive Functionals 47
- 2.7 Random Evolutions 50
  - 2.7.1 Continuous Random Evolutions 50
  - 2.7.2 Jump Random Evolutions 54
  - 2.7.3 Semi-Markov Random Evolutions 56
- 2.8 Extended Compensating Operators 59
- 2.9 Markov Additive Semimartingales 61
  - 2.9.1 Impulsive Processes 61
  - 2.9.2 Continuous Predictable Characteristics 63

3. **Stochastic Systems in the Series Scheme** — 67
   - 3.1 Introduction 67
   - 3.2 Random Evolutions in the Series Scheme 68
     - 3.2.1 Continuous Random Evolutions 68
     - 3.2.2 Jump Random Evolutions 72
   - 3.3 Average Approximation 74
     - 3.3.1 Stochastic Additive Functionals 74
     - 3.3.2 Increment Processes 79
   - 3.4 Diffusion Approximation 81
     - 3.4.1 Stochastic Integral Functionals 81
     - 3.4.2 Stochastic Additive Functionals 84
     - 3.4.3 Stochastic Evolutionary Systems 88
     - 3.4.4 Increment Processes 89
   - 3.5 Diffusion Approximation with Equilibrium 90
     - 3.5.1 Locally Independent Increment Processes 90
     - 3.5.2 Stochastic Additive Functionals with Equilibrium .. 93
     - 3.5.3 Stochastic Evolutionary Systems with Semi-Markov Switching 97

4. **Stochastic Systems with Split and Merging** — 103
   - 4.1 Introduction 103
   - 4.2 Phase Merging Scheme 104
     - 4.2.1 Ergodic Merging 104
     - 4.2.2 Merging with Absorption 110
     - 4.2.3 Ergodic Double Merging 112
   - 4.3 Average with Merging 116

    4.3.1 Ergodic Average 117
    4.3.2 Average with Absorption 119
    4.3.3 Ergodic Average with Double Merging 120
    4.3.4 Double Average with Absorption 121
  4.4 Diffusion Approximation with Split and Merging 122
    4.4.1 Ergodic Split and Merging 123
    4.4.2 Split and Merging with Absorption 126
    4.4.3 Ergodic Split and Double Merging 128
    4.4.4 Double Split and Merging 130
    4.4.5 Double Split and Double Merging 132
  4.5 Integral Functionals in Split Phase Space 134
    4.5.1 Ergodic Split 134
    4.5.2 Double Split and Merging 137
    4.5.3 Triple Split and Merging 138

5. **Phase Merging Principles**   **139**

  5.1 Introduction 139
  5.2 Perturbation of Reducible-Invertible Operators 140
    5.2.1 Preliminaries 140
    5.2.2 Solution of Singular Perturbation Problems . 141
  5.3 Average Merging Principle 150
    5.3.1 Stochastic Evolutionary Systems 151
    5.3.2 Stochastic Additive Functionals 152
    5.3.3 Increment Processes 154
    5.3.4 Continuous Random Evolutions 156
    5.3.5 Jump Random Evolutions 157
    5.3.6 Random Evolutions with Markov Switching ... 159
  5.4 Diffusion Approximation Principle 160
    5.4.1 Stochastic Integral Functionals 161
    5.4.2 Continuous Random Evolutions 165
    5.4.3 Jump Random Evolutions 169
    5.4.4 Random Evolutions with Markov Switching ... 172
  5.5 Diffusion Approximation with Equilibrium 173
    5.5.1 Locally Independent Increment Processes ... 174
    5.5.2 Stochastic Additive Functionals 175
    5.5.3 Stochastic Evolutionary Systems with
          Semi-Markov Switching 176
  5.6 Merging and Averaging in Split State Space 182
    5.6.1 Preliminaries 182

        5.6.2 Semi-Markov Processes in Split State Space 184
        5.6.3 Average Stochastic Systems 186
    5.7 Diffusion Approximation with Split and Merging 188
        5.7.1 Ergodic Split and Merging 188
        5.7.2 Split and Double Merging 189
        5.7.3 Double Split and Merging 190
        5.7.4 Double Split and Double Merging 191

6. Weak Convergence   193

    6.1 Introduction 193
    6.2 Preliminaries 193
    6.3 Pattern Limit Theorems 196
        6.3.1 Stochastic Systems with Markov Switching 196
        6.3.2 Stochastic Systems with Semi-Markov Switching . . 201
        6.3.3 Embedded Markov Renewal Processes 205
    6.4 Relative Compactness 209
        6.4.1 Stochastic Systems with Markov Switching 209
        6.4.2 Stochastic Systems with Semi-Markov Switching . . 212
        6.4.3 Compact Containment Condition 213
    6.5 Verification of Convergence 216

7. Poisson Approximation   219

    7.1 Introduction 219
    7.2 Stochastic Systems in Poisson Approximation Scheme . . . 220
        7.2.1 Impulsive Processes with Markov Switching 220
        7.2.2 Impulsive Processes in an Asymptotic
              Split Phase Space 225
        7.2.3 Stochastic Additive Functionals with Semi-Markov
              Switching 228
    7.3 Semimartingale Characterization 231
        7.3.1 Impulsive Processes as Semimartingales 232
        7.3.2 Stochastic Additive Functionals 237

8. Applications I   243

    8.1 Absorption Times 243
    8.2 Stationary Phase Merging 249
    8.3 Superposition of Two Renewal Processes 253
    8.4 Semi-Markov Random Walks 258

		8.4.1	Introduction	258
		8.4.2	The algorithms of approximation for SMRW	259
		8.4.3	Compensating Operators	262
		8.4.4	The singular perturbation problem	265
		8.4.5	Stationary Phase Merging Scheme	267
9.	Applications II			269
	9.1	Birth and Death Processes and Repairable Systems		269
		9.1.1	Introduction	269
		9.1.2	Diffusion Approximation	270
		9.1.3	Proofs of the Theorems	272
	9.2	Lévy Approximation of Impulsive Processes		276
		9.2.1	Introduction	276
		9.2.2	Lévy Approximation Scheme	278
		9.2.3	Proof of Theorems	282

Problems to Solve   287

Appendix A   Weak Convergence of Probability Measures   301

     A.1 Weak Convergence   301
     A.2 Relative Compactness   303

Appendix B   Some Limit Theorems for Stochastic Processes   305

     B.1 Two Limit Theorems for Semimartingales   305
     B.2 A Limit Theorem for Composed Processes   308

Appendix C   Some Auxiliary Results   311

     C.1 Backward Recurrence Time Negligibility   311
     C.2 Positiveness of Diffusion Coefficients   312

Bibliography   315

Notation   325

Index   329

Chapter 1

Markov and Semi-Markov Processes

1.1 Preliminaries

The aim of this chapter is to give a brief account of basic notions on Markov and semi-Markov processes, together with martingale characterization, needed throughout this book. Further introduction to semimartingales and stochastic linear operators is also given.

Let (E, r) be a complete, separable metric space, (that is, a *Polish space*), and let $\mathcal{E}$ be its Borel σ-algebra of subsets of E [52]. Throughout this book, we will call the measurable space $(E, \mathcal{E})$ a *standard state space*. The space $\mathbb{R}^d$, with the Euclidean metric, is a Polish space. We will denote by $\mathcal{B}_d$ its Borel σ-algebra, $d > 1$, with $\mathcal{B} := \mathcal{B}_1$.

In this book, the space where the trajectories of processes are considered is $\mathbf{D}[0, \infty)$, the space of right-continuous functions having left side limits with Skorokhod metric [16,45,70,132,153]. We call them cadlag trajectories and cadlag processes. Let also $\mathbf{C}[0, \infty)$ be the subspace of $\mathbf{D}[0, \infty)$, of continuous functions with the sup-norm, $\|x\| = \sup_{t \geq 0} |x(t)|$, $x \in \mathbf{C}[0, \infty)$. These two spaces are Polish spaces (see Appendix A).

Let $\mathbf{B}$ be the Banach space, that is a complete linear normed space, of all bounded real-valued measurable functions on E, with the sup-norm $\|\varphi\| = \sup_{x \in E} |\varphi(x)|$, $\varphi \in \mathbf{B}$.

Let us consider also a fixed *stochastic basis* $\Im = (\Omega, \mathcal{F}, \mathbf{F} = (\mathcal{F}_t, t \geq 0), \mathbb{P})$, where $(\mathcal{F}_t, t \geq 0)$ (for discrete time note that $\mathbf{F} = (\mathcal{F}_n, n \geq 0)$) is a filtration of sub-σ-algebras of $\mathcal{F}$, that is $\mathcal{F}_s \subseteq \mathcal{F}_t \subseteq \mathcal{F}$, for all $s < t$ and $t \geq 0$. The filtration $\mathbf{F} = (\mathcal{F}_t, t \geq 0)$ is said to be *complete*, if $\mathcal{F}_0$ contains all the $\mathbb{P}$-null sets. Set $\mathcal{F}_{t+} = \cap_{s > t} \mathcal{F}_s, t \geq 0$. If for any $t \geq 0$, $\mathcal{F}_t = \mathcal{F}_{t+}$, then the filtration $\mathcal{F}_t, t \geq 0$, is said to be *right-continuous*. If a filtration is complete and right-continuous, we say that it satisfies the *usual conditions*.

A mapping $T : \Omega \to [0, +\infty]$, such that $\{T \leq t\} \in \mathcal{F}_t$, is called a *stopping time*. If T is a stopping time, we denote by $\mathcal{F}_T$ the collection of all sets $A \in \mathcal{F}$ such that $A \cap \{T \leq t\} \in \mathcal{F}_t$.

An $(E, \mathcal{E})$-valued stochastic process $x(t), t \in I$, $(I = \mathbb{R}_+$ or $I = \mathbb{N})$, defined on the stochastic basis $\mathfrak{S}$, is *adapted*, if for any $t \in I$, $x(t)$ is $\mathcal{F}_t$-measurable. The set of values E is said to be the *state* (or *phase*) *space* of the process $x(t), t \geq 0$.

Given a probability measure μ on $(E, \mathcal{E})$, we define the probability measure $\mathbb{P}_\mu$, by

$$\mathbb{P}_\mu(B) = \mu \mathbb{P}(B) = \int_E \mu(dx)\mathbb{P}_x(B), \quad x \in E, B \in \mathcal{F},$$

where $\mathbb{P}_x(B) := \mathbb{P}(B \mid x(0) = x)$. We denote by $\mathbb{E}_\mu$ and $\mathbb{E}_x$ the expectations corresponding respectively to $\mathbb{P}_\mu$ and $\mathbb{P}_x$.

We will also consider the following spaces endowed by the corresponding sup-norms:

- **B** is the Banach space of real-valued measurable bounded functions $\varphi(u, x)$, $u \in \mathbb{R}^d$, $x \in E$;
- $\mathbf{B}^1 := C^1(\mathbb{R}^d \times E) \cap \mathbf{B}$ is the Banach space of continuously differentiable functions on $u \in \mathbb{R}^d$, uniformly on $x \in E$, with bounded first derivative;
- $\mathbf{B}^2 := C^2(\mathbb{R}^d \times E) \cap \mathbf{B}$ is the Banach space of twice continuously differentiable functions on $u \in \mathbb{R}^d$, uniformly on $x \in E$, with bounded first two derivatives.

1.2 Markov Processes

1.2.1 *Markov Chains*

Definition 1.1 A positive-valued function $P(x, B)$, $x \in E, B \in \mathcal{E}$, is called a *Markov transition function* or a *Markov kernel* or *transition (probability) kernel*, if

1) for any fixed $x \in E$, $P(x, \cdot)$ is a probability measure on $(E, \mathcal{E})$, and

2) for any fixed $B \in \mathcal{E}$, $P(\cdot, B)$ is a Borel measurable function, that is, an $(\mathcal{E}, \mathcal{B})$-measurable function.

If $P(x, E) \leq 1$, for a $x \in E$, then P is said to be a *sub-Markov kernel*. If for fixed $x \in E$, the $P(x, \cdot)$ is a *signed measure*, then it is said to be a *signed kernel*. In that case, we will suppose that the signed kernel P is of bounded variation, that is,

$$|P|(x, E) < +\infty. \tag{1.1}$$

If E is a finite or countable set, we take $\mathcal{E} = \mathcal{P}(E)$ (the set of all subsets of E), the Markov kernel is determined by the matrix $(P(i,j); \ i,j \in E)$, with $P(i, B) = \sum_{j \in B} P(i,j)$, $B \subset \mathcal{E}$.

Definition 1.2 A *time-homogeneous Markov chain* associated to a Markov kernel $P(x, B)$ is an adapted sequence of random variables $x_n, n \geq 0$, defined on some stochastic basis $\mathfrak{S}$, satisfying, for every $n \in \mathbb{N}$, $x \in E$, and $B \in \mathcal{E}$, the following relation

$$\mathbb{P}(x_{n+1} \in B \mid \mathcal{F}_n) = \mathbb{P}(x_{n+1} \in B \mid x_n) =: P(x_n, B), \quad \text{(a.s.)}, \tag{1.2}$$

which is called the *Markov property*.

In most cases we consider the Markov property with respect to $\mathcal{F}_n := \sigma(x_k, \ k \leq n)$, $n \geq 0$, the *natural filtration* generated by the chain $x_n, n \geq 0$. The Markov property (1.2) is satisfied for any finite $\mathcal{F}_n$-stopping time and it is called *strong Markov property* and the chain a *strong Markov chain*.

The product of two Markov kernels P and Q defined on $(E, \mathcal{E})$, is also a Markov kernel, defined by

$$PQ(x, B) = \int_E P(x, dy) Q(y, B). \tag{1.3}$$

Let us denote by $P^n(x, B) = \mathbb{P}(x_n \in B \mid x_0 = x) = \mathbb{P}(x_{n+m} \in B \mid x_m = x)$ the n-step transition probability which is defined inductively by (1.3). By the Markov property we get, for $n, m \in \mathbb{N}$,

$$P^{n+m}(x, B) = \int_E P^n(x, dy) P^m(y, B) = \int_E P^m(x, dy) P^n(y, B),$$

which is the *Chapman-Kolmogorov equation*.

A subset $B \in \mathcal{E}$, is called *accessible* from a state $x \in E$, if

$$\mathbb{P}_x(x_n \in B, \text{ for some } n \geq 1) > 0,$$

or equivalently $P^n(x, B) > 0$.

Definition 1.3

1) A Markov chain $x_n, n \geq 0$, is called *Harris recurrent* if there exists a σ-finite measure ψ on $(E, \mathcal{E})$, with $\psi(E) > 0$, such that

$$\mathbb{P}_x(\cup_{n \geq 1}\{x_n \in A\}) = 1, \quad x \in E, \tag{1.4}$$

for any $A \in \mathcal{E}$ with $\psi(A) > 0$.

2) If the probability (1.4) is positive, the Markov chain is called ψ-*irreducible*.

3) The Markov chain is said to be *uniformly irreducible* if, for any $A \in \mathcal{E}$,

$$\sup_x \mathbb{P}_x(\tau_A > N) \longrightarrow 0, \quad N \to \infty, \tag{1.5}$$

where $\tau_A := \inf\{n \geq 0 : x_n \in A\}$, is the hitting time of set $A \in \mathcal{E}$.

Definition 1.4

A Markov chain $x_n, n \geq 0$, is said to be d-*periodic* ($d > 1$), if there exists a cycle, that is a sequence $(C_1, ..., C_d)$ of sets, $C_i \in \mathcal{E}$, $1 \leq i \leq d$, with $P(x, C_{j+1}) = 1$, $x \in C_j$, $1 \leq j \leq d-1$, and $P(x, C_1) = 1$, $x \in C_d$, such that:

- the set $E \setminus \cup_{i=1}^d C_i$ is ψ-null;
- if $(C_1', ..., C_{d'}')$ is another cycle, then d' divides d and C_i' differs from a union of d/d' members of $(C_1, ..., C_d)$ only by a ψ-*null set* which is of type $\cup_{i \geq 1} V_i$, where, for any $i \geq 1$, $\mathbb{P}_x(\limsup\{x_n \in V_i\}) = 0$.

If $d = 1$ then the Markov chain is said to be *aperiodic*.

Definition 1.5

A probability measure ρ on $(E, \mathcal{E})$, is said to be a *stationary distribution* or *invariant probability* for the Markov chain x_n, $n \geq 0$, (or for the Markov kernel $P(x, B)$) if, for any $B \in \mathcal{E}$,

$$\rho(B) = \int_E \rho(dx) P(x, B).$$

Definition 1.6

1) If a Markov chain is ψ-irreducible and has an invariant probability, it is called *positive*, otherwise it is called *null*.

2) If a Markov chain is Harris recurrent and positive it is called *Harris positive*.

3) If a Markov chain is aperiodic and Harris positive it is called (Harris) *ergodic*.

Proposition 1.1

Let $x_n, n \geq 0$, be an ergodic Markov chain, then:

1) for any probability measure α on $(E, \mathcal{E})$, we have

$$\|\alpha P^n - \rho\| \to 0, \quad n \to \infty;$$

1.2. MARKOV PROCESSES

2) for any $\varphi \in \mathbf{B}$, we have

$$\lim_{n\to\infty} \frac{1}{n} \sum_{k=0}^{n-1} \varphi(x_k) = \int_E \rho(dx)\varphi(x), \quad \mathbb{P}_\mu\text{-}a.s.,$$

for any probability measure μ on $(E, \mathcal{E})$.

Let us denote by P the operator of transition probabilities on $\mathbf{B}$ defined by

$$P\varphi(x) = \mathbb{E}[\varphi(x_{n+1}) \mid x_n = x] = \int_E P(x, dy)\varphi(y),$$

and denote by P^n the n-step transition operator corresponding to $P^n(x, B)$.

The Markov property (1.2) can be represented in the following form

$$\mathbb{E}[\varphi(x_{n+1}) \mid \mathcal{F}_n] = P\varphi(x_n). \tag{1.6}$$

Definition 1.7 Let us denote by Π the *stationary projector* in $\mathbf{B}$ defined by the stationary distribution $\rho(B), B \in \mathcal{E}$ of the Markov chain x_n, as follows

$$\Pi\varphi(x) := \int_E \rho(dy)\varphi(y)\mathbf{1}(x) = \widehat{\varphi}\mathbf{1}(x), \quad \widehat{\varphi} := \int_E \rho(dx)\varphi(x),$$

where $\mathbf{1}(x) = 1$ for all $x \in E$. Of course, we have $\Pi^2 = \Pi$.

Definition 1.8 The Markov chain x_n is called *uniformly ergodic* if

$$\sup_{\|\varphi\|\leq 1} \|(P^n - \Pi)\varphi\| \longrightarrow 0, \quad n \to \infty, \tag{1.7}$$

Note that uniform ergodicity implies Harris recurrence [117,80,137,139].

Moreover, the convergence in (1.7) is of exponential rate (see, e.g. [137]). So the series

$$R_0 := \sum_{n=0}^{\infty} [P^n - \Pi],$$

is convergent and defines the *potential operator* of the Markov chain $x_n, n \geq 0$, satisfying the property (see Section 1.6)

$$R_0[I - P] = [I - P]R_0 = I - \Pi.$$

1.2.2 Continuous-Time Markov Processes

Let us consider a family of Markov kernels $(P_t = P_t(x,B),\ t \in \mathbb{R}_+)$ on $(E,\mathcal{E})$. Let an adapted $(E,\mathcal{E})$-valued stochastic process $x(t),\ t \geq 0$, be defined on some stochastic basis $\mathfrak{F}$.

Definition 1.9 A stochastic process $x(t),\ t \geq 0$, is said to be a *time-homogeneous Markov process*, if, for any fixed $s, t \in \mathbb{R}_+$ and $B \in \mathcal{E}$,

$$\mathbb{P}(x(t+s) \in B \mid \mathcal{F}_s) = \mathbb{P}(x(t+s) \in B \mid x(s)) = P_t(x(s), B), \text{ (a.s.)}. \tag{1.8}$$

When the Markov property (1.8) holds for any finite **F**-stopping time τ, instead of a deterministic time s, we say that the Markov process $x(t), t \geq 0$, satisfies the *strong Markov property*, and that the process $x(t)$ is a *strong Markov process*.

Definition 1.10 On the Banach space **B**, the operator P_t of transition probability, is defined by

$$P_t \varphi(x) = \mathbb{E}_x[\varphi(x(t))] = \int_E \varphi(y) P_t(x, dy), \quad \varphi \in \mathbf{B}. \tag{1.9}$$

This is a contractive operator (that is, $\|P_t \varphi\| \leq \|\varphi\|$).

The *Chapman-Kolmogorov equation* is equivalent to the following semigroup property of P_t,

$$P_t P_s = P_{t+s}, \text{ for all } t, s \in \mathbb{R}_+. \tag{1.10}$$

The Markov process $x(t),\ t \geq 0$, has a *stationary* (or invariant) *distribution*, π say, if, for any $B \in \mathcal{E}$,

$$\pi(B) = \int_E \pi(dx) P_t(x, B), \quad \pi(E) = 1, \quad t \geq 0.$$

Definition 1.11 The Markov process $x(t),\ t \geq 0$, is said to be *ergodic*, if for every $\varphi \in \mathbf{B}$, we have

$$\lim_{t \to \infty} \frac{1}{t} \int_0^t \varphi(x(s)) ds = \int_E \pi(dx) \varphi(x), \quad \mathbb{P}_\mu\text{-a.s.},$$

for any probability measure μ on $(E, \mathcal{E})$.

The stationary projector Π, of an ergodic Markov process with stationary distribution π, is defined as follows (see Definition 1.7)

$$\Pi \varphi(x) := \int_E \pi(dy) \varphi(y) \mathbf{1}(x) = \widehat{\varphi} \mathbf{1}(x), \quad \widehat{\varphi} := \int_E \pi(dx) \varphi(x),$$

1.2. MARKOV PROCESSES

where $\mathbf{1}(x) = 1$ for all $x \in E$. Of course, we have $\Pi^2 = \Pi$.

Let us consider a Markov process $x(t)$, $t \geq 0$, on the stochastic basis $\mathfrak{F}$, with trajectories in $\mathbf{D}[0, \infty)$, and semigroup $(P_t, t \geq 0)$.

There exists a linear operator Q acting on $\mathbf{B}$, defined by

$$\lim_{t \downarrow 0} \frac{1}{t}(P_t \varphi - \varphi) = Q\varphi, \tag{1.11}$$

where the limit exists in norm.

Let $\mathcal{D}(Q)$ be the subset of $\mathbf{B}$ for which the above limit exists, this is the *domain* of the operator Q. The operator Q is called a (strong) *generator* or (strong) *infinitesimal operator*.

Definition 1.12 A Markov semigroup $P_t, t \geq 0$, is said to be *uniformly continuous* on $\mathbf{B}$, if

$$\lim_{t \to 0} \|P_t - I\| = 0,$$

where I is the identity operator on $\mathbf{B}$.

A time-homogeneous Markov process is said to be (purely) *discontinuous* or of *jump type*, if its semigroup is uniformly continuous. In that case, the process stays in any state for a positive (strict) time, and after leaving a state it moves directly to another one. We will call it a *jump Markov process* [34,56,153,165].

Let $x(t)$, $t \geq 0$, be a time-homogeneous jump Markov process. Let $\tau_n, n \geq 0$ be the jump times for which we have $0 = \tau_0 \leq \tau_1 \leq \cdots \leq \tau_n \leq \cdots$. A Markov process is said to be *regular (non explosive)*, if $\tau_n \to \infty$, as $n \to \infty$ (see, e.g. [56]).

Definition 1.13 The stochastic process $x_n, n \geq 0$ defined by

$$x_n = x(\tau_n), \quad n \geq 0,$$

is called the *embedded Markov chain* of the Markov process $x(t)$, $t \geq 0$.

Let $P(x, B)$ be the transition probability of $x_n, n \geq 0$. The generator Q of the jump Markov process $x(t), t \geq 0$, is of the form (see, e.g. [56,34]),

$$Q\varphi(x) = q(x) \int_E P(x, dy)[\varphi(y) - \varphi(x)], \tag{1.12}$$

where the kernel $P(x, dy)$ is the transition kernel of the embedded Markov chain, and $q(x), x \in E$, is the intensity of jumps function.

Proposition 1.2 *(see, e.g.* [165,52]*) Let* $(P_t, t \geq 0)$ *be a uniformly continuous semigroup on* **B**, *and* Q *its generator with domain* $\mathcal{D}(Q) \subset \mathbf{B}$. *Then:*

1) the limit in (1.11) exists, and the operator Q is bounded with $\overline{\mathcal{D}}(Q) =$ **B**;

2) $dP_t/dt = QP_t = P_tQ$;

3) $P_t = \exp(tQ) = I + \sum_{k \geq 1}(tQ)^k/k!$.

If $x(t)$ has a *stationary distribution*, π, then x_n also has a stationary distribution, ρ, and we have

$$\pi(dx)q(x) = q\rho(dx), \quad q := \int_E \pi(dx)q(x).$$

Let us consider the *counting process*

$$\nu(t) = \max\{n : \tau_n \leq t\}, \qquad (1.13)$$

with $\max \emptyset = 0$. That gives the number of jumps of the Markov process in $(0, t]$.

▷ **Example 1.1.** The generator $\mathbb{L}$ of a Poisson process, with intensity $\lambda > 0$, is

$$\mathbb{L}\varphi(u) = \lambda[\varphi(u+1) - \varphi(u)], \quad u \in \mathbb{N}.$$

▷ **Example 1.2.** Let $\tau_n, n \geq 0$, be a renewal process on $\mathbb{R}_+$, $\tau_0 = 0$, with distribution function F, and hazard rate of inter-arrival times $\theta_n := \tau_n - \tau_{n-1}$,

$$\lambda(t) := -\frac{\overline{F}'(t)}{\overline{F}(t)}.$$

Let $\nu(t), t \geq 0$ be the corresponding counting process, that is $\nu(t) := \sup\{n : \tau_n \leq t\}$.

The generator of the Markov process $x(t) := t - \tau(t), t \geq 0$, $\tau(t) := \tau_{\nu(t)}$, is given by

$$\mathbb{L}\varphi(x) = \varphi'(x) + \lambda(x)[\varphi(0) - \varphi(x)], \quad u \in \mathbb{N}, \ t \in \mathbb{R}_+,$$

where $\overline{F}(t) := 1 - F(t)$. The domain of this generator is $\mathcal{D}(\mathbb{L}) = C^1(\mathbb{R})$.

▷ **Example 1.3.** Let $x(t), t \geq 0$, be a nonhomogeneous jump Markov process, the generator of the coupled Markov process $t, x(t), t \geq 0$, is defined as follows

$$\mathbb{L}\varphi(t, x) = \frac{\partial}{\partial t}\varphi(t, x) + Q\varphi(\cdot, x),$$

with $\mathcal{D}(\mathbb{L}) = C^{1,0}(\mathbb{R}_1 \times E)$.

▷ **Example 1.4.** Let $x(t), t \geq 0$, be a pure jump Markov process (that is, without drift and diffusion part) with state space E and generator Q, and let $\nu(t), t \geq 0$, be the corresponding counting process of jumps and $x_n, n \geq 0$ the embedded Markov chain. Let a be a real-valued measurable function on the state space E, and consider the *increment process*

$$\alpha(t) = \sum_{k=1}^{\nu(t)} a(x_k), \quad t \geq 0.$$

Then the generator of the coupled Markov process $\alpha(t), x(t), t \geq 0$ is

$$\mathbb{L} = Q + Q_0[\mathbb{\Gamma}(x) - I], \qquad (1.14)$$

where:

$$\mathbb{\Gamma}(x)\varphi(u) := \varphi(u + a(x))$$

$$Q_0\varphi(x) := q(x)\int_E P(x, dy)\varphi(y),$$

and I is the identity operator.

▷ **Example 1.5.** For the jump Markov process $x(t), t \geq 0$, as in the previous example, let us consider the process

$$\xi(t) := \int_0^t a(x(s))ds.$$

Then the generator of the coupled process $\xi(t), x(t), t \geq 0$ is

$$\mathbb{L} = Q + \mathbb{A}(x),$$

where $\mathbb{A}(x)\varphi(u) := a(x)\varphi'(u)$.

1.2.3 Diffusion Processes

A diffusion process is a strong Markov process in continuous time with almost surely continuous paths. This process is used in order to describe mathematically the physical phenomenon of diffusion, that is the movement of particles in a chaotic environment. In this book we will use diffusions as limit processes of stochastic systems switched by Markov and semi-Markov processes.

Definition 1.14 A Markov nonhomogeneous in time process $x(t), t \geq 0$, defined on a stochastic basis $\Im$, with values in the Euclidean space $\mathbb{R}^d$, $d \geq 1$, with transition function $P_{s,t}(x, B) := \mathbb{P}(x(t) \in B \mid x(s) = x)$, for $0 \leq s < t < \infty$, $x \in \mathbb{R}^d, B \in \mathcal{B}_d$, is said to be a *diffusion process*, with generator $\mathbb{L}_t$, if (a) it has continuous paths, and (b) for any $x \in \mathbb{R}^d$, and any $\varphi \in C^2(\mathbb{R}^d)$,

$$\int_{\mathbb{R}^d} P_{t,t+h}(x, dy)[\varphi(y) - \varphi(x)] = h\mathbb{L}_t\varphi(x) + o(h), \quad h \to 0.$$

Let a be a real-valued measurable function, defined on $\mathbb{R}^d \times \mathbb{R}_+$, and B a function defined on $\mathbb{R}^d \times \mathbb{R}_+$ with values in the space of symmetric positive operators from $\mathbb{R}^d$ to $\mathbb{R}^d$.

For any $t \geq 0$, the generator $\mathbb{L}_t$, of the diffusion $x(t), t \geq 0$, acts on functions φ in $C^2(\mathbb{R}^d)$ as follows

$$\mathbb{L}_t\varphi(x) = a(x,t)\varphi'(x) + \frac{1}{2}B(x,t)\varphi''(x). \quad (1.15)$$

The above means that the *drift coefficient* $a(x,t)$, and the *diffusion operator* (or coefficient) $B(x,t)$, apply as follows:

$$a(x,t)\varphi'(x) = \sum_{i=1}^{d} a_i(x,t)\frac{\partial}{\partial x_i}\varphi(x),$$

and

$$B(x,t)\varphi''(x) = \sum_{i,j=1}^{d} b_{ij}(x,t)\frac{\partial^2}{\partial x_i \partial x_j}\varphi(x).$$

For a *time-homogeneous diffusion* the generator does not depend on t, that means functions a and B are free of t.

If $a(x,t) \equiv 0$, and $B(x,t) = I$, the $x(t), t \geq 0$, is a *Wiener process* or a *Brownian motion*.

A diffusion process with a constant shift vector and a diffusion operator B has the following representation

$$x(t) = x(0) + ta + \sigma w(t),$$

where $w(t), t \geq 0$, is a Wiener process, with $\mathbb{E}w(t) = 0$, $\mathbb{E}(xw(t))^2 = t\|x\|^2$, $x \in \mathbb{R}^d$, and $\sigma := B^{1/2}$ ($B = \sigma\sigma^*$) is the positive symmetric square root of the matrix B.

For the existence and weak uniqueness of a diffusion with generator (1.15), we suppose that (see, e.g. [56,140,151,165])
(a) $a(x,t)$ and $B(x,t)$ are continuous, and
(b) $\|a(x,t)\| + \|B(x,t)\| \leq C(1+\|x\|)$, where C is some positive constant.
Weak uniqueness here means uniqueness of the transition function.

1.2.4 Processes with Independent Increments

Let $\mathfrak{F}$ be a stochastic basis, and $x(t), t \geq 0$, a stochastic process adapted to $\mathfrak{F}$ with values in $\mathbb{R}^d$.

Definition 1.15 The process $x(t), t \geq 0$, is said to be with *independent increments* (PII), if for any $s, t \in \mathbb{R}_+$, with $s < t$, $x(t) - x(s)$ is independent of $\mathcal{F}_s$.

This property is equivalent to the increment independence property of the process $x(t)$, when the filtration $(\mathcal{F}_t)$ is the natural filtration of the process. That is, for any $n \geq 1$, and any $0 \leq t_0 < t_1 < \cdots < t_n < \infty$, the random variables $x(t_0), x(t_1) - x(t_0), ..., x(t_n) - x(t_{n-1})$ are independent.

If, moreover, $x(t)$ has *stationary increments*, that is, the law of $x(t) - x(s)$ depends only on $t - s$, then the process $x(t)$ is said to be a *process with stationary independent increments* (PSII).

The main properties of PSII are that their distributions are infinitely divisible and have the Markov property. Let

$$\phi_t(\lambda) := \mathbb{E}\exp(i\lambda(x(t+s) - x(s)),$$

be the *characteristic function* of increments. Then it satisfies the semigroup property

$$\phi_{t+s}(\lambda) = \phi_t(\lambda)\phi_s(\lambda).$$

The characteristic function has the following natural representation

$$\phi_t(\lambda) = \exp[t\psi(\lambda)].$$

The *cumulant* $\psi(\lambda)$, has the well known *Lévy-Khintchine formula* [13,164]

$$\psi(\lambda) = i\lambda a - \frac{1}{2}\sigma^2\lambda^2 + \int_{\mathbb{R}}[e^{i\lambda z} - 1 - i\lambda z \mathbf{1}_{\{|z|\leq 1\}}]H(dz), \quad (1.16)$$

where H is the *spectral measure* and satisfies the following conditions

$$\int_{|z|\leq 1} z^2 H(dz) < \infty, \quad \int_{|z|>1} H(dz) < \infty.$$

In the case where the PSII process $x(t)$ is of finite variation (that is, it has trajectories of finite variation), the cumulant has the following form

$$\psi(\lambda) = i\lambda a + \int_{\mathbb{R}}[e^{i\lambda z} - 1]H(dz),$$

and the spectral measure satisfies the condition

$$\int_{|z|\leq 1} |z| H(dz) < \infty.$$

The most important PSII processes are Brownian motion, Poisson process, and Lévy process. [13]

▷ **Example 1.6.** The *Compound Poisson Process* is defined by

$$x(t) = \sum_{k=1}^{\nu(t)} \xi_k,$$

where $\nu(t), t \geq 0$, is a homogeneous Poisson process with intensity $\lambda > 0$, and $\xi_k, k \geq 1$, is an i.i.d. sequence of real random variables, independent of $\nu(t), t \geq 0$, with common distribution function F. Then we have $F(A) = H(A)/\lambda = \mathbb{P}(\xi_k \in A)$. And the cumulant of the compound Poisson process has the form

$$\psi(\lambda) = \lambda \int_{\mathbb{R}} [e^{i\lambda z} - 1]F(dz). \quad (1.17)$$

As stated above, the PSII satisfy the Markov property, that is, the transition probabilities are generated by the Markov semigroup

$$\Gamma_t \varphi(u) := \mathbb{E}\varphi(u + x(t)). \quad (1.18)$$

Lemma 1.1 ([163]) *The generator* Γ *of semigroup (1.18) has the following representation*

$$\Gamma\varphi(u) = \int_{\mathbb{R}} e^{i\lambda u}\psi(\lambda)\tilde{\varphi}(\lambda)d\lambda, \qquad (1.19)$$

for $\varphi(u) = \int_{\mathbb{R}} e^{i\lambda u}\tilde{\varphi}(\lambda)d\lambda$, *where* $\tilde{\varphi}(\lambda)$ *and* $\lambda^2\tilde{\varphi}(\lambda)$ *are integrable functions.*

PROOF. Let us consider the semigroup (1.18)

$$\Gamma_t\varphi(u) = \mathbb{E}\varphi(u + x(t)) = \mathbb{E}\int_{\mathbb{R}} e^{i\lambda(u+x(t))}\tilde{\varphi}(\lambda)d\lambda$$
$$= \int_{\mathbb{R}} e^{i\lambda u + t\psi(\lambda)}\tilde{\varphi}(\lambda)d\lambda.$$

Note that, according to the Lévy-Khintchine formula, the cumulant has the asymptotic form $\psi(\lambda) = O(\lambda^2)$, as $|\lambda| \to \infty$. Hence, the latter integral is convergent, uniformly on t. So, we get the derivative

$$\frac{d}{dt}\Gamma_t\varphi(u) = \int_{\mathbb{R}} e^{i\lambda u + t\psi(\lambda)}\psi(\lambda)\tilde{\varphi}(\lambda)d\lambda.$$

By the evolutionary equation for the semigroup, we have

$$\frac{d}{dt}\Gamma_t\varphi(u) = \Gamma_t\Gamma\varphi(u).$$

Comparing the two latter formulas, we get (1.19). $\square$

The meaning of representation (1.19) is that the cumulant $\psi(\lambda)$ of a PSII is the symbol of the generator Γ. In the particular case of a drift process $x(t) = at$, the corresponding generator is $\Gamma\varphi(u) = a\varphi'(u)$ since for the corresponding semigroup

$$\frac{d}{dt}\Gamma_t\varphi(u) = \int_{\mathbb{R}} e^{i\lambda u} i\lambda a\tilde{\varphi}(\lambda)d\lambda.$$

It is well-known that the standard Wiener process with the cumulant $\psi(\lambda) = -\sigma^2\lambda^2/2$ has the generator $\Gamma\varphi(u) = \sigma^2\varphi''(u)/2$.

Corollary 1.1 *The generator of the semigroup (1.18) with the cumulant (1.16) has the following representation*

$$\Gamma\varphi(u) = a\varphi'(u) - \frac{\sigma^2}{2}\varphi''(u) + \int_{\mathbb{R}}[\varphi(u+v) - \varphi(u) - v\varphi'(u)\mathbf{1}_{(|v|\leq 1)}]H(dv).$$

In the particular case of a compound Poisson process with the cumulant (1.17) the generator has the form

$$\mathbb{\Gamma}\varphi(u) = \lambda \int_{\mathbb{R}} [\varphi(u+v) - \varphi(u)] F(dv).$$

1.2.5 Processes with Locally Independent Increments

We consider now the Markov processes with *locally independent increments* (PLII). It is worth noticing that such processes include strictly the independent increment processes. Here we will restrict our interest to PLII without diffusion part. These processes are called also "Piecewise–deterministic Markov processes" [34], or "jump Markov process with drift", or "weak differentiable Markov processes" and have the same local structure as the PII (see [56]).

Roughly speaking, these processes are jump Markov processes with drift and without diffusion part. These processes are of increasing interest in the literature because of their importance in applications, for which they constitute an alternative to diffusion processes. For their detailed presentation and applications see [34].

These processes are defined by the generator $\mathbb{\Gamma}$ as follows

$$\mathbb{\Gamma}\varphi(u) = a(u)\varphi'(u) + \int_{\mathbb{R}^d} [\varphi(u+v) - \varphi(u)]\Gamma(u, dv), \quad (1.20)$$

with the intensity kernel $\Gamma(u, dv)$ satisfying the boundedness property: $\Gamma(u, \mathbb{R}^d) \in \mathbb{R}_+$.

We can also write the above generator in the following form, by extracting the drift, due to the jump part, and add it into the initial drift a to obtain the drift coefficient g, that is, $g(u) := a(u) + \int_{\mathbb{R}^d} v\Gamma(u, dv)$,

$$\mathbb{\Gamma}\varphi(u) = g(u)\varphi'(u) + \int_{\mathbb{R}^d} [\varphi(u+v) - \varphi(u) - v\varphi'(u)]\Gamma(u, dv). \quad (1.21)$$

Let us be given the Euclidean space $\mathbb{R}^d$ with the Borel σ-algebra $\mathcal{B}_d$ and the compact measurable space $(E, \mathcal{E})$. We consider the family of time-homogeneous Markov processes $\eta(t; x)$, $t \geq 0$, $x \in E$, with trajectories in $\mathbf{D}[0, \infty)$, with locally independent increments. These processes take values in the Euclidean space $\mathbb{R}^d$ ($d \geq 1$), and depend on the state $x \in E$, and their generators are given by

$$\Gamma(x)\varphi(u) = a(u;x)\varphi'(u) + \int_{\mathbb{R}^d} [\varphi(u+v) - \varphi(u) - v\varphi'(u)]\Gamma(u, dv; x). \quad (1.22)$$

These processes will be used in Chapter 2 as switched processes. A complete characterization of the above generators is given in [34]. Note also that $\eta(t;x)$ contains no diffusion part (see, e.g. [34,51,56,107]).

Of course, it is understood that when $d > 1$, we have

$$v\varphi'(u) = \sum_{k=1}^{d} v_k \frac{\partial \varphi}{\partial u_k}(u).$$

It is worth noting that slightly changed conditions allow one to include a locally compact space of values for the switched process.

The drift velocity $a(u;x)$ and the measure of the random jumps $\Gamma(u, dv; x)$ depend on the state $x \in E$. The time-homogeneous cadlag jump Markov process $x(t)$, $t \geq 0$, taking values in the state space $(E, \mathcal{E})$, is given by its generators (1.12)

$$Q\varphi(x) = q(x) \int_E P(x, dy)[\varphi(y) - \varphi(x)],$$

where q is the intensity of jumps, which is a nonnegative element of the Banach space $\mathbf{B}(E)$ of real bounded functions defined on the state space E, with the sup–norm, that is $\|\varphi\| := \sup_{x \in E} |\varphi(x)|$.

We will consider additive functionals of the process $\eta(t;x)$, of the following form

$$\xi(t) = \xi_0 + \int_0^t \eta(ds; x(s)) = \xi_0 + \sum_{k=1}^{\nu(t)-1} \eta(\theta_k; x_k) + \eta(t - \tau(t); x(t)),$$

with generator $\mathbb{L} = Q + \Gamma(x)$.

1.2.6 Martingale Characterization of Markov Processes

Let $\Im = (\Omega, \mathcal{F}, \mathbf{F} = (\mathcal{F}_t, t \geq 0), \mathbb{P})$ be a stochastic basis. Let $z(t), t \geq 0$, be a process, defined on $\Im$ and adapted to $\mathbf{F}$.

Definition 1.16 (see, e.g. [70,132,93]) A real-valued process $z(t), t \geq 0$, adapted to the filtration $\mathbf{F}$, is called an $\mathbf{F}$-*martingale (submartingale, su-*

permartingale), if $\mathbb{E}|z(t)| < \infty$, for all $t \geq 0$, and, for $s < t$,

$$\mathbb{E}[z(t) \mid \mathcal{F}_s] = z(s), \quad (\mathbb{E}[z(t) \mid \mathcal{F}_s] \geq z(s), \ \mathbb{E}[z(t) \mid \mathcal{F}_s] \leq z(s)), \quad \text{a.s.} \tag{1.23}$$

A martingale $\mu_t, t \geq 0$, is called *square integrable*, if

$$\sup_{t \geq 0} \mathbb{E}\mu_t^2 < \infty. \tag{1.24}$$

Then the process $\mu_t^2, t \geq 0$, is an $\mathcal{F}_t$-submartingale.

The *Doob-Meyer decomposition* [132] of a square integrable martingale $\mu_t, t \geq 0$, is as follows

$$\mu_t^2 = \langle \mu \rangle_t + z(t),$$

where $z(t), t \geq 0$, is a martingale, and the increasing process $\langle \mu \rangle_t, t \geq 0$, is called the *square characteristic* of the martingale μ_t.

Definition 1.17 (see, e.g. [132]) A process $z(t), t \geq 0$, adapted to **F** is called an **F**-*local martingale* if there exists an increasing to infinity sequence of *stopping times*, $\tau_n \to +\infty$, such that the stopped process $z^n(t) = z(t \wedge \tau_n), t \geq 0$, is a martingale for each $n \geq 1$.

It is clear that any martingale is a local martingale, by setting $\tau_n = n$, $n \geq 1$. The converse is false.

Let $x(t), t \geq 0$, be a Markov process with a standard state space $(E, \mathcal{E})$, defined on a stochastic basis $\mathfrak{F}$. Let $P_t(x, B), x \in E, B \in \mathcal{E}, t \geq 0$, be its transition function, and $P_t, t \geq 0$, its strongly continuous semigroup defined on the Banach space **B** of real-valued measurable functions defined on E, with the sup-norm, (see Definition 1.10). Let Q be the generator of the semigroup $P_t, t \geq 0$, with the dense domain of definition $\mathcal{D}(Q) \subset \mathbf{B}$.

For any function $\varphi \in \mathcal{D}(Q)$ and any $t > 0$, we have the *Dynkin formula*
[45,34]

$$P_t\varphi(x) = \varphi(x) + \int_0^t QP_s\varphi(x)ds. \tag{1.25}$$

From this formula, using conditional expectation, we get

$$\mathbb{E}_x\left[\varphi(x(t)) - \varphi(x) - \int_0^t Q\varphi(x(s))ds\right] = 0.$$

1.2. MARKOV PROCESSES

Thus, the process

$$\mu(t) := \varphi(x(t)) - \varphi(x) - \int_0^t Q\varphi(x(s))ds \qquad (1.26)$$

is an $\mathcal{F}_t^x = \sigma(x(s), s \leq t)$-martingale.

The following theorem gives the martingale characterization of Markov processes.

Theorem 1.1 ([45]) *Let $(E, \mathcal{E})$ be a standard state space and let $x(t), t \geq 0$, be a stochastic process on it, adapted to the filtration $\mathbf{F} = (\mathcal{F}_t, t \geq 0)$. Let Q be the generator of a strongly continuous semigroup $P_t, t \geq 0$, on the Banach space $\mathbf{B}$, with dense domain $\mathcal{D}(Q) \subset \mathbf{B}$. If for any $\varphi \in \mathcal{D}(Q)$, the process $\mu(t), t \geq 0$, defined by (1.26) is an $\mathcal{F}_t$-martingale, then $x(t), t \geq 0$, is a Markov process generated by the infinitesimal generator Q.*

The process $x(t), t \geq 0$, is said to solve the *martingale problem for the generator Q*.

The martingale (1.26) is a square integrable one whose the square integrable characteristic is the process:

Theorem 1.2 ([132]) *The square characteristic of the martingale $\mu(t), t \geq 0$, (see 1.26), denoted by $\langle\mu\rangle_t, t \geq 0$, is the process*

$$\langle\mu\rangle_t = \int_0^t [Q\varphi^2(x(s)) - 2\varphi(x(s))Q\varphi(x(s))]ds, \quad t \geq 0.$$

PROOF. Let us denote $\alpha_t := \int_0^t Q\varphi(x(s))ds$. Then, from the representation of the martingale $\mu(t)$, we have

$$\varphi^2 = (\mu + \alpha)^2 = \mu^2 + 2\mu\alpha + \alpha^2 = \mu^2 + L,$$

where $L := 2\mu\alpha + \alpha^2$. Differentiating L, we get

$$dL = 2d\mu\alpha + 2\mu Q\varphi ds + 2\alpha Q\varphi ds.$$

Now the martingale representation $\mu = \varphi - \alpha$, gives

$$dL = 2d\mu\alpha + 2\varphi Q\varphi ds,$$

and, by integration,

$$L = 2\int_0^t d\mu(s) \int_0^s Q\varphi dv + 2\int_0^t \varphi Q\varphi ds.$$

The first term is a martingale, since it is the integral with respect to a martingale $\mu(s)$. It is obvious that

$$\mu_1(t) := \varphi^2(x(t)) - \int_0^t Q\varphi^2(x(s))ds,$$

is a martingale. Hence

$$\mu^2 = \varphi^2 - L$$
$$= \varphi^2 - \int_0^t Q\varphi^2 ds + \int_0^t Q\varphi^2 ds - 2\int_0^t \varphi Q\varphi ds - \mu_1$$
$$= \int_0^t [Q\varphi^2 - 2\varphi Q\varphi]ds + \mu_2,$$

where μ_2 is a martingale. So, the latter relation gives the square characteristic of the martingale $\mu(t)$. □

Let $x_n, n \geq 0$, be a Markov chain on a measurable state space $(E, \mathcal{E})$ induced by a stochastic kernel $P(x, B)$, $x \in E, B \in \mathcal{E}$. Let P be the corresponding transition operator defined on the Banach space **B**.

Let us construct now the following martingale as a sum of martingale differences

$$\mu_n = \sum_{k=0}^{n-1} [\varphi(x_{k+1}) - \mathbb{E}(\varphi(x_{k+1}) \mid \mathcal{F}_k)]. \tag{1.27}$$

By using the Markov property and the rearrangement of terms in (1.27) the martingale takes the form

$$\mu_n = \varphi(x_n) - \varphi(x_0) - \sum_{k=1}^{n-1} [P - I]\varphi(x_k). \tag{1.28}$$

This representation of the martingale is associated with a Markov chain characterization.

Lemma 1.2 *Let $x_n, n \geq 0$, be a sequence of random variables taking values in a measurable space $(E, \mathcal{E})$ and adapted to the filtration $\mathcal{F}_n, n \geq 0$. Let P be a bounded linear positive operator on the Banach space **B** induced by a transition probability kernel $P(x, B)$ on $(E, \mathcal{E})$. If for every $\varphi \in \mathbf{B}$, the right hand side of (1.28) is a martingale $\mu_n, \mathcal{F}_n, n \geq 0$, then the sequence $x_n, n \geq 0$, is a Markov chain with transition probability kernel $P(x, B)$ induced by the operator P.*

PROOF. Using (1.28) we have

$$\mathbb{E}[\mu_n \mid \mathcal{F}_{n-1}] = \mathbb{E}[\varphi(x_n) \mid \mathcal{F}_{n-1}] - \varphi(x_0) - \sum_{k=1}^{n-1}[P - I]\varphi(x_k)$$
$$= \mathbb{E}[\varphi(x_n) \mid \mathcal{F}_{n-1}] - P\varphi(x_{k-1}) + \varphi(x_{n-1}) - \varphi(x_0)$$
$$= \mu_{n-1} + \mathbb{E}[\varphi(x_n) \mid \mathcal{F}_{n-1}] - P\varphi(x_{k-1}).$$

So, the martingale property $\mathbb{E}[\mu_n \mid \mathcal{F}_{n-1}] = \mu_{n-1}$, is equivalent to the Markov property

$$\mathbb{E}[\varphi(x_{n+1}) \mid \mathcal{F}_n] = \mathbb{E}[\varphi(x_{n+1}) \mid x_n] = P\varphi(x_n). \qquad (1.29)$$

□

By the definition of square characteristic of martingale it is easy to check that

$$\langle \mu \rangle_n = \sum_{k=0}^{n-1}[P\varphi^2(x_k) - (P\varphi(x_k))^2]. \qquad (1.30)$$

1.3 Semi-Markov Processes

The semi-Markov process is a generalization of the Markov and renewal processes. We will present shortly definitions and basic properties of semi-Markov process useful in the sequel of the book (see, e.g. [127,116]).

1.3.1 *Markov Renewal Processes*

Definition 1.18 A positive-valued function $Q(x, B, t)$, $x \in E$, $B \in \mathcal{E}$, $t \in \mathbb{R}_+$, is called a *semi-Markov kernel* on $(E, \mathcal{E})$ if

(i) $Q(x, B, \cdot)$, for $x \in E$, $B \in \mathcal{E}$, is a non-decreasing, right continuous real function, such that $Q(x, B, 0) = 0$;
(ii) $Q(\cdot, \cdot, t)$, for any $t \in \mathbb{R}_+$, is a sub-Markov kernel on $(E, \mathcal{E})$;
(iii) $P(\cdot, \cdot) = Q(\cdot, \cdot, \infty)$ is a Markov kernel on $(E, \mathcal{E})$.

For any fixed $x \in E$, the function $F_x(t) := Q(x, E, t)$ is a distribution function on $\mathbb{R}_+$. By Radon-Nikodym theorem, as $Q \ll P$ there exists a positive-valued function $F(x, y, \cdot)$, such that

$$Q(x, B, t) = \int_B F(x, y, t) P(x, dy), \quad B \in \mathcal{E}. \qquad (1.31)$$

We consider a special class of semi-Markov processes where $F(x, y, t)$ does not depend on the second argument y, we have $F(x, y, t) =: F_x(t)$. Nevertheless, any semi-Markov process can be transformed in one of the above kind, (see, e.g. [127]), by representing the semi-Markov kernel Q as follows

$$Q(x, B, t) = P(x, B)F_x(t). \tag{1.32}$$

Let us consider a $(E \times \mathbb{R}_+, \mathcal{E} \otimes \mathcal{B}_+)$-valued stochastic process $(x_n, \tau_n; n \geq 0)$, with $\tau_0 \leq \tau_1 \leq \cdots \leq \tau_n \leq \tau_{n+1} \leq \cdots$.

Definition 1.19 A *Markov renewal process* is a two component Markov chain, x_n, τ_n, $n \geq 0$, homogeneous with respect to the second component with transition probability defined by a semi-Markov kernel Q as follows,

$$\mathbb{P}(x_{n+1} \in B, \tau_{n+1} - \tau_n \leq t \mid \mathcal{F}_n) = \mathbb{P}(x_{n+1} \in B, \tau_{n+1} - \tau_n \leq t \mid x_n)$$
$$= Q(x_n, B, t) \quad (a.s.), \tag{1.33}$$

for any $n \geq 0$, $t \geq 0$, and $B \in \mathcal{B}_+$.

Let us define the counting process of jumps $\nu(t)$, $t \geq 0$, by

$$\nu(t) = \sup\{n \geq 0 : \tau_n \leq t\},$$

that gives the number of jumps of the Markov renewal process in the time interval $(0, t]$.

Definition 1.20 A stochastic process $x(t)$, $t \geq 0$, defined by the following relation

$$x(t) = x_{\nu(t)}, \quad t \geq 0,$$

is called a *semi-Markov process*, associated to the Markov renewal process x_n, τ_n, $n \geq 0$.

Remark 1.1. Markov jump processes are special cases of semi-Markov processes with semi-Markov kernel

$$Q(x, B, t) = P(x, B)[1 - e^{-q(x)t}].$$

Let $\theta_n := \tau_n - \tau_{n-1}$, that is $\tau_n = \tau_0 + \sum_{k=1}^n \theta_k$. The random variable $\theta_x, x \in E$, will denote the *sojourn time* in state x. The process $x_n, \theta_n, n \geq 0$, will be called also a *Markov renewal process*.

It is worth noticing that jump Markov and semi-Markov processes consider in this book are *regular*, that is [127,56],

$$\mathbb{P}(\nu(t) < \infty) = 1,$$

for every $t \geq 0$.

1.3.2 Markov Renewal Equation and Theorem

Let Q_1 and Q_2 be two semi-Markov kernels on $(E, \mathcal{E})$. Then their convolution, denoted by $Q_1 \star Q_2$, is defined by

$$(Q_1 \star Q_2)(x, B, t) = \int_E \int_0^t Q_1(x, dy, ds) Q_2(y, B, t-s), \quad (1.34)$$

where $x \in E$, $t \in \mathbb{R}_+$, $B \in \mathcal{E}$.

The function $Q_1 \star Q_2$ is also a semi-Markov kernel. For a semi-Markov kernel Q on $(E, \mathcal{E})$, we set by induction

$$Q^1 = Q, \quad Q^{n+1} = Q \star Q^n, \quad n \geq 0, \quad (1.35)$$

and

$$Q^0(x, B, t) = \begin{cases} 0 & \text{if } t \leq 0 \\ 1_B(x) & \text{if } t > 0. \end{cases} \quad (1.36)$$

We prove easily that

$$Q^{m+n} = Q^m \star Q^n. \quad (1.37)$$

Note that

$$Q^n(x, B, t) = \mathbb{P}(x_n \in B, \tau_n \leq t \mid x_0 = x).$$

Let us now consider two real-valued functions $U(x, t)$ and $V(x, t)$ defined on $E \times \mathbb{R}_+$.

Definition 1.21 The *Markov renewal equation* is defined as follows

$$U(x, t) - \int_E \int_0^t Q(x, dy, ds) U(y, t-s) = V(x, t), \quad x \in E, \quad (1.38)$$

where U is an unknown function and V a given function.

The above Markov renewal equation can be also written as follows

$$U(t) - \int_0^t Q(ds) U(t-s) = V(t),$$

which is the usual form in the scalar case of the classical renewal equation on the half-real line $t \geq 0$.

By using convolution $\star$, this equation can be written as follows

$$[I - Q] \star U = V \quad \text{or} \quad U = V + Q \star U,$$

where I is the identity operator: $I \star U = U \star I = U$.

Theorem 1.3 *(Markov renewal theorem [159]) Let the following conditions hold:*

C1: *the stochastic kernel $P(x, B) = Q(x, B, \infty)$ induces an irreducible ergodic Markov chain with the stationary distribution ρ,*

C2: *the mean sojourn times are uniformly bounded, that is:*

$$m(x) := \int_0^\infty \overline{F}_x(t)dt \leq C < +\infty,$$

and

$$m := \int_E \rho(dx) m(x) > 0,$$

C3: *the distribution functions $F_x(t) := Q(x, E, t)$, $x \in E$, are non arithmetic (that is, not concentrated on a set $\{na : n \in \mathbb{N}\}$, where $a > 0$, is a constant; the largest a is called the span of distribution).*

C4: *the nonnegative function $V(x,t)$ is direct Riemann integrable[47] on $\mathbb{R}_+$, so*

$$\int_E \rho(dx) \int_0^\infty V(x,t)dt < +\infty.$$

Then Equation (1.38) has a unique solution $U(x,t)$, and the following limit result holds

$$\lim_{t \to \infty} U(x,t) = \int_E \rho(dx) \int_0^\infty V(x,t)dt / m.$$

Let us apply the above theorem in order to obtain the limit distribution of the semi-Markov process.

The transition probabilities of the semi-Markov process

$$P_t(x, B) = \mathbb{P}(x(t) \in B \mid x(0) = x), \tag{1.39}$$

satisfy the following Markov renewal equation

$$P_t(x, B) - \int_E \int_0^t Q(x, dy, ds) P_{t-s}(y, B) = \mathbf{1}_B(x) \overline{F}_x(t). \tag{1.40}$$

1.3. SEMI-MARKOV PROCESSES

Now, applying the renewal limit theorem to the above equation, we get

$$\pi(B) := \lim_{t\to\infty} P_t(x, B) = \int_B \rho(dx) m(x)/m, \quad B \in \mathcal{E}.$$

The *limit distribution* $\pi(B)$ is said to be the *stationary distribution* of the semi-Markov process $x(t)$.

1.3.3 Auxiliary Processes

The following auxiliary processes will be used:

$$\tau(t) = \tau_{\nu(t)}, \quad \tau_+(t) = \tau_{\nu_+(t)}, \quad \nu_+(t) = \nu(t) + 1,$$
$$\gamma(t) = t - \tau(t), \quad \gamma_+(t) = \tau_+(t) - t. \tag{1.41}$$

The distributions of the auxiliary processes (1.41) satisfy the Markov renewal equation (1.38) with certain function $V(x,t)$. Let us consider the distribution function of the remaining sojourn time $\gamma_+(t)$, which will be used in the phase merging principle (see Chapter 4),

$$\Phi_x(u, t) := \mathbb{P}(\gamma_+(t) \le u \mid x(0) = x).$$

The right-hand side of the Markov renewal equation (1.38) is calculated as follows:

$$\begin{aligned} V_x(u, t) :&= \mathbb{E}[\mathbf{1}_{(\gamma_+(t) \le u, \tau_1 > t)} \mid x(0) = x] \\ &= \mathbb{E}[\mathbf{1}_{(\tau_1 - t \le u, \tau_1 > t)} \mid x(0) = x] \\ &= \mathbb{E}[\mathbf{1}_{(t < \tau_1 \le t + u)} \mid x(0) = x] \\ &= F_x(t + u) - F_x(t) \\ &= \overline{F}_x(t) - \overline{F}_x(t + u). \end{aligned}$$

Now, the Markov renewal theorem yields the following limit result:

$$\begin{aligned} \lim_{t\to\infty} \Phi_x(u, t) &= \int_E \rho(dx) \int_0^\infty [\overline{F}_x(t) - \overline{F}_x(t + u)] dt/m \\ &= \int_E \rho(dx) \left[\int_0^\infty \overline{F}_x(t) dt - \int_u^\infty \overline{F}_x(t) dt \right]/m \\ &= \int_E \rho(dx) \int_0^u \overline{F}_x(t) dt/m \\ &= \int_0^u \overline{F}(t) dt/m, \end{aligned}$$

where, by definition,

$$\overline{F}(t) := \int_E \rho(dx)\overline{F}_x(t).$$

It is worth noticing that the limit remaining sojourn time γ_+ distribution, naturally called the *stationary renewal time distribution*, is defined by

$$F_+(u) := \mathbb{P}(\gamma_+ \leq u) = \int_0^u \overline{F}(t)dt/m,$$

with density

$$f_+(u) = \overline{F}(u)/m.$$

1.3.4 Compensating Operators

The compensating operator is a basic important device in our analysis of stochastic systems switched by semi-Markov processes.

Let us consider a Markov renewal process

$$x_n, \quad \tau_n, \quad n \geq 0, \tag{1.42}$$

where $x_n = x(\tau_n)$, and $\tau_{n+1} = \tau_n + \theta_{n+1}$, $n \geq 0$, and

$$\mathbb{P}(\theta_{n+1} \leq t \mid x_n = x) = F_x(t) = \mathbb{P}(\theta_x \leq t).$$

Definition 1.22 ([174]) The *compensating operator* $\mathbb{L}$ of the Markov renewal process (1.42) is defined by the following relation

$$\mathbb{L}\varphi(x_0, \tau_0) = q(x_0)\mathbb{E}[\varphi(x_1, \tau_1) - \varphi(x_0, \tau_0) \mid \mathcal{F}_0], \tag{1.43}$$

where $q(x) = 1/m(x)$, $m(x) = \mathbb{E}\theta_x = \int_0^\infty \overline{F}_x(t)dt$, and

$$\mathcal{F}_t := \sigma(x(s), \tau(s); \, 0 \leq s \leq t).$$

Of course, by the homogeneity property of the Markov renewal process, we have

$$q(x)\mathbb{E}[\varphi(x_{n+1}, \tau_{n+1}) - \varphi(x_n, \tau_n) \mid \mathcal{F}_n] = \mathbb{L}\varphi(x_n, \tau_n). \tag{1.44}$$

Proposition 1.3 *The compensating operator (1.43) of the Markov renewal process (1.42) can be defined by the relation*

$$\mathbb{L}\varphi(x,t) = q(x) \int_0^\infty \int_E Q(x,dy,ds)\left[\varphi(y,t+s) - \varphi(x,t)\right]$$
$$= q(x)\left[\int_0^\infty F_x(ds) \int_E P(x,dy)\varphi(y,t+s) - \varphi(x,t)\right]. \quad (1.45)$$

If $q(x) = 0$, then $\mathbb{L}\varphi(x,t) = 0$.

The claim of Proposition 1.3 follows directly from Definition 1.22.

1.3.5 Martingale Characterization of Markov Renewal Processes

Let $x(t), t \geq 0$, be a semi-Markov process and let $x_n, \tau_n, n \geq 0$, be the corresponding Markov renewal process, and $\mathbb{L}$ be the compensating operator. Define the process $\xi_n, n \geq 0$, by

$$\xi_n := \varphi(x_n, \tau_n) - \sum_{i=1}^n (\tau_i - \tau_{i-1})\mathbb{L}\varphi(x_{i-1}, \tau_{i-1}), \quad n \geq 0,$$

and $\mathcal{G}_n := \sigma(x_k, \tau_k; k \leq n), n \geq 0$.

Proposition 1.4 ([174]) *The process $\xi_n, n \geq 0$, is a $\mathcal{G}_n$-martingale sequence for any function $\varphi \in \mathbf{B}(E \times \mathbb{R}_+)$, such that $\mathbb{E}_x |\varphi(x_1, \tau_1)| < +\infty$.*

1.4 Semimartingales

Let us consider an adapted stochastic process $x(t)$, $t \geq 0$, on the stochastic basis $\mathfrak{F}$, with trajectories in the space $\mathbf{D}[0, +\infty)$. Set $\mathbb{R}_0 = \mathbb{R} \setminus \{0\}$.

Definition 1.23 ([70]) *The stochastic process $x(t)$, $t \geq 0$, is said to be:*

(1) *a* semimartingale, *if it has the following representation*

$$x(t) = x_0 + \mu(t) + \alpha(t), \quad t \geq 0, \quad (1.46)$$

where $x_0 = x(0)$ is a finite $\mathcal{F}_0$-measurable random variable, $\mu(t)$ is a local martingale, with $\mu_0 = 0$, and $\alpha(t)$ is a bounded variation process, with $\alpha(0) = 0$;

(2) *a* special semimartingale, *if it has the representation (1.46) with $\alpha(t)$ a predictable process.*[70]

Then we note $x \in \mathcal{S}$ and $x \in \mathcal{S}_p$ respectively.

The representation (1.46), for a special semimartingale is unique. A semimartingale with bounded jumps is a special semimartingale. While the representation (1.46) for a semimartingale is not unique, the continuous martingale part is unique.

Let h be a *truncation function*, that is, $h : \mathbb{R}_0^d \to \mathbb{R}_0^d$, bounded with compact support and $h(x) = x$ in a neighborhood of 0. Let $x(t)$, $t \geq 0$, be a d-dimensional semimartingale.

For a fixed truncation function h, let us consider the processes:

$$\check{x}_h(t) = \sum_{s \leq t}[\Delta x(s) - h(\Delta x(s))] = \int_0^t \int_{\mathbb{R}_0} (u - h(u))\mu(ds, du), \quad t \geq 0$$

$$x_h(t) = x(t) - \check{x}_h(t), \quad t \geq 0,$$

where $\Delta x(s) := x(s) - x(s-)$, and $\mu(ds, du)$ is the measure of jumps of $x(t)$.

Since $\check{x}_h(t)$ is of bounded variation, the process $x_h(t)$ is a semimartingale and has bounded jumps, consequently, it is a special semimartingale, and has the canonical representation

$$x_h(t) = x_0 + M_h(t) + B_h(t), \quad t \geq 0, \tag{1.47}$$

where $M_h(t)$ is a local martingale, with $M_h(0) = 0$ and $B_h(t)$ is a predictable process of bounded variation.

Definition 1.24 ([70]) For a fixed truncation function h, we call a triple of *predictable characteristics*, with respect to h, the triplet $T = (B, C, \nu)$, of which the semimartingale $x(t)$, $t \geq 0$, has the following representation [70,132]

$$x(t) = x_0 + B_h(t) + x^c(t) + \int_0^t \int_{\mathbb{R}_0} h(u)(\mu(ds, du) - \nu(ds, du))$$
$$+ \int_0^t \int_{\mathbb{R}^*} (u - h(u))\mu(ds, du),$$

where:

- $B = B_h$, is the predictable process in (1.47),
- $C = (c^{ij})$ is a continuous process of bounded variation, with $c^{ij} = <x^{ic}, x^{jc}>$, which is the predictable process of (x^{ic}, x^{jc}), that is, the process $x^{ic}x^{jc} - <x^{ic}, x^{jc}>$ is a local martingale;

1.4. SEMIMARTINGALES

- ν is the *compensator of the measure* μ of jumps of $x(t)$, that is a predictable measure on $\mathbb{R}_+ \times \mathbb{R}_0^d$.

It is convenient to introduce the second modified characteristic $\tilde{C} = (\tilde{c}^{ij})$, by

$$(\tilde{c}^{ij}) = \ll M_h^i, M_h^j \gg.$$

We will use the semimartingales as a tool in order to establish Poisson approximation results (see Chapter 7).

▷ **Example 1.7.** *Brownian motion.* Let $w(t), t \geq 0$ be a Wiener process with $w(0) = 0$. This is a local martingale with $\langle w, w \rangle_t = \sigma^2(t)$. Its predictable characteristics are $(B, C, \nu) = (0, \sigma^2(t), 0)$.

▷ **Example 1.8.** *Gaussian process.* Let $x(t), t \geq 0$, be a Gaussian process. We have $(B, C, \nu) = (\mathbb{E}x(t), \mathbb{E}(x(t) - \mathbb{E}x(t))^2, 0)$.

▷ **Example 1.9.** *Generalized diffusion* ([62]). Let us consider Borel functions $a \geq 0$ and b defined on $\mathbb{R}_+ \times \mathbb{R}$, and a family of transition kernels K_t, $t \geq 0$, on $(\mathbb{R}, \mathcal{B})$, satisfying the following conditions:

$$K_t(x, \{0\}) = 0,$$
$$\int_{\mathbb{R}} (1 \wedge y)^2 K_t(x, dy) < +\infty.$$

Let $x(t), t \geq 0$, be a semimartingale with predictable characteristics (B, C, ν) given by:

$$B_h(t) = \int_0^t b(s, x(s)) ds,$$
$$C(t) = \int_0^t a(s, x(s)) ds,$$
$$\nu(dt, dx) = K_t(x(t), dx) dt.$$

In that case, the semimartingale $x(t), t \geq 0$, is said to be a *generalized diffusion*. If $a(t, x)$, $b(t, x)$ and $K_t(x, B)$ do not depend upon t, them it is called a *time-homogeneous generalized diffusion* (compare with PLII).

For a time-homogeneous generalized diffusion $x(t), t \geq 0$, the infinitesimal generator $\mathbb{L}$ acts on functions $\varphi \in C^2(\mathbb{R})$, as follows

$$\mathbb{L}\varphi(x) = b(x)\varphi'(x) + \frac{1}{2}a(x)\varphi''(x)$$
$$+ \int_{\mathbb{R}} K(x, dy)[\varphi(x+y) - \varphi(x) - h(y)\varphi'(y)]. \quad (1.48)$$

The triplet (b, a, K) is called the infinitesimal characteristics of the generalized time-homogeneous diffusion.

▷ **Example 1.10.** *Processes with stationary independent increments.* For the processes with stationary independent increments given in Section 1.2.4, with cumulant function $\psi(\lambda)$, given in (1.16), we have $(B_t, C_t, \nu_t(dx)) = (at, \sigma^2 t, tH(dx))$.

1.5 Counting Markov Renewal Processes

In this section, we consider counting processes as semimartingales.

Let $x_n, \theta_n, n \geq 0$, be a Markov renewal process taking values in $E \times [0, +\infty)$, and defined by the semi-Markov kernel

$$Q(x, B, t) = P(x, B)F_x(t).$$

So, the components x_{n+1} and θ_{n+1} are conditionally independent

$$\mathbb{P}(x_{n+1} \in B, \theta_{n+1} \leq t \mid x_n = x)$$
$$= \mathbb{P}(x_{n+1} \in B \mid x_n = x)\mathbb{P}(\theta_{n+1} \leq t \mid x_n = x)$$
$$= P(x, B)F_x(t).$$

The renewal moments are defined by

$$\tau_n = \sum_{k=1}^{n} \theta_k, \quad n \geq 1, \quad \tau_0 = 0.$$

The counting process is defined by

$$\nu(t) = \max\{n \geq 1 : \tau_n \leq t\}.$$

Definition 1.25 (see, e.g. [21,70,133]) An integer-valued *random measure*

1.5. COUNTING MARKOV RENEWAL PROCESSES

for the Markov renewal process $x_n, \tau_n, n \geq 0$, is defined by the relation

$$\mu(dx, dt) = \sum_{n \geq 1} \delta_{(x_n, \tau_n)}(dx, dt) \mathbf{1}_{(\tau_n < +\infty)}, \quad (1.49)$$

where δ_a is the *Dirac measure* concentrated at point a.

It is worth noticing that the random measure (1.49) defines the multivariate point process $x_n, \tau_n, n \geq 0$, (see, e.g. [70]). By Theorem III.1.26 [70], there exists a unique predictable random measure $\bar{\nu}(dx, dt)$ which is the compensator of the measure $\mu(dx, dt)$, that is, for any nonnegative continuous function $w(x, t)$,

$$\int_0^t \int_E w(x, s)[\mu(dx, dt) - \bar{\nu}(dx, ds)]$$

is a local martingale (Section 1.2.6) with respect to the natural filtration

$$\mathcal{F}_t^x := \sigma(x(s), s \leq t), \quad t \geq 0, \quad (1.50)$$

of the corresponding semi-Markov process $x(t), t \geq 0$.

Certainly, there exists a unique predictable random measure $\bar{\nu}(t)$ which is a compensator of the counting process $\nu(t), t \geq 0$, that is, the process $\mu(t) = \nu(t) - \bar{\nu}(t), t \geq 0$, is a local martingale with respect to the filtration (1.50). Note that

$$\bar{\nu}(dt) = \int_E \bar{\nu}(dx, dt). \quad (1.51)$$

Moreover, it is possible to give the constructive representation of the compensator (1.51) for the Markov renewal process, at any rate, where the family of distributions $F_x(t), x \in E$, is absolutely continuous, with respect to the Lebesgue measure on $\mathbb{R}_+$, and has the following representation

$$\bar{F}_x(t) := 1 - F_x(t) = \exp[-\Lambda(x,t)], \quad \Lambda(x,t) = \int_0^t \lambda(x,s)ds. \quad (1.52)$$

Proposition 1.5 ([118]) *The compensator $\bar{\nu}(t), t \geq 0$, for the counting process $\nu(t), t \geq 0$, of a Markov renewal process can be represented as follows*

$$\bar{\nu}(t) = \int_0^t \lambda(x(s), \gamma(s))ds,$$

where $\gamma(s) := s - \tau(s), \tau(s) := \tau_{\nu(s)}$.

It is worth noticing that the compensator of the counting Markov renewal process is a *stochastic integral functional* of the Markov process $x(t), \gamma(t), t \geq 0$ (see Section 2.2).

PROOF. Introduce the conditional distributions of the Markov renewal process $x_n, \tau_n, n \geq 0$:

$$F_n(dx, dt) = \mathbb{P}(x_{n+1} \in dx, \tau_{n+1} \in dt \mid \mathcal{F}_n) \qquad (1.53)$$
$$= \mathbb{P}(x_{n+1} \in dx, \tau_{n+1} \in dt \mid x_n, \tau_n)$$
$$= P(x_n, dx) F_{x_n}(dt - \tau_n). \qquad (1.54)$$

By Theorem III.1.33 [70], the compensating measure of the multivariate point process (1.49) can be represented as follows

$$\bar{\nu}(dx, dt) = \sum_{n \geq 0} \mathbf{1}_{(\tau_n < t \leq \tau_{n+1})} F_n(dx, dt) / \overline{H}_n(t),$$

where

$$\overline{H}_n(t) = \int_t^\infty \int_E F_n(dx, ds) = \overline{F}_{x_n}(t - \tau_n).$$

Therefore, the compensator of the counting process $\nu(t), t \geq 0$, is represented by

$$\bar{\nu}(t) = \sum_{n \geq 0} \int_{\tau_n}^{t \wedge \tau_{n+1}} H_n(ds) / \overline{H}_n(s).$$

Now, calculate

$$\int_{\tau_n}^{\tau_{n+1}} H_n(ds) / \overline{H}_n(s) = \int_{\tau_n}^{\tau_{n+1}} F_{x_n}(ds - \tau_n) / \overline{F}_{x_n}(s)$$
$$= \int_0^{\theta_{n+1}} F_{x_n}(ds) / \overline{F}_{x_n}(s)$$
$$= \int_0^{\theta_{n+1}} \lambda(x_n, s) ds, \quad \text{(by using (1.52))}$$
$$= \Lambda(x_n, \theta_{n+1}).$$

Similarly, for $\tau_n < t \leq \tau_{n+1}$, we get

$$\int_{\tau_n}^t H_n(ds) / \overline{H}_n(s) = \Lambda(x_n, t - \tau_n) = \Lambda(x_n, \gamma(t)).$$

Hence,

$$\bar{\nu}(t) = \sum_{n=0}^{\nu(t)} \Lambda(x_n, \theta_{n+1}) + \Lambda(x(t), \gamma(t))$$
$$= \int_0^t \lambda(x(s), \gamma(s))ds.$$

□

Corollary 1.2 *The compensator of the counting process for a Markov jump process with the intensity function $q(x), x \in E$, is represented as follows*

$$\bar{\nu}(t) = \int_0^t q(x(s))ds.$$

1.6 Reducible-Invertible Operators

Let **B** be the Banach space of real-valued measurable functions, with $\|\cdot\|$ the sup-norm, defined on the state space E.

Let $Q \colon \mathbf{B} \to \mathbf{B}$ be a linear operator acting on **B**, and denote by

$$\mathcal{D}_Q := \{\varphi \colon \varphi \in \mathbf{B}, Q\varphi \in \mathbf{B}\}, \quad \text{the domain of } \mathbf{Q},$$

by

$$\mathcal{R}_Q := \{\psi \colon \psi = Q\varphi, \varphi \in \mathbf{B}\}, \quad \text{the range of } \mathbf{Q},$$

and by

$$\mathcal{N}_Q := \{\varphi \colon Q\varphi = 0, \varphi \in \mathbf{B}\}, \quad \text{the null-space kernel of } \mathbf{Q}.$$

An operator Q is said to be bounded if there exist constants $C > 0$, such that

$$\|Q\varphi\| \leq C \|\varphi\|, \quad \varphi \in \mathcal{D}_Q.$$

The least of these constants is called the norm of the operator Q, and is denoted by $\|Q\|$. The operator norm is

$$\|Q\| = \sup_{\varphi \in \mathbf{B}} \frac{\|Q\varphi\|}{\|\varphi\|}. \tag{1.55}$$

Let $Q\colon \mathbf{B} \to \mathbf{B}$ be a linear operator that maps $\mathcal{D}_Q$ to $\mathcal{R}_Q$ one-to-one. Thus a linear operator Q^{-1} is defined as a map $\mathcal{R}_Q$ onto $\mathcal{D}_Q$, which satisfies the following conditions:

$$Q^{-1}Q\varphi = \varphi, \quad \varphi \in \mathcal{D}_Q, \quad QQ^{-1}\psi = \psi, \quad \psi \in \mathcal{R}_Q.$$

The operator Q^{-1} is defined uniquely and is called an *inverse operator*.

Definition 1.26 ([116])

(1) An operator Q is said to be *densely defined* if its domain of definition is dense in $\mathbf{B}$, that is, $\overline{\mathcal{D}_Q} = \mathbf{B}$, ($\overline{\mathcal{D}_Q}$ is the closure of $\mathcal{D}_Q$).

(2) An operator Q is said to be *closed* if for every convergent sequence $x_n \to x$, and $Qx_n \to y$, as $n \to \infty$, it follows that $x \in \mathcal{D}_Q$ and $Qx = y$.

(3) A bounded linear operator Q is said to be *reducible-invertible* if the Banach space $\mathbf{B}$ can be decomposed in a direct sum of two subspaces, that is

$$\mathbf{B} = \mathcal{N}_Q \oplus \mathcal{R}_Q, \qquad (1.56)$$

where the null-space has nontrivial dimension, $\dim \mathcal{N}_Q \geq 1$.

(4) A densely defined operator $Q\colon \mathbf{B} \to \mathbf{B}$ is said to be *normally solvable* if its range of values $\mathcal{R}_Q$ is closed.

Remark 1.2.
1) A reducible-invertible operator is normally solvable.
2) The decomposition (1.56) generates the *projector* Π on the subspace $\mathcal{N}_Q$

$$\Pi\varphi := \begin{cases} \varphi, & \varphi \in \mathcal{N}_Q, \\ 0, & \varphi \in \mathcal{R}_Q. \end{cases}$$

The operator $I - \Pi$ is the projector on the subspace $\mathcal{R}_Q$

$$(I - \Pi)\varphi := \begin{cases} 0, & \varphi \in \mathcal{N}_Q, \\ \varphi, & \varphi \in \mathcal{R}_Q. \end{cases}$$

where I is the *identity operator* in $\mathbf{B}$.

Lemma 1.3 ([100]) *If the linear operator Q is normally resolvable, then the operator $Q + \Pi$ has an inverse.*

1.6. REDUCIBLE-INVERTIBLE OPERATORS

PROOF. Applying the projector Π to both sides of the equation

$$[Q + \Pi]\varphi = \psi, \tag{1.57}$$

and since $\Pi Q\varphi = Q\Pi\varphi = 0$, we get

$$\Pi\varphi = \Pi\psi.$$

On the other hand, rewriting (1.57), we have

$$Q\varphi = [I - \Pi]\psi \in \mathcal{R}_Q.$$

This equation, as Q is a normally solvable operator, has a solution which is the solution of (1.57). □

Definition 1.27 ([116]) Let Q be a reducible-invertible operator. The operator

$$R_0 := \Pi - (Q + \Pi)^{-1} \tag{1.58}$$

is called the *potential operator* or the *potential* of Q.

It is easy to check that the potential can also written as follows

$$R_0 = (\Pi - Q)^{-1} - \Pi. \tag{1.59}$$

Proposition 1.6 *The following equalities hold:*

$$QR_0 = R_0 Q = \Pi - I, \tag{1.60}$$

$$\Pi R_0 = R_0 \Pi = 0, \tag{1.61}$$

$$QR_0^n = R_0^n Q = R_0^{n-1}, \quad n \geq 1, \tag{1.62}$$

$$\|R_0\| = \|Q_0^{-1}\|. \tag{1.63}$$

Equation (1.60) is called *Poisson equation*. The right hand side of this equation is sometimes defined as $I - \Pi$ (see, e.g. [116]).

Proposition 1.7 *Let $Q: \mathbf{B} \to \mathbf{B}$ be a reducible-invertible operator. Then the equation*

$$Q\varphi = \psi, \tag{1.64}$$

under the solvability condition $\Pi\psi = 0$, has a general solution with representation

$$\varphi = -R_0\psi + \varphi_0, \quad \varphi_0 \in \mathcal{N}_Q.$$

If moreover the condition $\Pi\varphi = 0$ holds, Equation (1.64) has a unique solution represented by

$$\varphi = -R_0\psi. \tag{1.65}$$

For a *uniformly ergodic Markov chain*, the operator $Q := P - I$ (where I is the identity operator), is *reducible-invertible* [100], that is,

$$\mathbf{B} = \mathcal{N}_Q \oplus \mathcal{R}_Q,$$

with $dim\mathcal{N}_Q = 1$.

For a uniformly ergodic Markov process with generator Q, and semigroup $P_t, t \geq 0$, the potential R_0 is a bounded operator defined by

$$R_0 = \int_0^\infty (P_t - \Pi)dt, \tag{1.66}$$

where the projector operator Π is defined as follows

$$\Pi\varphi(u) = \int_E \rho(dx)\varphi(x)\mathbf{1}(x),$$

with $\rho(B), B \in \mathcal{E}$, the stationary distribution of the Markov chain and the indicator function $\mathbf{1}(x) = 1$, for any $x \in E$.

Definition 1.28 Let $Q_\varepsilon \colon \mathbf{B} \to \mathbf{B}$, $\varepsilon > 0$, be a family of linear operators. We say that it *converges in the strong sense*, as $\varepsilon \to 0$, to the operator Q, if

$$\lim_{\varepsilon \to 0} \|Q_\varepsilon \varphi - Q\varphi\| = 0 \quad \varphi \in \mathbf{B}.$$

And we note $s - \lim_{\varepsilon \to 0} Q_\varepsilon = Q$.

Chapter 2

Stochastic Systems with Switching

2.1 Introduction

This chapter deals with *switched-switching processes*. When real systems are studied an important problem arise, connected to the fact that the local characteristics of the systems are not fixed but depend upon random factors. Concerning this problem, one describes the random changes of local characteristics by a stochastic process, called a *switching process*, or *driving* or *modulation process* [2,3]. In applications, switching processes could represent the environment [100,146], or, in the particular case of dynamic reliability, they represent the structure of the system [36].

Usually, the switching process is assumed to be an ergodic process. Nevertheless, in many practical problems switching non ergodic stochastic processes have to be considered, for example when the system is observed up to the hitting time to some subset of the state space.

Switched processes were first considered by Ezhov and Skorokhod [46]. Starting with a basic process, say $\eta(t;x), t \geq 0, x \in E$, where x is a parameter, one considers an additional process, say $x(t), t \geq 0$, with values in the set E. Then one considers the composed process [169] $\eta(t;x(t)), t \geq 0$, meaning that the process $\eta(t;x)$ depends on the current state of the process $x(t)$ in time t. The process $\eta(t;x)$ is said to be a *switched process*, and $x(t)$ a *switching process*. The switched process is also called a *modulated process* or a *process driven* by the process $x(t)$.

Such a composition of stochastic processes is very useful for studying real problems. A switching process can represent the variation of environment, the structure evolution of systems, etc. A switched process can represent the level of performance for stochastic systems, the flux of materials or information, the parameter values, as temperature, pressure, etc. A very large literature exists concerning this kind of problems. It includes additive stochastic functionals, stopped processes, hidden processes in statistics, etc. In our case, the switching process $x(t)$ will be a Markov or a semi-Markov process.

The present chapter includes stochastic evolutionary systems, stochastic additive functionals, increment processes, Markov additive semimartingales, impulsive processes, and finally, *random evolution* which generalize all the previous processes.

The main object of this chapter is to introduce generators and compensating operators of the coupled switched-switching processes, which are the main tools for the proofs of weak convergence presented in Chapters 3-6.

2.2 Stochastic Integral Functionals

A stochastic model of systems with switching is constructed using a real-valued measurable bounded function $a(x), x \in E$, and a regular cadlag semi-Markov process $x(t), t \geq 0$, on the measurable space $(E, \mathcal{E})$, given by a semi-Markov kernel (see Section 1.3.1)

$$Q(x, B, t) = \mathbb{P}(x_{n+1} \in B, \theta_{n+1} \leq t \mid x_n = x) = P(x, B) F_x(t),$$

for $x \in E$, $B \in \mathcal{E}$, $t \geq 0$.

Definition 2.1 A *Stochastic integral functional* with semi-Markov is represented by

$$U(t) = u + \int_0^t a(x(s)) ds, \quad t \geq 0, u \in \mathbb{R}. \tag{2.1}$$

Using the Markov renewal process $x_n, \tau_n, n \geq 0$ (see Section 1.3.1), the stochastic integral functional (2.1) can be represented as follows

$$U(t) = u + \sum_{k=0}^{\nu(t)-1} a(x_k) \theta_{k+1} + \gamma(t) a(x(t)), \quad t \geq 0, u \in \mathbb{R}, \tag{2.2}$$

where $\gamma(t) := t - \tau(t)$ and $\tau(t) := \tau_{\nu(t)}$.

Let us introduce the family of associated evolutionary equations

$$\begin{vmatrix} \frac{d}{dt} U_x(t; u_0) = a(x), & x \in E, \\ U_x(0; u_0) = u_0. \end{vmatrix} \tag{2.3}$$

2.2. STOCHASTIC INTEGRAL FUNCTIONALS

The trajectories of (2.3) are

$$U_x(t; u_0) = u_0 + a(x)t, \quad t \geq 0, \tag{2.4}$$

and satisfy the semigroup property

$$U_x(t + t'; u_0) = U_x(t; U_x(t'; u_0)). \tag{2.5}$$

This property can be described by the family of semigroup operators

$$A_t(x)\varphi(u) := \varphi(U_x(t; u)), \quad t \geq 0, x \in E, \tag{2.6}$$

defined on the Banach space $C(\mathbb{R})$ of real-valued bounded continuous functions $\varphi(u), u \in \mathbb{R}$, with the sup norm $\|\varphi\| := \sup_{u \in \mathbb{R}} |\varphi(u)|$.

The semigroup property of the operators in (2.6) take the form

$$A_{t+t'}(x) := A_t(x)A_{t'}(x), \quad t, t' \geq 0, x \in E, \tag{2.7}$$

which can be verified by using (2.5).

It is worth noticing that the semigroup (2.6) is continuous and contractive. That is, for any $x \in E$,

$$\lim_{t \to 0} A_t(x) = I, \quad \text{and} \quad \|A_t(x)\| \leq 1.$$

Hence, the generators (Section 1.2.2)

$$\mathbb{A}(x)\varphi(u) := \lim_{t \downarrow 0} t^{-1}[A_t(x) - I]\varphi(u), \quad x \in E, \tag{2.8}$$

exist on the test functions $\varphi \in C^1(\mathbb{R})$, the real-valued functions with bounded continuous first derivatives.

Lemma 2.1 *The generators (2.8) of the semigroups (2.6) are represented by*

$$\mathbb{A}(x)\varphi(u) = a(x)\varphi'(u). \tag{2.9}$$

PROOF. Indeed, by definitions (2.6) and (2.4), we get

$$\mathbb{A}(x)\varphi(u) = \lim_{t \downarrow 0} t^{-1}[\varphi(u + a(x)t) - \varphi(u)]$$
$$= a(x)\varphi'(u).$$

□

Relation (2.6) determines the evolution

$$\Phi_t^x(u) := A_t(x)\varphi(u) = \varphi(U_x(t; u)), \quad t \geq 0, x \in E, u \in \mathbb{R}, \tag{2.10}$$
$$\Phi_0^x(u) = \varphi(u),$$

in the Banach space $C(\mathbb{R})$. The operators $A_t(x)$ transform an initial test function $\varphi(u)$ into $\Phi_t^x(u)$. The evolution (2.10) can be determined by a solution of the evolution equation

$$\begin{cases} \frac{d}{dt}\Phi_t^x(u) = \mathbb{A}(x)\Phi_t^x(u), \\ \Phi_0^x(u) = \varphi(u), \end{cases} \qquad (2.11)$$

or, in another form (see (2.9))

$$\begin{cases} \frac{d}{dt}\Phi_t^x(u) = a(x)\frac{d}{du}\Phi_t^x(u), \\ \Phi_0^x(u) = \varphi(u). \end{cases} \qquad (2.12)$$

Definition 2.2 The *random evolution* for integral functional is defined on a test function $\varphi \in C(\mathbb{R})$ by the relation

$$\Phi_t(u) := \varphi(U(t)), \quad t \geq 0, u \in \mathbb{R}, \qquad (2.13)$$

where $U(t), t \geq 0$, is the integral functional (2.1).

Lemma 2.2 *The random evolution (2.13) can be represented in the following form*

$$\Phi_t(u) = A_{\gamma(t)}(x(t)) \prod_{k=0}^{\nu(t)-1} A_{\theta_{k+1}}(x_k)\varphi(u), \quad t \geq 0, u \in \mathbb{R}, \qquad (2.14)$$

and satisfies the evolution equation

$$\begin{cases} \frac{d}{dt}\Phi_t(u) = \mathbb{A}(x(t))\Phi_t(u), \quad t \geq 0, \\ \Phi_0(u) = \varphi(u). \end{cases}$$

Indeed, if $\nu(t) = 0$, then $\gamma(t) = t, x(t) = x$, hence in (2.14)

$$\Phi_t(u) = A_t(x)\varphi(u) = \varphi(U(t)).$$

Next, the formula (2.14) can be proved by induction using (2.2).

The characterization of the stochastic integral functional with semi-Markov switching is realized by using the *compensating operator* for the extended Markov renewal process

$$U_n := U(\tau_n), \quad x_n := x(\tau_n), \quad \tau_n, \quad n \geq 0, \qquad (2.15)$$

where $\tau_n, n \geq 0$, are the renewal jump times of the semi-Markov process $x(t), t \geq 0$.

2.2. STOCHASTIC INTEGRAL FUNCTIONALS

Definition 2.3 The *compensating operator* of the *extended Markov renewal process* (2.15) is defined by the relation

$$\mathbb{L}\varphi(u,x,t) = \mathbb{E}[\varphi(U_1,x_1,\tau_1) - \varphi(u,x,t) \mid U_0 = u, x_0 = x, \tau_0 = t]/m(x). \tag{2.16}$$

It is easy to verify that the compensating operator has the following homogeneous property

$$\mathbb{L}\varphi(u,x,t) = \mathbb{E}[\varphi(U_{n+1},x_{n+1},\tau_{n+1}) - \varphi(u,x,t) \mid U_n = u, x_n = x, \tau_n = t]/m(x),$$

where $m(x) := \mathbb{E}\theta_x = \int_0^\infty [1 - F_x(t)]dt$.

Lemma 2.3 *The compensating operator (2.16) can be represented in the following form*

$$\mathbb{L}\varphi(u,x,t) = q(x)\left[\int_0^\infty F_x(ds)A_s(x)\int_E P(x,dy)\varphi(u,y,t+s) - \varphi(u,x,t)\right],$$

where $A_s(x), s \geq 0, x \in E$, are the semigroups defined in (2.6) by the generators $\mathbb{A}(x), x \in E$ in (2.9), and $q(x) = 1/m(x)$.

The transformation of the compensating operator can be realized as follows. By definition the compensating operator, acting on functions $\varphi(u,x)$, is given by the following relation

$$\mathbb{L}\varphi(u,x) = q(x)[\int_0^\infty F_x(ds)A_s(x)\int_E P(x,dy)\varphi(u,y) - \varphi(u,x)],$$

or, in a symbolic form

$$\mathbb{L} = q[\mathbb{F}(x)P - I],$$

where

$$\mathbb{F}(x) := \int_0^\infty F_x(ds)A_s(x).$$

The first step of transformation is the following

$$\mathbb{L} = Q + [\mathbb{F}(x) - I]Q_0, \tag{2.17}$$

where

$$Q := q[P - I],$$

is the generator of the associated Markov process, and

$$Q_0\varphi(x) = q(x)\int_E P(x,dy)\varphi(y).$$

The second step of transformation consists in using the integral equation for the semigroup

$$A_s(x) - I = \mathbb{A}(x)\int_0^s A_v(x)dv.$$

For the second term in (2.17), we obtain

$$\mathbb{F}(x) - I = \int_0^\infty F_x(ds)[A_s(x) - I]$$
$$= \mathbb{A}(x)\int_0^\infty F_x(ds)\int_0^s A_v(x)dv$$
$$= \mathbb{A}(x)\int_0^\infty \overline{F}_x(s)A_s(x)ds.$$

So, we get the equivalent representation

$$\mathbb{F}(x) - I = \mathbb{A}(x)F^{(1)}(x),$$

where, by definition,

$$F^{(1)}(x) := \int_0^\infty \overline{F}_x(s)A_s(x)ds.$$

So doing, we have proven the following result.

Lemma 2.4 *The compensating operator of the extended Markov renewal process (2.15) is represented as follows*

$$\mathbb{L} = Q + \mathbb{A}(x)F^{(1)}(x)Q_0. \qquad (2.18)$$

2.3 Increment Processes

The discrete analogue of the integral functional considered in Section 2.2 is the increment process defined by the sum on the embedded Markov chain $x_n, n \geq 0$,

$$\alpha(t) := \sum_{k=1}^{\nu(t)} a(x_k) = \int_0^t a(x(s))d\nu(s), \quad t \geq 0, \qquad (2.19)$$

2.3. INCREMENT PROCESSES

with the given real-valued measurable bounded function $a(x), x \in E$. The counting process

$$\nu(t) := \max\{n \geq 0 : \tau_n \leq t\}, \quad t \geq 0, \tag{2.20}$$

is defined by the renewal moments $\tau_n, n \geq 0$, of the switching semi-Markov process $x(t), t > 0$.

Introduce the family of shift linear operators on the Banach space $C(\mathbb{R})$

$$\mathbb{D}(x)\varphi(u) = \varphi(u + a(x)), \quad x \in E, u \in \mathbb{R}. \tag{2.21}$$

Definition 2.4 The random evolution associated with the increment process $\alpha(t), t \geq 0$, is defined by the relation

$$\Phi_t(u) := \varphi(\alpha(t)), \quad \alpha(0) = u. \tag{2.22}$$

Clearly, the random evolution (2.22) can be represented in the following form

$$\Phi_t(u) = \prod_{k=1}^{\nu(t)} \mathbb{D}(x_k)\varphi(u), \quad t \geq 0, \quad \Phi_0(u) = \varphi(u). \tag{2.23}$$

Indeed, for $t < \tau_1$, by definition

$$\Phi_t(u) = \varphi(u).$$

Next, for $\tau_n \leq t < \tau_{n+1}$, from (2.23) and (2.21)

$$\Phi_t(u) = \Phi_{\tau_n}(u) = \prod_{k=1}^{n} \mathbb{D}(x_k)\varphi(u)$$

$$= \varphi(u + \sum_{k=1}^{n} a(x_k))$$

$$= \varphi(\alpha(\tau_n))$$

$$= \varphi(\alpha(t)),$$

that is Equation (2.22).

The recursive relation for the random evolution (2.23)

$$\Phi_{\tau_{n+1}}(u) - \Phi_{\tau_n}(u) = [\mathbb{D}(x_{n+1}) - I]\Phi_{\tau_n}(u), \tag{2.24}$$

provides the following additive representation of the random evolution (2.23)

$$\Phi_t(u) = \varphi(u) + \sum_{k=1}^{\nu(t)} [\mathbb{D}(x_k) - I]\Phi_{\tau_{k-1}}(u), \quad t \geq 0. \tag{2.25}$$

In what follows it will be useful to characterize the increment process by the generator of the coupled increment process

$$\alpha(t), \quad x(t), \quad t \geq 0. \tag{2.26}$$

Let the switching process $x(t), t \geq 0$, be Markovian and defined by the generator

$$Q\varphi(x) = q(x) \int_E P(x, dy)[\varphi(y) - \varphi(x)]. \tag{2.27}$$

Proposition 2.1 *The coupled increment process (2.26) is also Markovian and can be defined by the generator*

$$\mathbb{L}\varphi(u, x) = [Q + Q_0(\mathbb{D}(x) - I)]\varphi(u, x), \tag{2.28}$$

where $Q_0 = q(x) \int_E P(x, dy)\varphi(y)$.

Let the switching semi-Markov process $x(t), t \geq 0$, associated to the Markov renewal process $x_n, \tau_n, n \geq 0$, be given by the semi-Markov kernel

$$Q(x, B, t) = P(x, B)F_x(t), \quad x \in E, B \in \mathcal{E}, t \geq 0. \tag{2.29}$$

Introduce the extended Markov renewal process

$$\alpha_n := \alpha(\tau_n), \quad x_n, \quad \tau_n, \quad n \geq 0. \tag{2.30}$$

Proposition 2.2 *The compensating operator of the extended Markov renewal process (2.30) can be represented as follows*

$$\mathbb{L}\varphi(u, x, t) = q(x)\left[\int_0^\infty F_x(ds)\int_E P(x, dy)\mathbb{D}(y)\varphi(u, y, t+s) - \varphi(u, x, t)\right]$$

PROOF. Let $\mathbb{P}_{u,x,t}$ be the conditional probability on $(\alpha_0 = u, x_0 = x, \tau_0 = t)$, and $\mathbb{E}_{u,x,t}$ the corresponding expectation.

Then we have

$$\mathbb{E}_{u,x,t}[\varphi(\alpha_1, x_1, \tau_1)] =$$
$$\int_{\mathbb{R}\times E\times \mathbb{R}_+} \mathbb{P}_{u,x,t}(a(x_1) \in dv, x_1 \in dy, \theta_1 \in ds)\varphi(u+v, y, t+s).$$

But

$$\mathbb{P}_{u,x,t}(a(x_1) \in dv, x_1 \in dy, \theta_1 \in ds) = \delta_{a(y)}(dv)P(x, dy)F_x(ds).$$

So,

$$\mathbb{E}_{u,x,t}[\varphi(\alpha_1, x_1, \tau_1)] = \int_0^t F_x(ds) \int_E P(x, dy)\mathbb{D}(y)\varphi(u, y, t+s),$$

and the conclusion follows from Definition 2.3. □

It is easy to verify the following result.

Corollary 2.1 *The compensating operator* $\mathbb{L}$ *acts on test functions* $\varphi(u, x)$ *as follows*

$$\mathbb{L}\varphi(u, x) = [Q + Q_0(\mathbb{D}(x) - I)]\varphi(u, x).$$

(Compare with (2.28)).

2.4 Stochastic Evolutionary Systems

Various stochastic systems can be described by evolutionary processes with Markov or semi-Markov switching.

Definition 2.5 *The* evolutionary switched process $U(t)$, $t \geq 0$, *in* $\mathbb{R}^d$, *is defined as a solution of the evolutionary equation*

$$\begin{vmatrix} \frac{d}{dt}U(t) = a(U(t); x(t)), \\ U(0) = u. \end{vmatrix} \qquad (2.31)$$

The local velocity is given by the $\mathbb{R}^d$-valued continuous function $a(u; x)$, $u \in \mathbb{R}^d, x \in E$. The switching regular semi-Markov process $x(t), t \geq 0$, is considered in the standard phase space $(E, \mathcal{E})$, given by the semi-Markov kernel (see Section 1.3.1)

$$Q(x, B, t) = P(x, B)F_x(t).$$

The integral form of the evolutionary equation is

$$U(t) = u + \int_0^t a(U(s); x(s))ds. \tag{2.32}$$

In what follows, it is assumed that the velocity $a(u; x)$ satisfies the condition of the unique global solvability of the deterministic problems (2.33). That is Lipschitz condition on $u \in \mathbb{R}^d$, with a constant which is independent of $x \in E$. In order to emphasize the dependence of $U(t)$ on the initial condition u, let us write

$$\left| \begin{array}{l} \frac{d}{dt} U(t; x, u) = a(U(t; x, u); x), \\ U(0; x, u) = u, \quad \text{for all } x \in E. \end{array} \right. \tag{2.33}$$

The well-posedness of the stochastic process $U(t)$, $t \geq 0$, by the solution of Equation (2.31) follows from the fact that this solution can be represented in the following recursive form by using the solution of Problem (2.33)

$$U(t) = U(\gamma(t); x(t), U(\tau(t))), \quad t \geq 0. \tag{2.34}$$

The initial values for Problem (2.33) are defined by the following recursive relation

$$U(\tau_{n+1}) = U(\theta_{n+1}; x_n, U(\tau_n)), \quad n \geq 0. \tag{2.35}$$

The recursive relation (2.34) can be represented in the following form

$$U(t) = U(t - \tau_n; x_n, U(\tau_n)), \quad \tau_n \leq t < \tau_{n+1}, \quad n \geq 0.$$

The existence of a global solution $U(t)$, $t \geq 0$, for arbitrary time-interval $[0, T]$ follows from the regular property of the switching semi-Markov process $x(t)$, $t \geq 0$ (see Section 1.3).

It is well known (see, e.g. [100]) that the solution of the deterministic problem (2.33) under fixed value of $x \in E$ has a semigroup property which can be expressed as follows

$$U(t + t'; x, u) = U(t'; x, U(t; x, u)). \tag{2.36}$$

It means that the trajectory at time $t + t'$ with initial value u can be obtained by extending the trajectory at time t' of the trajectory with initial value $U(t; x, u)$.

2.4. STOCHASTIC EVOLUTIONARY SYSTEMS

The semigroup property (2.36) can be reformulated for the semigroup operators in abstract form by the relation

$$\Gamma_t(x)\varphi(u) := \varphi(U(t;x,u)), \quad t \geq 0, \tag{2.37}$$

in the Banach space $C(\mathbb{R}^d)$ of continuous bounded real-valued functions $\varphi(u)$, $u \in \mathbb{R}^d$.

It is easy to see that the operators $\Gamma_t(x)$, $t \geq 0$, satisfy a semigroup property

$$\Gamma_{t+t'}(x) = \Gamma_{t'}(x)\Gamma_t(x).$$

Indeed:

$$\begin{aligned}
\Gamma_{t'}(x)\Gamma_t(x)\varphi(u) &= \Gamma_{t'}(x)\varphi(U(t;x,u)) \\
&= \varphi(U(t';x,U(t;x,u))) \\
&= \varphi(U(t+t';x,u)) \quad \text{by (2.36)} \\
&= \Gamma_{t+t'}(x)\varphi(u).
\end{aligned}$$

Definition (2.37) of the semigroup $\Gamma_t(x)$, $t \geq 0$, implies the contraction property [118]

$$\|\Gamma_t(x)\| \leq 1,$$

and their uniform continuity

$$\lim_{t \to 0} \|\Gamma_t(x) - I\| = 0.$$

Proposition 2.3 *The generator $\mathbf{\Gamma}(x)$ of the semigroup $\Gamma_t(x)$, $t \geq 0$, is defined by the following relation*

$$\mathbf{\Gamma}(x)\varphi(u) = a(u;x)\varphi'(u). \tag{2.38}$$

PROOF. We have

$$\begin{aligned}
\mathbf{\Gamma}(x)\varphi(u) &= \lim_{t \to 0} t^{-1}\left[\Gamma_t(x) - I\right]\varphi(u) \\
&= \lim_{t \to 0} t^{-1}\left[\varphi(u + \int_0^t a(U(s);x)ds) - \varphi(u)\right] \\
&= \lim_{t \to 0} t^{-1} \int_0^t a(U(s);x)ds\varphi'(u) \\
&= a(u;x)\varphi'(u).
\end{aligned}$$

For sake of simplicity, we have written $U(s) := U(s;x,u)$. $\square$

Remark 2.1. In the vector case we have to consider the scalar product, that is $a\varphi'(u) = \sum a_k \frac{\partial}{\partial u_k}\varphi(u)$.

Note that the domain of definition $\mathcal{D}(\mathbf{\Gamma}(x))$ of the generator $\mathbf{\Gamma}(x)$ contains $C^1(\mathbb{R}^d)$, the continuously differentiable functions $\varphi(u)$ with bounded first derivative.

2.5 Markov Additive Processes

Markov additive processes (MAP) constitute a very large family of processes including semi-Markov processes as a particular case. Of course, since the MAP are a generalization of the Markov renewal (semi-Markov) processes, of Markov processes, and of renewal processes, the field of their applications is very large: reliability, survival analysis, queuing theory, risk process, etc.

Definition 2.6 An $\mathbb{R}^d \times E$-valued coupled stochastic process $\xi(t), x(t)$, $t \geq 0$ is called a MAP if:
 1) the coupled process $\xi(t), x(t), t \geq 0$ is a Markov process;
 2) and, on $\{\xi(t) = u\}$, we have, (a.s.),

$$\mathbb{P}(\xi(t+s) \in A, x(t+s) \in B \mid \mathcal{F}_t) = \mathbb{P}(\xi(t+s) - \xi(t) \in A - u, x(t+s) \in B \mid x(t)),$$

for all $A \in \mathcal{B}(\mathbb{R}^d)$, $B \in \mathcal{E}$, $t \geq 0$ and $s \geq 0$, where $\mathcal{F}_t := \sigma(\xi(s), x(s); s \leq t)$, $t \geq 0$.

From 2., it is clear that $x(t), t \geq 0$, is a Markov process. A typical example of a MAP is the Markov renewal process when the time t is discrete and $\xi(t), t \geq 0$, is an increasing sequence of $\mathbb{R}_+$-valued random variables.

Let us define also the transition function $P_t(x, A, B), A \in \mathcal{B}_d, B \in \mathcal{E}, t \geq 0$, by

$$P_t(x, A, B) := \mathbb{P}(\xi(t+s) - \xi(s) \in A, x(t+s) \in B \mid x(s) = x).$$

The semigroup property is written

$$P_{t+s}(x, A, B) = \int_{\mathbb{R}^d} \int_E P_t(x, du, dy) P_s(x, A - u, B) =: P_t P_s(x, A, B).$$

In the discrete case, the MAP $\xi_n, x_n, n \geq 0$, can be represented as follows

$$\xi_n = \xi_0 + y_1 + \cdots + y_n,$$

where

$$\mathbb{P}(\xi_1 \in A, y_1 \in B \mid \xi_0 = u, x_0 = x) = \mathbb{P}(\xi_1 \in A, y_1 \in B \mid x_0 = x)$$
$$=: P(x, A, B).$$

Let us present some examples of Markov additive processes.

▷ **Example 2.1.** Sums of i.i.d. random vectors in $\mathbb{R}^d$.

▷ **Example 2.2.** *The increment process.* Define $\xi_n = \sum_{k=1}^n a(x_k)$, where $x_n, n \geq 0$, is a Markov chain on $(E, \mathcal{E})$ and $a : E \to \mathbb{R}^d$ a measurable function. In applications, ξ_n can be considered in the following form $\xi_n = \sum_{k=1}^n a(x_{k-1}, x_k)$.

▷ **Example 2.3.** A Markov renewal process $\xi_n, x_n, n \geq 0$, is a MAP with additive component $\xi_n = \tau_n, n \geq 0$.

▷ **Example 2.4.** *The integral functional.* Define $\xi(t) = \int_0^t a(x(s))ds$, where $x(t), t \geq 0$ is a cadlag Markov process on $(E, \mathcal{E})$, and $a : E \to \mathbb{R}^d$ a measurable function. When $x(t), t \geq 0$, is a semi-Markov process, then the extended Markov renewal process $\xi_n = \xi(\tau_n), \tau_n, x_n, n \geq 0$, is a MAP with additive component the coupled process $\xi_n, \tau_n, n \geq 0$.

2.6 Stochastic Additive Functionals

An evolutionary system with switching considered in Section 2.3 is characterized by a deterministic trajectory with velocity depending on the state of the switching process between successive renewal moments.

Various stochastic systems can be characterized by a stochastic trajectory under fixed state of switching process. The Markovian locally independent increment processes (see Section 1.2.4) constitute a wide class of such processes in $\mathbb{R}^d$ with semigroup property.

Definition 2.7 The Markov additive process $\xi(t), x(t), t \geq 0$, in $\mathbb{R}^d \times E$, with Markov switching $x(t), t \geq 0$, defined by the relation

$$\xi(t) = \xi_0 + \int_0^t \eta(ds; x(s)), \quad t \geq 0, \tag{2.39}$$

is called a *stochastic additive functional*.

The cadlag Markov processes $\eta(t;x)$, $t \geq 0, x \in E$, are of locally independent increment processes determined by its generators

$$\mathbb{\Gamma}(x)\varphi(u) = a(u;x)\varphi'(u)$$
$$+ \int_{\mathbb{R}^d} [\varphi(u+v) - \varphi(u) - v\varphi'(u)]\Gamma(u, dv; x), \quad (2.40)$$

where the positive kernels $\Gamma(u, dv; x), x \in E$, satisfy the following conditions: the functions

$$\Lambda(u;x) := \Gamma(u, \mathbb{R}^d; x),$$
$$b(u;x) := \int_{\mathbb{R}^d} v\Gamma(u, dv; x),$$
$$C(u;v) := \int_{\mathbb{R}^d} vv^*\Gamma(u, dv; x),$$

are continuous and bounded on $u \in \mathbb{R}^d$, uniformly on $x \in E$.

Note that the common domain of definition $\mathcal{D}(\mathbb{\Gamma}) = \cap_{x \in E}\mathcal{D}(\mathbb{\Gamma}(x))$ contains the full and complete class of functions $C^2(\mathbb{R}^d)$ [45].

The generators of locally independent increment processes $\eta(t;x)$, $t \geq 0, x \in E$, can be represented in the following form

$$\mathbb{\Gamma}(x)\varphi(u) = a_0(u;x)\varphi'(u) + \int_{\mathbb{R}^d} [\varphi(u+v) - \varphi(u)]\Gamma(u, dv; x), \quad (2.41)$$

with $a_0(u;x) = a(u;x) - b(u;x)$.

According to (2.41) the *Markov additive process* $\xi(t), x(t), t \geq 0$, has a deterministic drift defined by a solution of the evolutionary equation

$$\frac{d}{dt}U(t;x) = a_0(U(t;x);x),$$

and a pure jump part defined by the generators

$$\mathbb{\Gamma}_0(x)\varphi(u) = \Lambda(u;x)\int_{\mathbb{R}^d} [\varphi(u+v) - \varphi(u)]F(u, dv; x),$$

with intensity of jump moments $\Lambda(u;x)$ and with distribution functions of jump values

$$F(u, dv; x) = \Gamma(u, dv; x)/\Lambda(u;x).$$

In the particular important case where

$$a(u;x) = a(x) \quad \text{and} \quad \Gamma(u, dv; x) = \Gamma(dv; x),$$

2.6. STOCHASTIC ADDITIVE FUNCTIONALS

the family of Markov processes $\eta(t;x)$, $t \geq 0, x \in E$, are processes with independent increments. This is why such a process is said to be with *locally independent increments* [100].

In this case the pure jump part is defined by the compound Poisson process with generators

$$\mathbb{\Gamma}_0(x)\varphi(u) = \Lambda(x) \int_{\mathbb{R}^d} [\varphi(u+v) - \varphi(u)] F(dv;x).$$

It is well-known that the compound Poisson process [100] can be represented by a sum of i.i.d. random variables

$$\eta(t;x) = \sum_{k=1}^{\nu(t;x)} \alpha_k(x), \quad t \geq 0, x \in E, \qquad (2.42)$$

where $\nu(t;x)$, $t \geq 0, x \in E$, are Poisson processes with intensity $\Lambda(x)$, and $\alpha_k(x)$, $k \geq 1$, $x \in E$, is a sequence of i.i.d. r.v., under fixed x, with the distribution function $F(dv;x)$.

The switching jump Markov process $x(t)$, $t \geq 0$, in (2.39), is defined by its generator

$$Q\varphi(x) = q(x) \int_E P(x,dy)[\varphi(y) - \varphi(x)].$$

Lemma 2.5 *The Markov additive process $\xi(t), x(t)$, $t \geq 0$, is determined by the generator*

$$\mathbb{L}\varphi(u,x) = Q\varphi(u,x) + \mathbb{\Gamma}(x)\varphi(u,x),$$

where $\mathbb{\Gamma}(x)$, $x \in E$, is the family of generators (2.40)

PROOF. The proof of Lemma 2.5 is based on the asymptotic representation of the conditional expectation

$$\mathbb{E}\left[\varphi(u + \Delta\xi(t), x(t+\Delta t)) \mid \xi(t) = u, x(t) = x\right]$$
$$= \mathbb{E}\left[\varphi(u + \Delta\eta(t;x), x)\mathbf{1}(\theta_x > \Delta t)\right]$$
$$\quad + \mathbb{E}\left[\varphi(u + \Delta\eta(t;x), x(t+\Delta t)) \mid x(t) = x\right]\mathbf{1}(\theta_x \leq \Delta t) + o(\Delta t)$$
$$= \mathbb{E}\varphi(u + \Delta\eta(t;x), x)(1 - \Delta t q(x))$$
$$\quad + \mathbb{E}\left[\varphi(u, x(t+\Delta t))\right]\Delta t q(x) + o(\Delta t)$$
$$= \mathbb{E}\varphi(u + \Delta\eta(t;x), x) + \Delta t Q\varphi(u,x) + o(\Delta t)$$
$$= \varphi(u,x) + \Delta t[\mathbb{\Gamma}(x) + Q]\varphi(u,x) + o(\Delta t).$$

The last equality leads to the desired result.   □

2.7 Random Evolutions

The stochastic systems considered in the above Sections 2.2-2.6, can be described by the abstract mathematical model in the Banach space $\mathbf{B}(\mathbb{R}^d)$ of functions $\varphi(u), u \in \mathbb{R}^d$, called *random evolution model*, introduced Griego and Hersh [59,63,64].

Some particular examples of random evolution models were considered in Sections 2.2 and 2.3.

Now we will introduce two models of random evolution: *continuous* and *jump random evolution*.

2.7.1 *Continuous Random Evolutions*

Let the family of continuous semigroup operators $\Gamma_t(x), t \geq 0, x \in E$, be given on the Banach space $C(\mathbb{R}^d)$, of real-valued continuous bounded (with respect to the sup-norm) functions $\varphi(u), u \in \mathbb{R}^d$.

Let the switching semi-Markov process $x(t), t \geq 0$, be given by the semi-Markov kernel

$$Q(x, B, t) = \mathbb{P}(x_{n+1} \in B, \theta_{n+1} \leq t \mid x_n = x),$$

which determines the Markov renewal process $x_n, \tau_n, n \geq 0$, associated with the semi-Markov process $x(t), t \geq 0$, and the Markov renewal process $x_n, \theta_n, n \geq 0$, associated with the semi-Markov (or Markov) process $x(t), t \geq 0$.

Definition 2.8 The *semi-Markov continuous random evolution* on the Banach space $C(\mathbb{R}^d)$ of continuous bounded functions $\varphi(u), u \in \mathbb{R}^d$, is determined by the relation

$$\Phi(t) = \Gamma_{\gamma(t)}(x(t)) \prod_{k=1}^{\nu(t)-1} \Gamma_{\theta_k}(x_{k-1}), \quad t > 0, \quad \Phi(0) = I. \quad (2.43)$$

Particularly, in the renewal moments

$$\Phi(\tau_n) = \prod_{k=1}^{n} \Gamma_{\theta_k}(x_{k-1}), \quad n \geq 0. \quad (2.44)$$

2.7. RANDOM EVOLUTIONS

Proposition 2.4 *The random evolution (2.43) can be determined by a solution of the evolutionary equation*

$$\left| \begin{array}{l} \frac{d}{dt}\Phi(t) = \mathbb{\Gamma}(x(t))\Phi(t), \quad t > 0, \\ \Phi(0) = I, \end{array} \right. \quad (2.45)$$

where $\mathbb{\Gamma}(x), x \in E$, *is the family of the generators for semigroups* $\Gamma_t(x)$, $t \geq 0, x \in E$. *The integral equivalent of (2.45) is*

$$\Phi(t) = I + \int_0^t \mathbb{\Gamma}(x(s))\Phi(s)ds. \quad (2.46)$$

PROOF. The semigroups $\Gamma_t(x), t \geq 0, x \in E$, satisfy the integral equation

$$\Gamma_t(x) = I + \mathbb{\Gamma}(x) \int_0^t \Gamma_s(x)ds. \quad (2.47)$$

By Definition 2.8, the random evolution (2.43), on the interval $[0, \tau_1]$ is represented by

$$\Phi(t) = \Gamma_t(x_0), \quad 0 \leq t < \tau_1, \quad (2.48)$$

and, in general form by

$$\Phi(t) = \Gamma_{t-\tau_n}(x_n)\Phi(\tau_n), \quad \tau_n \leq t < \tau_{n+1}. \quad (2.49)$$

By substitution of (2.47) into (2.48) we get

$$\Phi(t) = I + \int_0^t \mathbb{\Gamma}(x(s))\Phi(s)ds, \quad 0 \leq t \leq \tau_1, \quad (2.50)$$

since $\mathbb{\Gamma}(x(s)) = \mathbb{\Gamma}(x_0)$ for $0 \leq s < \tau_1$.

Now, by induction, let relation (2.46) hold for $t \leq \tau_n$. Then, from (2.49), for $\tau_n \leq t < \tau_{n+1}$, we get:

$$\Phi(t) = \Gamma_{t-\tau_n}(x_n)\Phi(\tau_n), \quad \tau_n \leq t < \tau_{n+1}$$

$$= \Phi(\tau_n) + \int_{\tau_n}^t \mathbb{\Gamma}(x(s))\Phi(s)ds, \quad \text{(by substitution of (2.47))}$$

$$= I + \int_0^{\tau_n} \mathbb{\Gamma}(x(s))\Phi(s)ds + \int_{\tau_n}^t \mathbb{\Gamma}(x(s))\Phi(s)ds,$$

(because $\mathbb{\Gamma}(x(s)) = \mathbb{\Gamma}(x_n)$, for $\tau_n \leq s < \tau_{n+1}$)

$$= I + \int_0^t \mathbb{\Gamma}(x(s))\Phi(s)ds.$$

Definition 2.9 The *coupled random evolution* is determined on the Banach space $C(\mathbb{R}^d \times E)$, by the relation

$$\Phi(t, x(t)) := \Phi(t)\varphi(u, x(t)), \quad t \geq 0. \tag{2.51}$$

Proposition 2.5 *The mean value of the coupled random evolution (2.51), defined by the family of semigroups $\Gamma_t(x), t \geq 0, x \in E$, on the Banach space $C(\mathbb{R}^d \times E)$ of $\varphi(u, x), u \in \mathbb{R}^d, x \in E$,*

$$U(t, x) := \mathbb{E}_x[\Phi(t, x(t))] := \mathbb{E}[\Phi(t, x(t)) \mid x(0) = x]$$

is determined by a solution of the Markov renewal equation

$$U(t, x) - \int_0^t F_x(ds)\Gamma_s(x) \int_E P(x, dy)U(t - s, y) = \overline{F}_x(t)\Gamma_t(x)\varphi(u, x), \tag{2.52}$$

PROOF. Let us use the following recursive relation for random evolution (see (2.49) and (2.48))

$$\Phi(t) = \Gamma_{t-\tau_1}(x)\Phi(\tau_1)\mathbf{1}_{(\tau_1 \leq t)} + \Gamma_t(x)\mathbf{1}_{(\tau_1 > t)},$$

where τ_1 is the first renewal moment of the switching semi-Markov process. Now calculate the first term by using the Markov property in the renewal moments

$$\begin{aligned}
U(t, x) &= \mathbb{E}_x \Phi(t, x(t)) = \mathbb{E}_x[\Phi(t)\varphi(u, x(t))] \\
&= \mathbb{E}_x[\Gamma_{t-\tau_1}(x)\Phi(\tau_1, x_1)\mathbf{1}_{(\tau_1 \leq t)}] + \mathbb{E}_x[\Gamma_t(x)\varphi(u, x)\mathbf{1}_{(\tau_1 > t)}] \\
&= \int_0^t F_x(ds)\Gamma_{t-s}(x) \int_E P(x, dy)U(s, y) + \overline{F}_x(t)\Gamma_t(x)\varphi(u, x) \\
&\equiv \int_0^t F_x(ds)\Gamma_s(x) \int_E P(x, dy)U(t - s, y) + \overline{F}_x(t)\Gamma_t(x)\varphi(u, x).
\end{aligned}$$

Hence, the Markov renewal equation (2.52) holds. □

Let the switching process $x(t), t \geq 0$, be Markovian and defined by the generator

$$Q\varphi(x) = q(x) \int_E P(x, dy)[\varphi(y) - \varphi(x)]. \tag{2.53}$$

The stochastic evolutionary switched process $U(t), t \geq 0$, is defined by a solution of the evolutionary equation (2.31) (see Section 2.4).

2.7. RANDOM EVOLUTIONS

Then the coupled process

$$U(t), \quad x(t), \quad t \geq 0, \qquad (2.54)$$

is also Markovian.

Remark 2.2. The coupled random evolution is

$$\Phi(t)\varphi(u, x(t)) = \varphi(U(t), x(t)), \quad U(0) = u.$$

Lemma 2.6 *The coupled Markov process (2.54) can be determined by the generator*

$$\mathbb{L}\varphi(u, x) = Q\varphi(u, x) + \Gamma(x)\varphi(u, x), \qquad (2.55)$$

on the test functions $\varphi \in C^1(\mathbb{R} \times E)$.

PROOF. See proof of Lemma 2.5 which includes the present case. □

Corollary 2.2 *The mean value of the coupled Markov random evolution is determined by a solution of the evolutionary equation*

$$\left| \begin{array}{l} \frac{d}{dt} U(t, x) = [Q + \Gamma(x)]U(t, x), \\ U(0, x) = \varphi(u, x). \end{array} \right. \qquad (2.56)$$

So, the mean value of the Markov random evolution is characterized by the generator

$$\mathbb{L} = Q + \Gamma(x). \qquad (2.57)$$

PROOF. The mean value of the Markov random evolution satisfies the Markov renewal equation (2.52) with the distribution function of jumps

$$F_x(ds) = q(x)e^{-q(x)s}ds, \quad x \in E,$$

So,

$$U(t, x) - q(x) \int_0^t e^{-q(x)s} \Gamma_s(x) ds \int_E P(x, dy) U(t - s, y)$$
$$= e^{-q(x)t} \Gamma_t(x)\varphi(u, x),$$

or

$$U(t,x) - q(x)\int_0^t e^{-q(x)(t-s)}\Gamma_{t-s}(x)ds \int_E P(x,dy)U(s,y)$$
$$= e^{-q(x)t}\Gamma_t(x)\varphi(u,x).$$

Now, differentiating out, we get

$$\frac{d}{dt}U(t,x) - q(x)\int_E P(x,dy)U(t,y) + [q(x) - \mathbb{\Gamma}(x)]J(t,x)$$
$$= [-q(x) + \mathbb{\Gamma}(x)]e^{-q(x)t}\Gamma_t(x)\varphi(u,x),$$

where

$$J(t,x) := U(t,x) - e^{-q(x)t}\Gamma_t(x)\varphi(u,x).$$

Hence,

$$\frac{d}{dt}U(t,x) - q(x)\int_E P(x,dy)U(t,y) + [q(x) - \mathbb{\Gamma}(x)]U(t,x) = 0, \quad (2.58)$$

that is exactly (2.56). □

It is worth noticing that the generator $\mathbb{L}$ in (2.55) characterizes the coupled Markov process $U(t), x(t), t \geq 0$, (see Section 2.2), and also the switched evolutionary stochastic system $U(t)$ defined in Section 2.5.

2.7.2 Jump Random Evolutions

The random evolution model for the increment process (see Section 2.3) is constructed similarly by using the family of bounded operators $\mathbb{D}(x), x \in E$, characterizing jumps of stochastic system in state $x \in E$, and the Markov renewal process $x_n, \tau_n, n \geq 0$, associated with the switching semi-Markov process $x(t), t \geq 0$.

Definition 2.10 The *semi-Markov jump random evolution* on the Banach space $C(\mathbb{R}^d)$ of continuous bounded functions $\varphi(u), u \in \mathbb{R}^d$, is determined by the relation

$$\Phi(t) = \prod_{k=1}^{\nu(t)} \mathbb{D}(x_k), \quad t \geq 0, \quad \Phi(0) = I. \quad (2.59)$$

2.7. RANDOM EVOLUTIONS

Particularly, in the renewal moments, we have

$$\Phi(\tau_n) = \prod_{k=1}^{n} \mathbb{D}(x_k), \quad n \geq 1. \tag{2.60}$$

Proposition 2.6 *The jump random evolution can be represented in the following form*

$$\Phi(t) = I + \sum_{k=1}^{\nu(t)} [\mathbb{D}(x_k) - I]\Phi(\tau_{k-1}), \quad t \geq 0. \tag{2.61}$$

PROOF. From (2.59) it follows that

$$\Phi(\tau_k) - \Phi(\tau_{k-1}) = [\mathbb{D}(x_k) - I]\Phi(\tau_{k-1}). \tag{2.62}$$

The sum in (2.61) gives (compare with (2.25), (2.46) and (2.63) instead of relation (2.60))

$$\Phi(\tau_n) - I = \sum_{k=1}^{n} [\mathbb{D}(x_k) - I]\Phi(\tau_{k-1}), \tag{2.63}$$

that is an equivalent form of (2.60). □

The linear forms of the relations (2.46) and (2.61) provide the most effective asymptotic analysis of stochastic systems in the series scheme considered in the next Chapter 3.

Proposition 2.7 *The mean value of the coupled random evolution defined by the family of bounded operators $\mathbb{D}(x), x \in E$, is determined by a solution of the Markov renewal equation*

$$U(t, x) - \int_0^t F_x(ds) \int_E P(x, dy)\mathbb{D}(y)U(t - s, y) = \overline{F}_x(t)\varphi(u, x). \tag{2.64}$$

PROOF. We use the following recursive relation for the jump random evolution

$$\Phi(t, x(t)) = \mathbb{D}(x_1)\Phi(t - \tau_1, x(t - \tau_1))\mathbf{1}_{(\tau_1 \leq t)} + \varphi(u, x)\mathbf{1}_{(\tau_1 > t)}. \tag{2.65}$$

The mean value of (2.65) gives the Markov renewal equation (2.64). □

Corollary 2.3 *The mean value of the Markov jump random evolution is determined by a solution of the evolutionary equation*

$$\left|\begin{array}{l} \frac{d}{dt}U(t,x) = [Q + Q_0(\mathbb{D}(x) - I)]U(t,x), \\ U(0,x) = \varphi(u,x). \end{array}\right.$$

So, the mean value of the Markov jump random evolution is characterized by the generator

$$\mathbb{L}_0 = Q + Q_0[\mathbb{D}(x) - I].$$

It is worth noticing that the generator $\mathbb{L}_0$ characterizes the coupled Markov process $\alpha(t), x(t), t \geq 0$, (see Section 2.3), with the switched increment process $\alpha(t), t \geq 0$. The operator Q_0, defined in Section 2.3, is

$$Q_0 \varphi(x) = q(x) \int_E P(x, dy) \varphi(y).$$

2.7.3 Semi-Markov Random Evolutions

The semi-Markov random evolution can be characterized by the compensating operator.

Definition 2.11 The *compensating operator* of the continuous coupled random evolution (2.51) is determined by the relation

$$\mathbb{L}\Phi(t,x) := \{\mathbb{E}[\Phi(\tau_1, x_1) \mid \tau_0 = t, x_0 = x] - \Phi(t,x)\}/\mathbb{E}[\theta_x]. \quad (2.66)$$

It is worth noticing that the compensating operator (2.66) satisfies the homogeneous condition

$$\mathbb{L}\Phi(t,x) := \{\mathbb{E}[\Phi(\tau_{n+1}, x_{n+1}) \mid \tau_n = t, x_n = x] - \Phi(t,x)\}/\mathbb{E}[\theta_{n+1} \mid x_n = x]. \quad (2.67)$$

Proposition 2.8 *The compensating operator of the continuous coupled random evolution $\Phi(t, x(t))$, (2.51), can be represented as follows*

$$\mathbb{L}\Phi(t,x) = q(x) \Big[\int_0^\infty F_x(ds) \Gamma_s(x) \int_E P(x,dy) \Phi(t,y) - \Phi(t,x) \Big], \quad (2.68)$$

where $q(x) := 1/m(x) := 1/\mathbb{E}\theta_x$.

PROOF. Let us calculate by using (2.51)

$$\mathbb{E}_{t,x}\Phi(\tau_1, x_1) = \mathbb{E}_{t,x} \Gamma_{\theta_1}(x) \Phi(t) \varphi(u, x_1)$$

$$= \int_0^\infty F_x(ds) \Gamma_s(x) \mathbb{E}_{t,x} \Phi(t, x_1)$$

$$= \int_0^\infty F_x(ds) \Gamma_s(x) \int_E P(x, dy) \Phi(t, y).$$

Hence, (2.68) follows from definition (2.67). □

2.7. RANDOM EVOLUTIONS

Proposition 2.9 *The compensating operator (2.68) acting on the test functions $\varphi \in C^k(\mathbb{R}^d \times E), k \geq 3$, can be transformed into*

$$\mathbb{L}\varphi(u,x) = Q\varphi(\cdot,x) + \mathbb{\Gamma}(x)\mathbb{F}_1(x)Q_0\varphi(u,x) \quad (2.69)$$

$$\mathbb{L}\varphi(u,x) = Q\varphi(\cdot,x) + \mathbb{\Gamma}(x)P\varphi(u,x) + \mathbb{\Gamma}^2(x)\mathbb{F}_2(x)Q_0\varphi(u,x) \quad (2.70)$$

and

$$\mathbb{L}\varphi(u,x) = Q\varphi(\cdot,x) + \mathbb{\Gamma}(x)P\varphi(u,x) + \mu_2(x)\mathbb{\Gamma}^2(x)P\varphi(u,x)$$
$$+ \mathbb{\Gamma}^3(x)\mathbb{F}_3(x)Q_0\varphi(u,x). \quad (2.71)$$

The operator

$$Q\varphi(x) := q(x)\int_E P(x,dy)[\varphi(y) - \varphi(x)],$$

is the generator of the associated Markov process $x^0(t), t \geq 0$, with the same embedded Markov chain $x_n, n \geq 0$, as in the semi-Markov process, and the intensity of sojourn times

$$q(x) = 1/m(x), \quad m(x) = \mathbb{E}\theta_x := \int_0^\infty \overline{F}_x(t)dt.$$

As usual, $Q_0\varphi(x) := q(x)\int_E P(x,dy)\varphi(y)$. The bounded operators $\mathbb{F}_k(x), x \in E, k = 1, 2, 3$, are given by the relation

$$\mathbb{F}_k(x) := \int_0^\infty \overline{F}_x^{(k)}(s)\Gamma_s(x)ds, \quad k \geq 1,$$

where

$$\overline{F}_x^{(k)}(s) := \int_s^\infty \overline{F}_x^{(k-1)}(t)dt, \quad k \geq 2, \quad \overline{F}_x^{(1)}(t) = \overline{F}_x(t),$$

and

$$\mu_2(x) := m_2(x)/[2m(x)], \quad m_2(x) := 2\int_0^\infty \overline{F}_x^{(2)}(t)dt.$$

PROOF. Integration by parts gives

$$\mathbb{F}(x) := \int_0^\infty F_x(ds)\Gamma_s(x)ds = I + \int_0^\infty \overline{F}_x(s)d\Gamma_s(x). \quad (2.72)$$

Using the differential equation for semigroups

$$d\Gamma_s(x) = \mathbb{\Gamma}(x)\Gamma_s(x)ds,$$

we get
$$\mathbb{F}(x) = I + \mathbf{\Gamma}(x)\mathbb{F}_1(x),$$
where
$$\mathbb{F}_1(x) = \int_0^\infty \overline{F}_x(s)\Gamma_s(x)ds.$$
In the same way, we get
$$\mathbb{F}_1(x) = m(x)I + \mathbf{\Gamma}(x)\mathbb{F}_2(x), \qquad (2.73)$$
where
$$\mathbb{F}_2(x) = \int_0^\infty \overline{F}_x^{(2)}(s)\Gamma_s(x)ds,$$
that is
$$\mathbb{F}_2(x) = \frac{m_2(x)}{2}I + \mathbf{\Gamma}(x)\mathbb{F}_3(x), \qquad (2.74)$$
where
$$\mathbb{F}_3(x) := \int_0^\infty \overline{F}_x^{(3)}(s)\Gamma_s(x)ds.$$

Successively putting the $\mathbb{F}_k(x), k = 1, 2, 3$, given by formulas (2.73) and (2.74), into (2.72), we get the representations (2.69)-(2.71).

Proposition 2.10 *The compensating operator for the jump random evolution (2.59) can be represented as follows*

$$\mathbb{L}\Phi(t, x) = q(x)\left[\int_E P(x, dy)\mathbb{D}(y)\Phi(t, y) - \Phi(t, x)\right]. \qquad (2.75)$$

PROOF. From Definition 2.9, we calculate:
$$\mathbb{E}_{t,x}\Phi(\tau_1, x_1) = \mathbb{E}_{t,x}\mathbb{D}(x_1)\Phi(t, x_1) \qquad (2.76)$$
$$= \int_E P(x, dy)\mathbb{D}(y)\Phi(t, y).$$

Now, by Definition 2.11 we obtain (2.75). □

It is worth noticing that the compensating operator for the random evolution can be considered directly on the test functions $\varphi(u, x)$ as represented in Proposition 2.9.

Corollary 2.4 *The compensating operator of the jump random evolution can be represented as follows*

$$\mathbb{L}\varphi(u,x) = Q\varphi(u,x) + Q_0[\mathbb{D}(x) - I]\varphi(u,x).$$

2.8 Extended Compensating Operators

The semi-Markov continuous random evolution can be characterized by the extended compensating operator which is constructed by using the extended Markov renewal process

$$\xi_n = \xi(\tau_n), \quad x_n, \tau_n, \quad n \geq 0, \tag{2.77}$$

where $x_n, \tau_n, n \geq 0$, is the Markov renewal process associated with the switching semi-Markov process $x(t), t \geq 0$, defined by the semi-Markov kernel (see Section 1.3.1)

$$Q(x,B,t) = \mathbb{P}(x_{n+1} \in B, \theta_{n+1} \leq t \mid x_n = x) = P(x,B)F_x(t),$$

with $\theta_{n+1} := \tau_{n+1} - \tau_n, n \geq 0$.

The first component in (2.77) is the continuous part of the random evolution generated by the values of semigroup $\Gamma_{\theta_{n+1}}(x_n), t \geq 0$, such that

$$\Gamma_{\theta_{n+1}}(x_n)\varphi(u) = \varphi(u + \Delta \xi_n) \tag{2.78}$$

where $\Delta \xi_n := \xi_{n+1} - \xi_n, n \geq 0$.

For example, the evolutionary stochastic system defined in Section 2.3 is generated by the first component of the extended Markov renewal process (2.77) by the semigroup

$$\Gamma_t(x)\varphi(u) = \varphi(u + \Delta_x U(t)),$$

where $U(t), t \geq 0$, is a solution of the evolutionary equation

$$\begin{vmatrix} \frac{d}{dt}U_x(t) = a(U_x(t); x), \\ U_x(0) = u, \end{vmatrix}$$

and

$$\Delta U_x(t) := U_x(t) - U_x(0) = U_x(t) - u.$$

Analogously the first component of the extended Markov renewal process (2.77) can be defined for other stochastic systems considered in above Sections 2.2-2.4.

Definition 2.12 The *extended compensating operator* of the Markov renewal process (2.77) is defined by the relation

$$\mathbb{L}\varphi(u,x,t) := \mathbb{E}\left[\varphi(\xi_1, x_1, \tau_1) - \varphi(u,x,t) \mid x_0 = x, \tau = t\right]/m(x),$$

where $m(x) := \mathbb{E}\theta_x = \int_0^\infty \overline{F}_x(t)dt$.

Proposition 2.11 *The extended compensating operator of the Markov renewal process (2.77) is represented as follows*

$$\mathbb{L}\varphi(u,x,t) = \left[\int_0^\infty F_x(ds)\Gamma_s(x)\int_E P(x,dy)\varphi(u,y,t+s) - \varphi(u,x,t)\right]/m(x), \tag{2.79}$$

where $\Gamma_s(x)$, $s \geq 0, x \in E$, is the family of semigroups with generators $\Gamma(x)$, $x \in E$, defined in (2.40).

The proof of Proposition 2.11 is based on representation (2.78) for the increments of the first component and on the homogeneous property of the compensating operator

$$\mathbb{L}\varphi(u,x,t) = \mathbb{E}\left[\varphi(\xi_{n+1}, x_{n+1}, \tau_{n+1}) - \varphi(\xi_n, x_n, \tau_n) \mid \xi_n = u, x_n = x, \tau_n = t\right].$$

Proposition 2.12 *The compensating operator (2.79) acting on the test functions* $\varphi \in C^k(\mathbb{R}^d \times E), k \geq 3$, *does not depend on time t and can be transformed into:*

$$\mathbb{L}\varphi(u,x) = Q\varphi + \Gamma(x)F^{(1)}(x)Q_0\varphi,$$

$$\mathbb{L}\varphi(u,x) = Q\varphi + \Gamma(x)P\varphi + \Gamma^2(x)F^{(2)}(x)Q_0\varphi,$$

$$\mathbb{L}\varphi(u,x) = Q\varphi + \Gamma(x)P\varphi + \mu_2(x)\Gamma^2(x)P\varphi + \Gamma^3(x)F^{(3)}(x)Q_0\varphi,$$

where:

$$F^{(k)}(x) := \int_0^\infty \overline{F}_x^{(k)}(x)\Gamma_s(x)ds, \quad k = 1,2,3, \quad \overline{F}_x^{(1)}(x) = \overline{F}_x(x),$$

$$Q_0\varphi(x) = q(x)\int_E P(x,dy)\varphi(y),$$

$$\mu_2(x) := m_2(x)/2m(x) = \int_0^\infty \overline{F}_x^{(2)}(s)ds/m(x), \quad m_2(x) = \int_0^\infty s^2 F_x(ds),$$

$$\overline{F}_x^{(k+1)}(t) := \int_t^\infty \overline{F}_x^{(k)}(s)ds, \quad k = 1, 2, \ldots$$

2.9 Markov Additive Semimartingales

2.9.1 *Impulsive Processes*

Impulsive processes constitute a very active area of research since they are involved in many applications, especially in risk theory[7,44,149,55]. They are in some respects related to stochastic additive functionals as considered in [107]. The the increment processes and the compound Poisson processes constitute very particular cases of impulsive processes. These are also a typical case of stopped processes[169]. A large literature exists concerning risk processes, but few papers concerning functional type limit theorems, that we are interested in here (see, e.g.[53,7]).

The impulsive process $\zeta(t), t \geq 0$, considered here (see Definition 2.13 below), is the stochastic part of a risk process switched by a Markov process. To be specific, if we define the process $\zeta_0(t) \equiv ct - \zeta(t)$ (for some constant c), this process is the classical risk process used in insurance, reliability, queueing theory, etc. [7,44,100,149,106,111]. The switching process is a useful means of taking into account the environment[2,107]. For example, if the impulsive process gives the amount of damages for an insurance company in the interval of time $[0, t]$, that is, if $\alpha_n(x, s)$ is the amount of the n-th damage, $1 \leq n \leq \nu(t)$, under $x_n = x$, where x denotes the state of the environment, for example weather, time, etc., and s the time elapsed from the last damage, then the amount is strongly dependent of the environment.

The particular class of switched semimartingale processes is defined over a switching Markov renewal process.

Definition 2.13 The *impulsive process* with semi-Markov switching is defined by

$$\zeta(t) = \sum_{n=1}^{\nu(t)} \alpha_n(x_n, \theta_n), \quad t \geq 0, \tag{2.80}$$

where $\theta_{n+1} = \tau_{n+1} - \tau_n$, $n \geq 0$, and $x_n, \tau_n, n \geq 0$, is a Markov renewal process defined by the semi-Markov kernel

$$Q(x, B, t) = P(x, B) F_x(t) = \mathbb{P}(x_{n+1} \in B, \theta_{n+1} \leq t \mid x_n = x).$$

The counting process of jump times is

$$\nu(t) = \max\{n : \tau_n \leq t\}, \quad t \geq 0.$$

The family of random variables

$$\alpha_n(x, t), \quad t \geq 0, \ n \geq 1, \ x \in E,$$

is supposed to be i.i.d. under fixed x and t, defined by the distribution function

$$\Phi(B; x, t) = \mathbb{P}(\alpha_n(x, t) \in B),$$

and such that for every fixed sequence $z_n \in E$, $s_n \in \mathbb{R}_+$, $n \geq 0$, $\alpha_n(z_n, s_n)$, $n \geq 1$, are independent r.v.

Lemma 2.7 *The impulsive process (2.80) can be characterized by its predictable characteristics represented in the following form:*

- *the predictable process is*

$$B(t) = \sum_{n=1}^{\nu(t)} b(x_{n-1}), \quad t \geq 0,$$

where:

$$\begin{aligned} b(x) &= \int_0^\infty F_x(ds) \int_E P(x, dy) a(y, s) \\ &= \int_0^\infty \int_E Q(x, dy, ds) a(y, s), \end{aligned} \quad (2.81)$$

$$a(y, s) = \mathbb{E}\alpha_n(y, s) = \int_\mathbb{R} z \Phi(dz; y, s);$$

- *the second modified characteristic is*

$$\widetilde{C}(t) = \sum_{n=1}^{\nu(t)} C(x_{n-1}), \quad t \geq 0,$$

where:

$$C(x) = \int_0^\infty F_x(ds) \int_E P(x,dy) C_0(y,s) = \int_0^\infty \int_E Q(x,dy,ds) C_0(y,s),$$

$$C_0(y,s) = \mathbb{E}\alpha_n^2(y,s) = \int_\mathbb{R} z^2 \Phi(dz; y,s);$$

- the compensating measure of jumps is

$$\gamma_g(t) = \sum_{n=1}^{\nu(t)} \Gamma_g(x_{n-1}), \quad t \geq 0,$$

where

$$\Gamma_g(x) = \int_0^\infty \int_E Q(x,dy,ds) \int_\mathbb{R} g(z) \Phi(dz; y,s).$$

PROOF. The proof of Lemma 2.6 implies the standard formulas for predictable characteristics of increment processes [70]. For $\mathcal{F}_n := \sigma(x_k, \tau_k; 0 \leq k \leq n)$, we have:

$$B(t) = \sum_{n=1}^{\nu(t)} \mathbb{E}[\alpha_n(x_n, \theta_n) \mid \mathcal{F}_{n-1}]$$

$$= \sum_{n=1}^{\nu(t)} \int_0^\infty \int_E Q(x_{n-1}, dy, ds) \int_\mathbb{R} z\Phi(dz; y,s)$$

$$= \sum_{n=1}^{\nu(t)} b(x_{n-1}),$$

where $b(x)$ is defined by (2.81). □

2.9.2 Continuous Predictable Characteristics

The real-valued semimartingales $\xi(t), t \geq 0$, defined for a Markov process $x(t), t \geq 0$, on a stochastic basis $\mathfrak{F} = (\Omega, \mathcal{F}, \mathbf{F} = (\mathcal{F}_t, t \geq 0), \mathbb{P})$, with some standard state space $(E, \mathcal{E})$ constitute the general class of stochastic processes considered here as mathematical models of real stochastic systems. Let $\vartheta_s, s \geq 0$ be the shift operator.

Definition 2.14 ([27]) A *Markov additive semimartingale* $\xi(t)$, $t \geq 0$, defined over $\mathfrak{F}$, is characterized by the additivity property

(Θ) $\xi(0) = 0$ and $\vartheta_s \xi(t) = \xi(t) - \xi(s)$ (a.s.) for all $0 \leq s \leq t$.

It is well known that if $(\mathcal{F}_t,\ t \geq 0)$ is a strong Markov filtration, the process $\xi(t),\ t \geq 0$, satisfies the strong additive property, that is,

$(\Theta_S)\quad \xi(0) = 0$ and $\vartheta_S \xi(t) = [\xi(t) - \xi(S)]\mathbf{1}_{[S,\infty)}(t)$ (a.s.) for any $\mathcal{F}_t$-stopping time S.

Remark 2.3. The exceptional null set in the additivity properties (Θ) or (Θ_S) can be chosen in order not to depend on s, t.

The switching Markov process $x(t),\ t \geq 0$, considered here is defined by the following generator

$$Q\varphi(x) = q(x) \int_E P(x, dy)[\varphi(y) - \varphi(x)].$$

The factorization theorem (see [27]) implies the following canonical representation of the Markov additive semimartingale

$$\xi(t) = \xi_0 + \int_0^t b(\xi_{[0,s)}; x(s))ds + \int_0^t c(\xi_{[0,s)}; x(s))dw(s)$$
$$+ \int_0^t \int_{\mathbb{R}} v[\mu(ds, dv) - h(v)\Gamma(\xi_{[0,s)}, dv; x(s))ds]. \qquad (2.82)$$

Here $w(s),\ s \geq 0$, is a standard Wiener process, $\mu(ds, dv)$ is the measure of jumps of the semimartingale $\xi(t)$, and $h(v) := \mathbf{1}(|v| \leq 1)$ is a truncation function, and $\xi_{[0,s)} := (\xi_u, 0 \leq u < s)$.

The canonical representation (2.82) is characterized by the triplet of characteristics:

- the predictable process

$$B(t) = \int_0^t b(\xi_{[0,s)}; x(s))ds; \qquad (2.83)$$

- the second predictable characteristic

$$C(t) = \int_0^t c(\xi_{[0,s)}; x(s))ds; \qquad (2.84)$$

- and the compensating measure

$$\Gamma_g(t) = \int_0^\infty \int_{\mathbb{R}} g(v)\Gamma(\xi_{[0,s)}, dv; x(s))ds, \quad g \in C^1(\mathbb{R}). \qquad (2.85)$$

2.9. MARKOV ADDITIVE SEMIMARTINGALES

The Markov additive semimartingale (2.82) has the following representation

$$\xi(t) = \xi_0 + \int_0^t \eta(ds; x(s)), \quad t \geq 0, \qquad (2.86)$$

where

$$\eta(ds, x(s)) = b(\xi(s), x(s))ds + c(\xi(s), x(s))dw(s)$$
$$+ \int_{\mathbb{R}} v[\mu - h(v)]\Gamma(\xi(s), dv; x(s))]ds.$$

In the particular case where $\eta(t;x)$, $t \geq 0$, $x \in E$, are locally independent increment processes defined by their generators (2.40), the predictable characteristics are represented as follows:

$$B_x(t) = \int_0^t b(\eta(s;x);x)ds,$$

$$C_x(t) = \int_0^t C(\eta(s;x);x)ds, \qquad (2.87)$$

$$\Gamma_g(t;x) = \int_0^t \int_{\mathbb{R}} g(v)\Gamma(\eta(s;x), dv; x)ds.$$

The predictable characteristics (2.87) of the Markov additive semimartingale (2.86), with switching Markov process $x(t), t \geq 0$, can be characterized by the tripled Markov processes

$$\xi(t), \ A(t), \ x(t), \ t \geq 0,$$

defined by the generator

$$\mathbb{L}\varphi(u,v,x) = [Q + \mathbb{\Gamma}(x) + \mathbb{A}(x)]\varphi(u,v,x),$$

where $A(t)$ is one of the predictable characteristics (2.83)-(2.85) defined by the generator

$$\mathbb{A}(x)\varphi(u) := a(u;x)\varphi'(u).$$

The function $a(u;x)$ is one of the local characteristics $b(u;x)$, $c(u;x)$ and

$$\Gamma_g(u;x) := \int_{\mathbb{R}} g(v)\Gamma(u, dv; x).$$

Chapter 3

Stochastic Systems in the Series Scheme

3.1 Introduction

This chapter deals with the stochastic systems presented in Chapter 2, in a series scheme. That is, for a process $\xi(t), t \geq 0$, as in the previous chapter, we will consider here a family of processes, $\xi^\varepsilon(t), t \geq 0, \varepsilon > 0$, where $0 < \varepsilon < \varepsilon_0$ is the series parameter, defined on a stochastic basis $\Im = (\Omega, \mathcal{F}, \mathbf{F} = (\mathcal{F}_t, t \geq 0), \mathbb{P})$. We are interested in the weak convergence of the probability measures $\mathbb{P} \circ (\xi^\varepsilon)^{-1}$, as $\varepsilon \to 0$. Here ε is supposed to be a sequence $\varepsilon_n \to 0$, as $n \to \infty$. Instead of a common probability space, we could consider different spaces for each ε.

Two different schemes are considered here, the average approximation and the diffusion approximation. The switching semi-Markov process is considered with fast time-scale "ε^{-1}" for average approximation, and "ε^{-2}" for diffusion approximation. The ergodic property of the switching process is used in the average and diffusion approximation algorithms.

The main results presented in this chapter concern the asymptotic representation of compensating operators of the coupled switched-switching processes. First we give results for random evolution (Propositions 3.1-3.5), on which the average and diffusion approximation results will be based. The average approximation is presented for stochastic additive functionals and increment processes (Theorems 3.1-3.2). The diffusion approximation is presented in two different schemes. The first one is the usual one whose equilibrium point is the average limit fixed point (Theorems 3.3-3.5), and the second one is one whose equilibrium point is a deterministic function (Theorems 3.6-3.8). In the next chapter we will also present results for a diffusion approximation whose equilibrium point is a random process.

3.2 Random Evolutions in the Series Scheme

The characterization of random evolutions in the series schemes is considered with two different switching processes: semi-Markov and Markov processes, with the different algorithms.

3.2.1 *Continuous Random Evolutions*

The continuous random evolution with semi-Markov switching in the average scheme with the small series parameter $\varepsilon > 0, \varepsilon \to 0$, is given by a solution of the evolutionary equation (compare with Proposition 2.4),

$$\begin{vmatrix} \frac{d}{dt}\Phi^\varepsilon(t) = \mathbb{\Gamma}(x(t/\varepsilon))\Phi^\varepsilon(t), & t \geq 0, \\ \Phi^\varepsilon(0) = I. \end{vmatrix} \quad (3.1)$$

Here $\mathbb{\Gamma}(x), x \in E$, is the family of generators of the semigroup operators $\Gamma_t(x), t \geq 0, x \in E$, which determines the random evolution in the following form (compare with Definition 2.8)

$$\Phi^\varepsilon(t) = \Gamma_{\varepsilon\gamma(t/\varepsilon)}(x(t/\varepsilon)) \prod_{k=1}^{\nu(t/\varepsilon)} \Gamma_{\varepsilon\theta_k}(x_k), \quad t > 0, \quad \Phi^\varepsilon(0) = I. \quad (3.2)$$

The semi-Markov continuous random evolution $\Phi^\varepsilon(t), t \geq 0$, in the average series scheme can be characterized by the compensating operator on the test functions $\varphi \in C(\mathbb{R}^d \times E)$, given by the following relation (compare with Proposition 2.8)

$$\mathbb{L}^\varepsilon\varphi(u,x) = \varepsilon^{-1}q(x)\Big[\int_0^\infty F_x(ds)\Gamma_{\varepsilon s}(x)\int_E P(x,dy)\varphi(u,y) - \varphi(u,x)\Big]. \quad (3.3)$$

The normalized factor "ε^{-1}" corresponds to the fast time-scaling of the switching semi-Markov process in (3.1). The small time-scaling "ε" in the semigroup $\Gamma_{\varepsilon s}(x)$ provides the representation (3.2) for the random evolution in the series scheme.

As usual, we will suppose that the domain $\mathcal{D}_{\mathbb{\Gamma}(x)}$ contains the Banach space $C^1(\mathbb{R}^d)$.

Proposition 3.1 *The compensating operator (3.3) in the average scheme on the test functions $\varphi \in C^{2,0}(\mathbb{R}^d \times E)$ has the following asymptotic*

3.2. RANDOM EVOLUTIONS IN THE SERIES SCHEME

representation (compare with Proposition 2.9):

$$\mathbb{L}^\varepsilon\varphi(u,x) = [\varepsilon^{-1}Q + \mathbb{\Gamma}(x)P + \varepsilon\theta_2^\varepsilon(x)]\varphi$$
$$= [\varepsilon^{-1}Q + \theta_1^\varepsilon(x)]\varphi, \tag{3.4}$$

where:

$$\theta_k^\varepsilon(x) := \mathbb{\Gamma}^h(x)\mathbb{F}_\varepsilon^{(l_0)}(x)Q_0, \quad \mathbb{F}_\varepsilon^{(k)}(x) := \int_0^\infty \overline{F}_x^{(k)}(s)\Gamma_{\varepsilon s}(x)ds, \tag{3.5}$$

for $k = 1, 2$, and, as usual,

$$Q\varphi(x) = q(x)\int_E P(x, dy)[\varphi(y) - \varphi(x)].$$

PROOF. The same transformation as in the proof of Proposition 2.9 is used with one essential difference. The equation for semigroup is now

$$\Gamma_{\varepsilon s}(x) = I + \varepsilon\mathbb{\Gamma}(x)\int_0^s \Gamma_{\varepsilon v}(x)dv,$$

that is, in differential form,

$$d\Gamma_{\varepsilon s}(x) = \varepsilon\mathbb{\Gamma}(x)\Gamma_{\varepsilon s}(x)ds.$$

□

The continuous random evolution in the *diffusion approximation scheme* with accelerated switching is represented by a solution of the evolutionary equation

$$\left| \begin{array}{l} \frac{d\Phi^\varepsilon}{dt}(t) = \mathbb{\Gamma}^\varepsilon(x(t/\varepsilon^2))\Phi^\varepsilon(t), \quad t \geq 0, \\ \Phi^\varepsilon(0) = I. \end{array} \right. \tag{3.6}$$

The family of generators $\mathbb{\Gamma}^\varepsilon(x), x \in E$, has the following representation

$$\mathbb{\Gamma}^\varepsilon(x) = \varepsilon^{-1}\mathbb{\Gamma}(x) + \mathbb{\Gamma}_1(x). \tag{3.7}$$

Note that the generalization of the average scheme in such a way would not be productive.

The compensating operator of the random evolution (3.6) on the test functions $\varphi \in C(\mathbb{R}^d \times E)$ is given by the relation (in symbolic form, see Section 2.8)

$$\mathbb{L}^\varepsilon\varphi(u,x) = \varepsilon^{-2}q(x)[\mathbb{F}_\varepsilon(x)P - I]\varphi, \tag{3.8}$$

where
$$\mathbb{F}_\varepsilon(x) := \int_0^\infty F_x(ds)\Gamma^\varepsilon_{\varepsilon^2 s}(x). \qquad (3.9)$$

□

Proposition 3.2 *The compensating operator (3.8)-(3.9), in the diffusion approximation scheme, acting on the test functions $\varphi \in C^3(\mathbb{R}^d \times E)$ has the following asymptotic representation:*

$$\begin{aligned}
\mathbb{L}^\varepsilon \varphi(u,x) &= [\varepsilon^{-2}Q + \varepsilon^{-1}\mathbb{\Gamma}(x)P + Q_2(x)P + \varepsilon\theta_3^\varepsilon(x)]\varphi \\
&= [\varepsilon^{-2}Q + \varepsilon^{-1}\mathbb{\Gamma}(x)P + \varepsilon\theta_2^\varepsilon(x)]\varphi \\
&= [\varepsilon^{-2}Q + \varepsilon^{-1}\theta_1^\varepsilon(x)]\varphi,
\end{aligned} \qquad (3.10)$$

where
$$Q_2(x) := \mathbb{\Gamma}_1(x) + \mu_2(x)\mathbb{\Gamma}^2(x), \quad \mu_2(x) := m_2(x)/[2m(x)], \qquad (3.11)$$

and the remaining terms are:
$$\theta_1^\varepsilon(x) := \mathbb{\Gamma}_\varepsilon(x)F_\varepsilon^{(1)}(x)Q_0. \qquad (3.12)$$

$$\theta_2^\varepsilon(x) := [\mathbb{\Gamma}^2(x)F_\varepsilon^{(2)}(x) + \mathbb{\Gamma}_1(x)P]Q_0, \qquad (3.13)$$

$$\theta_3^\varepsilon(x) := \mathbb{\Gamma}_\varepsilon(x)[\mathbb{\Gamma}^2(x)F_\varepsilon^{(3)}(x) + \mathbb{\Gamma}_1(x)F_\varepsilon^{(2)}(x)]Q_0. \qquad (3.14)$$

Here, by definition $\mathbb{\Gamma}_\varepsilon(x) := \mathbb{\Gamma}(x) + \varepsilon\mathbb{\Gamma}_1(x)$.

PROOF. The starting point is the integral equation for semigroup
$$\Gamma^\varepsilon_{\varepsilon^2 s}(x) - I = \varepsilon^2 \mathbb{\Gamma}^\varepsilon(x) \int_0^s \Gamma^\varepsilon_{\varepsilon^2 v}(x)dv, \qquad (3.15)$$

or, in differential form,
$$d\Gamma^\varepsilon_{\varepsilon^2 s}(x) = \varepsilon \mathbb{\Gamma}_\varepsilon(x)\Gamma^\varepsilon_{\varepsilon^2 s}(x)ds. \qquad (3.16)$$

There we use the following relation
$$\varepsilon^2 \mathbb{\Gamma}^\varepsilon(x) = \varepsilon \mathbb{\Gamma}_\varepsilon(x).$$

The initial representation of the compensating operator is
$$\mathbb{L}^\varepsilon = \varepsilon^{-2}Q + \varepsilon^{-2}[\mathbb{F}_\varepsilon(x) - I]Q_0,$$

3.2. RANDOM EVOLUTIONS IN THE SERIES SCHEME

where

$$\mathbb{F}_\varepsilon(x) = \int_0^\infty F_x(ds)\Gamma^\varepsilon_{\varepsilon^2 s}(x)ds \qquad (3.17)$$

is transformed, by using (3.15), into

$$\mathbb{F}_\varepsilon(x) - I = \varepsilon\mathbb{\Gamma}_\varepsilon(x)\mathbb{F}^{(1)}_\varepsilon(x),$$

with

$$\mathbb{F}^{(1)}_\varepsilon(x) := \int_0^\infty \overline{F}_x(s)\Gamma^\varepsilon_{\varepsilon^2 s}(x)ds.$$

Now, by using (3.16), an integration by parts gives

$$\mathbb{F}^{(1)}_\varepsilon(x) = m(x)I + \varepsilon\mathbb{\Gamma}_\varepsilon(x)\mathbb{F}^{(2)}_\varepsilon(x) \qquad (3.18)$$
$$\mathbb{F}^{(2)}_\varepsilon(x) = \frac{1}{2}m_2(x)I + \varepsilon\mathbb{\Gamma}_\varepsilon(x))\mathbb{F}^{(3)}_\varepsilon(x),$$

where, by definition:

$$\mathbb{F}^{(k)}_\varepsilon(x) := \int_0^\infty \overline{F}^{(k)}_x(s)\Gamma^\varepsilon_{\varepsilon^2 s}(x)ds, \quad k = 1, 2, ..., \qquad (3.19)$$
$$\overline{F}^{(1)}_x(s) = \overline{F}_x(s),$$

and

$$m_2(x) := \int_0^\infty s^2 F_x(ds).$$

Now by putting (3.18) and (3.19) into (3.15) and then into (3.8) and by using (3.7), we get (3.10). □

The following result concerns the *coupled random evolution* defined in Definition 2.9.

Proposition 3.3 *The coupled Markov random evolution, with the switching Markov process $x(t), t \geq 0$, in the average scheme can be characterized by the generator*

$$\mathbb{L}^\varepsilon\varphi(u,x) = \varepsilon^{-1}Q\varphi + \mathbb{\Gamma}(x)\varphi.$$

The coupled Markov random evolution in the diffusion approximation scheme can be characterized by the generator

$$\mathbb{L}^\varepsilon\varphi(u,x) = \varepsilon^{-2}Q\varphi + \varepsilon^{-1}\mathbb{\Gamma}(x)\varphi + \mathbb{\Gamma}_1(x)\varphi.$$

It is worth noticing that the characterization of the Markov random evolution is comparatively simpler than the characterization of the semi-Markov random evolution (see Propositions 3.1 and 3.2).

3.2.2 Jump Random Evolutions

The jump random evolution in the average series scheme is represented with fast-scaling

$$\Phi^\varepsilon(t) = \prod_{k=1}^{\nu(t/\varepsilon)} \mathbb{D}^\varepsilon(x_k^\varepsilon), \quad t \geq 0, \quad \Phi^\varepsilon(0) = I. \tag{3.20}$$

The family of bounded operators $\mathbb{D}^\varepsilon(x), x \in E$, is supposed to have the following asymptotic representation

$$\mathbb{D}^\varepsilon(x) = I + \varepsilon \mathbb{D}(x) + \mathbb{D}_1^\varepsilon(x), \quad x \in E, \tag{3.21}$$

on the space $\mathbf{B}_0$ dense in $C_0^2(\mathbb{R}^d \times E)$, with the negligible term

$$\|\mathbb{D}_1^\varepsilon(x)\varphi\| \to 0, \quad \varepsilon \to 0, \quad \varphi \in \mathbf{B}_0. \tag{3.22}$$

The compensating operator of the semi-Markov jump random evolution in the average scheme is represented as follows (see Proposition 2.10)

$$\mathbb{L}^\varepsilon \varphi(u, x) = \varepsilon^{-1} q(x) \Big[\int_E P(x, dy) \mathbb{D}^\varepsilon(y) \varphi(u, y) - \varphi(u, x) \Big]. \tag{3.23}$$

Proposition 3.4 *The compensating operator (3.23) has the following asymptotic representations:*

$$\begin{aligned}\mathbb{L}^\varepsilon \varphi(u, x) &= [\varepsilon^{-1} Q + Q_0 \mathbb{D}(x) + Q_0 \mathbb{D}_1^\varepsilon(x)] \varphi(u, x) \\ &= [\varepsilon^{-1} Q + Q_0 \mathbb{D}_0^\varepsilon(x)] \varphi(u, x),\end{aligned} \tag{3.24}$$

with

$$\mathbb{D}_0^\varepsilon(x) := \mathbb{D}(x) + \mathbb{D}_1^\varepsilon(x)$$

and the negligible term

$$\|Q_0 \mathbb{D}_1^\varepsilon(x)\varphi\| \to 0, \quad \varepsilon \to 0, \quad \varphi \in \mathbf{B}_0, \tag{3.25}$$

where, as usual, $Q_0 \varphi(x) = q(x) \int_E P(x, dy) \varphi(y)$.

3.2. RANDOM EVOLUTIONS IN THE SERIES SCHEME

PROOF. The proof is obtained by putting the expansion (3.21) in (3.23). □

The *jump random evolution* in the diffusion approximation scheme is considered in the accelerated fast-scaling scheme:

$$\Phi^\varepsilon(t) = \prod_{k=1}^{\nu(t/\varepsilon^2)} \mathbb{D}^\varepsilon(x_k^\varepsilon), \quad t > 0, \quad \Phi^\varepsilon(0) = I. \quad (3.26)$$

The family of bounded operators $\mathbb{D}^\varepsilon(x), x \in E$, has the following asymptotic expansion

$$\mathbb{D}^\varepsilon(x) = I + \varepsilon \mathbb{D}(x) + \varepsilon^2 \mathbb{D}_1(x) + \varepsilon^2 \mathbb{D}_2^\varepsilon(x), \quad (3.27)$$

on the test functions $\varphi \in \mathbf{B}_0$, a dense subset of $C^2(\mathbb{R}^d)$, with the negligible term

$$\|\mathbb{D}_2^\varepsilon(x)\varphi\| \to 0, \quad \varepsilon \to 0, \quad \varphi \in \mathbf{B}_0. \quad (3.28)$$

The compensating operator acting on the test functions $\varphi(u,x)$ is

$$\mathbb{L}^\varepsilon \varphi(u,x) = \varepsilon^{-2} q(x) \left[\int_E P(x, dy) \mathbb{D}^\varepsilon(y) \varphi(u, y) - \varphi(u, x) \right]. \quad (3.29)$$

Proposition 3.5 *The compensating operator of the jump random evolution in the diffusion approximation scheme has the following asymptotic representation*

$$\mathbb{L}^\varepsilon \varphi(u,x) = [\varepsilon^{-2} Q + \varepsilon^{-1} Q_0 \mathbb{D}(x) + Q_0 \mathbb{D}_1(x) + Q_0 \mathbb{D}_2^\varepsilon(x)] \varphi(u,x). \quad (3.30)$$

PROOF. The proof is obtained by putting (3.27) in (3.29). □

The Markov jump random evolutions in the average and diffusion approximation schemes are respectively characterized by the generators $\mathbb{L}^\varepsilon$ represented in (3.24) and (3.30), with the generator Q of the switching Markov process.

It is worth noticing that the semi-Markov random evolution is characterized by the compensating operators in asymptotic forms (3.24) and (3.30) with the generator Q of the associated Markov process $x(t), t \geq 0$. The intensity function of the renewal times is $q(x) = 1/m(x)$, where $m(x) := \mathbb{E}\theta_x$, is the mean value of renewal times of the switching semi-Markov process.

3.3 Average Approximation

The phase merging effect for stochastic systems can be achieved under different scaling of the stochastic system and of the switching semi-Markov process. Let us consider the main model of stochastic systems, presented in the previous Chapter 2, that is the stochastic additive functionals model.

3.3.1 *Stochastic Additive Functionals*

Stochastic additive functionals are considered in the following scaling scheme

$$\xi^\varepsilon(t) = \xi^\varepsilon(0) + \int_0^t \eta^\varepsilon(ds; x(s/\varepsilon)), \quad t \geq 0. \tag{3.31}$$

The switching semi-Markov process $x(t), t \geq 0$, on the standard phase space $(E, \mathcal{E})$ is given by the semi-Markov kernel $Q(x, B, t)$,

$$Q(x, B, t) = P(x, B)F_x(t), \quad x \in E, B \in \mathcal{E}, t \geq 0. \tag{3.32}$$

The family of Markov processes with locally independent increments $\eta^\varepsilon(t; x), t \geq 0, x \in E$, with values in the Euclidean space $\mathbb{R}^d, d \geq 1$, is given by the generators

$$\mathbb{\Gamma}_\varepsilon(x)\varphi(u) = a(u, x)\varphi'(u) + \varepsilon^{-1} \int_{\mathbb{R}^d} [\varphi(u + \varepsilon v) - \varphi(u)]\Gamma(u, dv; x), \tag{3.33}$$

defined on the Banach space $C^1(\mathbb{R}^d)$.

The fast time-scaling for the switching process in (3.31) corresponds to the scale factor ε for the increments εv of the switched processes $\eta^\varepsilon(t; x), t \geq 0$. This explains why the large-scale intensity of the switching process is compensated by the small-scale of increments of the switched processes.

By subtracting the first moment of the jump values in (3.33), the generator takes the form

$$\mathbb{\Gamma}_\varepsilon(x)\varphi(u) = \mathbb{\Gamma}(x)\varphi(u) + \gamma_\varepsilon(x)\varphi(u), \tag{3.34}$$

where:

$$\mathbb{\Gamma}(x)\varphi(u) := g(u; x)\varphi'(u), \tag{3.35}$$

$$\gamma_\varepsilon(x)\varphi(u) := \varepsilon^{-1} \int_{\mathbb{R}^d} [\varphi(u + \varepsilon v) - \varphi(u) - \varepsilon v \varphi'(u)]\Gamma(u, dv; x). \tag{3.36}$$

3.3. AVERAGE APPROXIMATION

Here

$$g(u;x) := a(u;x) + b(u;x), \quad b(u;x) := \int_{\mathbb{R}^d} v\Gamma_\varepsilon(u, dv; x). \quad (3.37)$$

Let us consider the following assumptions.

A1: The switching semi-Markov process $x(t), t \geq 0$, is uniformly ergodic with stationary distribution $\pi(B), B \in \mathcal{E}$.

A2: The function $g(u;x), u \in \mathbb{R}^d, x \in E$, is (globally) Lipschitz continuous on $u \in \mathbb{R}^d$, with common Lipschitz constant L for all $x \in E$. So, there exists a global solution to the evolutionary systems

$$\frac{d}{dt}U_x(t) = g(U_x(t), x), \quad x \in E.$$

A3: The operators $\gamma_\varepsilon(x)$ are negligible for $\varphi \in C_0^2(\mathbb{R}^d)$, that is,

$$\|\gamma_\varepsilon(x)\varphi\| \to 0, \quad \varepsilon \to 0.$$

A4: The initial value condition is

$$\xi^\varepsilon(0) \xrightarrow{P} \xi(0), \quad \mathbb{E}|\xi^\varepsilon(0)| \leq c < +\infty.$$

The *average phase merging principle* is formulated as follows.

Theorem 3.1 *Under Assumptions A1-A4, the stochastic additive functional (3.31) converges weakly, as $\varepsilon \to 0$, to the average evolutionary deterministic system $\widehat{U}(t), t \geq 0$, determined by a solution of the evolutionary equation*

$$\left| \begin{array}{l} \frac{d}{dt}\widehat{U}(t) = \widehat{g}(\widehat{U}(t)), \\ \\ \widehat{U}(0) = \xi_0, \end{array} \right. \quad (3.38)$$

where the average velocity $\widehat{g}$ is given by

$$\widehat{g}(u) = \widehat{a}(u) + \widehat{b}(u), \quad (3.39)$$

where:

$$\widehat{a}(u) = \int_E \pi(dx)a(u;x), \quad \widehat{b}(u) = \int_E \pi(dx)b(u;x). \quad (3.40)$$

Remark 3.1. The weak convergence

$$\xi^{\varepsilon}(t) \Longrightarrow \widehat{U}(t), \quad \varepsilon \to 0, \qquad (3.41)$$

means, in particular, that for every finite time $T > 0$,

$$\sup_{0 \le t \le T} \left| \xi^{\varepsilon}(t) - \widehat{U}(t) \right| \xrightarrow{P} 0, \quad \varepsilon \to 0.$$

The verification of the average merging principle (3.38)-(3.40) is made in Chapter 5. The weak convergence (3.41) is investigated in Chapter 6.

The proof of Theorem 3.1 is based on the representation of the stochastic additive functional (3.31) by the associated continuous random evolution (3.1). The corresponding family of generators $\mathbb{\Gamma}_{\varepsilon}(x), x \in E, \varepsilon > 0$, is represented in (3.34). Setting (3.34) in the asymptotic representation (3.4) of the compensating operators (3.3) (see Proposition 3.1) we get the following form of the compensating operator for the stochastic additive functional:

$$\begin{aligned} \mathbb{L}^{\varepsilon}\varphi(u,x) &= [\varepsilon^{-1}Q + \mathbb{\Gamma}(x)P + \varepsilon \theta_1^{\varepsilon}(x)]\varphi \\ &= [\varepsilon^{-1}Q + \theta_1^{\varepsilon}(x)]\varphi, \end{aligned} \qquad (3.42)$$

where

$$\theta_1^{\varepsilon}(x) := \gamma_{\varepsilon}(x)P + \varepsilon \theta_2^{\varepsilon}(x), \qquad (3.43)$$

and the remaining terms $\theta_k^{\varepsilon}(x), k = 1, 2$, are given in (3.19) with the generators $\mathbb{\Gamma}_{\varepsilon}(x)$ and the semigroups $\Gamma_{\varepsilon s}^{\varepsilon}(x)$, depending on the parameter series $\varepsilon > 0$.

The family of generators $\mathbb{\Gamma}(x), x \in E$, is represented in (3.35), and the negligible term $\gamma_{\varepsilon}(x), x \in E$, is represented in (3.36).

Let us now give some heuristic explanation about the phase merging effect of Theorem 3.1. The average algorithm in Theorem 3.1 is evident from the ergodic theorem point of view. The problem is how does the ergodicity principle works?

In order to explain this, let us consider the Markov additive process $\xi^{\varepsilon}(t), x(t/\varepsilon), t \ge 0$, which can be characterized by the following generator,

$$\mathbb{L}^{\varepsilon}\varphi(u,x) = [\varepsilon^{-1}Q + \mathbb{\Gamma}(x)]\varphi(u,x),$$

on the Banach space $C^1(\mathbb{R}^d \times E)$ of $\varphi(u,x)$, where the generator $\mathbb{\Gamma}(x)$ is defined in (3.35), and the negligible operator (3.36) is neglected. The

3.3. AVERAGE APPROXIMATION

uniform ergodicity of the switching Markov process with the generator Q provides the definition of the projection operator Π (see Section 1.6), which satisfies the following property

$$\Pi Q = Q\Pi = 0.$$

The projector Π acts on the functions $\varphi \in \mathbf{B}(E)$ as follows

$$\Pi\varphi(x) = \int_E \pi(dx)\varphi(x)\mathbf{1}(x) = \widehat{\varphi}\mathbf{1}(x),$$

where $\mathbf{1}(x) = 1$, for all $x \in E$, and

$$\widehat{\varphi} = \int_E \pi(dx)\varphi(x).$$

Since $\Pi\varphi(u) = \varphi(u)$, the generator $\mathbb{L}^\varepsilon$ of the Markov additive process acts on a function $\varphi \in C^1(\mathbb{R}^d)$, which does not depend on $x \in E$, as follows,

$$\mathbb{L}^\varepsilon \varphi(u) = \mathbb{\Gamma}(x)\varphi(u).$$

Note that, since $\Pi Q = 0$,

$$\Pi\mathbb{L}^\varepsilon \varphi(u,x) = [\varepsilon^{-1}\Pi Q + \Pi\mathbb{\Gamma}(x)]\varphi(u,x) = \Pi\mathbb{\Gamma}(x)\varphi(u,x).$$

Hence, we have

$$\Pi\mathbb{L}^\varepsilon \Pi\varphi(u,x) = \Pi\mathbb{\Gamma}(x)\Pi\varphi(u,x) = \Pi\mathbb{\Gamma}(x)\Pi\widehat{\varphi}(u),$$

where $\widehat{\varphi}(u) := \int_E \pi(dx)\varphi(u;x)$.

The average evolution in Theorem 3.1 is characterized by the main part of the average generator

$$\widehat{\mathbb{\Gamma}}\Pi = \Pi\mathbb{\Gamma}(x)\Pi.$$

Note that the problem of verification of such a scheme is still open (see Chapters 5 and 6).

The *stochastic homogeneous additive functional* in series scheme

$$\xi^\varepsilon(t) = \xi_0 + \int_0^t \zeta^\varepsilon(ds; x(s/\varepsilon)), \quad t \geq 0, \tag{3.44}$$

is given by the family of Markov processes with independent increments $\zeta^\varepsilon(t;x), t \geq 0, x \in E$, defined by the generators

$$\mathbb{\Gamma}_\varepsilon(x)\varphi(u) = a(x)\varphi'(u) + \varepsilon^{-1}\int_{\mathbb{R}^d}[\varphi(u+\varepsilon v) - \varphi(u)]\Gamma(dv;x), \quad x \in E.$$

The intensity kernel $\Gamma(dv; x), x \in E$, is supposed to satisfy the average condition

$$\int_{\mathbb{R}^d} v\Gamma(dv; x) = b(x), \qquad (3.45)$$

with $b(x), x \in E$, a bounded function.

The average algorithm for the stochastic functional $\xi^\varepsilon(t), t \geq 0$, given in (3.44)-(3.45), is formulated as follows.

Corollary 3.1 *Under Assumptions A1-A4, the weak convergence*

$$\xi^\varepsilon(t) \Longrightarrow \widehat{U}(t) = \xi_0 + \widehat{g}t, \quad t \geq 0, \quad \varepsilon \to 0,$$

holds true. The average deterministic velocities of the drift are defined by the relations:

$$\widehat{g} = \widehat{a} + \widehat{b}, \quad \widehat{a} = \int_E \pi(dx)a(x), \quad \widehat{b} = \int_E \pi(dx)b(x).$$

Corollary 3.2 *The stochastic integral functional in the series scheme*

$$\xi^\varepsilon(t) = \xi_0 + \int_0^t a(x(s/\varepsilon))ds, \quad t \geq 0, \qquad (3.46)$$

converges weakly, as $\varepsilon \to 0$, to the deterministic linear drift

$$\widehat{U}(t) = \xi_0 + \widehat{a}t, \quad t \geq 0,$$

where $\widehat{a} = \int_E \pi(dx)a(x)$.

Corollary 3.3 *Under Assumptions A1, A2, the stochastic evolutionary system, defined by a solution of the evolutionary equation*

$$\left| \begin{array}{l} \frac{d}{dt}U^\varepsilon(t) = g(U^\varepsilon(t); x(t/\varepsilon)), \\ \\ U^\varepsilon(0) = u_0 \end{array} \right. \qquad (3.47)$$

converges weakly to the solution of the average equation

$$\left| \begin{array}{l} \frac{d}{dt}\widehat{U}(t) = \widehat{g}(\widehat{U}(t)), \\ \\ \widehat{U}(0) = u_0. \end{array} \right.$$

3.3.2 Increment Processes

The increment process considered here in the series scheme has the following form

$$\xi^\varepsilon(t) = \xi_0 + \varepsilon \sum_{n=1}^{\nu(t/\varepsilon)} a(x_n), \quad t \geq 0, \tag{3.48}$$

where $\nu(t) := \max\{n \geq 0 : \tau_n \leq t\}, t \geq 0$, is the counting process of renewal moments $\tau_n, n \geq 0$, of the switching semi-Markov process $x(t), t \geq 0$, given by the semi-Markov kernel

$$Q(x, B, t) = P(x, B) F_x(t), \quad x \in E, B \in \mathcal{E}, t \geq 0.$$

The average algorithm for the increment process (3.48) is formulated as follows.

Theorem 3.2 *Let the switching semi-Markov process $x(t), t \geq 0$, be uniformly ergodic with stationary distribution $\pi(B), B \in \mathcal{E}$, satisfying the relation*

$$\pi(dx) = \rho(dx) m(x)/m,$$

where $\rho(dx)$ is the stationary distribution of the embedded Markov chain $x_n, n \geq 0$, defined by the stochastic kernel $P(x, B)$.

The increment function $a(x), x \in E$ is supposed to be bounded, that is, $a \in \mathbf{B}(E)$.

Then the weak convergence

$$\xi^\varepsilon(t) \Longrightarrow \xi_0 + \widehat{a} t, \quad t \geq 0, \quad \varepsilon \to 0,$$

holds true. The average velocity is

$$\widehat{a} = \int_E \rho(dx) a(x)/m. \tag{3.49}$$

The proof of Theorem 3.2 is based on the representation of the increment process (3.48) by the associated jump random evolution (3.20).

The family of shift operators $\mathbb{D}^\varepsilon(x), x \in E$, is given by the relation

$$\mathbb{D}^\varepsilon(x)\varphi(u) = \varphi(u + \varepsilon a(x)).$$

By using Taylor's expansion we get the following asymptotic representation on $\varphi \in C_0^2(\mathbb{R}^d)$

$$\mathbb{D}^\varepsilon(x)\varphi(u) = [I + \varepsilon \mathbb{D}(x) + \varepsilon \mathbb{D}_1^\varepsilon(x)]\varphi(u),$$

where

$$\mathbb{D}(x)\varphi(u) = a(x)\varphi'(u), \quad \mathbb{D}_1^\varepsilon(x)\varphi(u) = \varepsilon^{-1}[\mathbb{D}^\varepsilon(x) - I - \varepsilon\mathbb{D}(x)]\varphi(u),$$

and the negligible term $\mathbb{D}_1^\varepsilon(x)$ satisfies the condition

$$\|\mathbb{D}_1^\varepsilon(x)\varphi\| \to 0, \quad \varepsilon \to 0, \quad \varphi \in C_0^2(\mathbb{R}^d).$$

The compensating operator of the jump random evolution associated with the increment process (3.48) is represented in asymptotic form (3.24) in Proposition 3.4.

Remark 3.2. The formula (3.49) can be explained as follows. The increment process (3.48) takes its increments in the renewal moments $\tau_n, n \geq 0$, connected by the recursive relation

$$\tau_n = \tau_{n-1} + \theta_n, \quad n \geq 0.$$

The sojourn times $\theta_n, n \geq 1$, are defined by the distribution functions

$$F_x(t) = \mathbb{P}(\theta_n \leq t \mid x_{n-1} = x).$$

Hence, it is almost evident that the average value of jumps have to be calculated by using the stationary distribution $\rho(dx)$ of the embedded Markov chain. The average approximation process $\widehat{\xi}(t) = \xi_0 + \widehat{a}t, t \geq 0$, is continuous in time. Hence, the average velocity $\widehat{a}$ of the limit process have to be normalized by the average mean value of the sojourn times

$$m = \int_E \rho(dx)m(x),$$

that is according to (3.49).

The increment process can be considered in more general forms, for example, as follows

$$\xi^\varepsilon(t) = \xi_0 + \varepsilon \sum_{n=1}^{\nu(t/\varepsilon)} a(x_{n-1}, x_n), \quad t \geq 0. \tag{3.50}$$

It is easy to formulate the average algorithm for the increment process (3.50) by taking into account that the coupled sequence $x_{n-1}, x_n, n \geq 0$, is also a Markov chain with the stationary distribution

$$\rho'(dx, dx') = \rho(dx)P(x, dx').$$

Corollary 3.4 *Under the conditions of Theorem 3.2 for the bounded increment function $a(x, x')$, $x, x' \in E$, the weak convergence*

$$\xi^\varepsilon(t) \Longrightarrow \xi_0 + \widehat{a}t, \quad t \geq 0, \quad \varepsilon \to 0,$$

holds. The average velocity of the limit linear drift $\widehat{a}$ is calculated by

$$\widehat{a} = \int_E \rho(dx) \int_E P(x, dx')a(x, x')/m. \tag{3.51}$$

An additional interesting form of the increment process is

$$\xi^\varepsilon(t) = \xi_0 + \varepsilon \sum_{k=1}^{\nu(t/\varepsilon)} a(x_k, \theta_{k+1}), \quad t \geq 0, \tag{3.52}$$

for which we have the following result.

Corollary 3.5 *Let us consider the process (3.52). Then, under the conditions of Theorem 3.2, and the additional condition that the function*

$$\widetilde{a}(x) := \int_0^\infty F_x(dt)a(x, t), \quad x \in E,$$

is bounded, the average velocity of the limit linear drift $\widehat{a}$, is calculated by

$$\widehat{a} = \int_E \rho(dx) \int_0^\infty F_x(dt)a(x, t).$$

3.4 Diffusion Approximation

3.4.1 *Stochastic Integral Functionals*

The diffusion approximation scheme applies to the integral functionals in the series scheme with accelerated time-scaling of the switching semi-Markov process in the following form

$$\alpha^\varepsilon(t) = \alpha_0 + \int_0^t a^\varepsilon(x(s/\varepsilon^2))ds, \quad t \geq 0. \tag{3.53}$$

The velocity function $a^\varepsilon(x)$ is supposed to depend on the parameter series as follows

$$a^\varepsilon(x) = \varepsilon^{-1}a(x) + a_1(x). \tag{3.54}$$

The first term in (3.54) satisfies the *balance condition*

$$\int_E \pi(dx)a(x) = 0. \tag{3.55}$$

The accelerated scaling of the switching process and the balance condition (3.55) provide the diffusion approximation of fluctuation of the integral functionals (3.53). Under the balance condition the average approximation scheme (Section 3.3.1) in Corollary 3.2 provides the weak convergence

$$\int_0^t a(x(s/\varepsilon))ds \Longrightarrow 0, \quad \varepsilon \to 0.$$

Let us state the following conditions.

D1: The switching semi-Markov process $x(t), t \geq 0$, is uniformly ergodic with the stationary distribution $\pi(B), B \in \mathcal{E}$.

D2: The second moments of the sojourn times are uniformly bounded:

$$m_2(x) = \int_0^\infty t^2 F_x(dt) \leq M < +\infty,$$

and

$$\sup_{x \in E} \int_T^\infty t^2 F_x(dt) \longrightarrow 0, \quad T \to \infty.$$

D3: The velocity functions $a(x)$ and $a_1(x)$ are bounded and $a(x)$ satisfies the balance condition (3.55).

Let us denote by $w(t), t \geq 0$, the standard Wiener process, that is, $\mathbb{E}w(t) = 0$ and $\mathbb{E}w(s)w(t) = s \wedge t$.

Theorem 3.3 *Under Conditions D1-D3 the following weak convergence holds*

$$\alpha^\varepsilon(t) \Longrightarrow \alpha^0(t) := \alpha_0 + a_1 t + \sigma w(t), \quad \varepsilon \to 0,$$

provided that $\sigma^2 > 0$. The variance σ^2 is calculated by

$$\sigma^2 = \sigma_0^2 + \sigma_\mu,$$

where

$$\sigma_0^2 = 2 \int_E \pi(dx)a_0(x), \quad a_0(x) = a(x)R_0 a(x),$$

3.4. DIFFUSION APPROXIMATION

$$\sigma_\mu = \int_E \pi(dx)\mu(x)a^2(x), \quad \mu(x) := [m_2(x) - 2m^2(x)]/m(x),$$

and the velocity of the drift is

$$a_1 = \int_E \pi(dx)a_1(x).$$

The potential operator R_0 (see Section 1.6) corresponds to the generator Q associated to the Markov process

$$Q\varphi(x) = q(x)\int_E P(x,dy)[\varphi(y) - \varphi(x)],$$

where $q(x) := 1/m(x)$, $m(x) := \int_0^\infty \overline{F}(t)dt$.

Remark 3.3. The function $\mu(x)$ is positive, if the density f_x (with respect to Lebesgue measure on $\mathbb{R}_+$) of F_x is a completely monotone function. That means if the derivatives of $f_x^{(n)}$, for $n = 1, 2, ...$, exist and $(-1)^n f_x^{(n)}(x) \geq 0$. This class of distribution function is included in the class of decreasing failure rate distribution functions. (See [84]). In the case where f_x is of Polya frequency function of infinite order (PF$_\infty$), we have $\mu(x) \leq 0$. This class of distribution functions is a subset of the class of increasing failure rate distribution functions. We have $\mu(x) = 0$ for exponential distributed renewal times, that is, for switching Markov processes.

Corollary 3.6 *Under Conditions D1-D3, the integral functionals (3.53) with the switching Markov process, converge weakly*

$$\xi^\varepsilon(t) \Longrightarrow \alpha^0(t) := \alpha_0 + a_1 t + \sigma_0 w(t), \quad \varepsilon \to 0,$$

where σ_0^2 is defined as in Theorem 3.3.

Let us give here some heuristic explanation of the diffusion approximation of the integral functional.

By using the representation (3.54), the integral functional takes the form:

$$\int_0^t a_\varepsilon(x(s/\varepsilon^2))ds = \varepsilon^2 \int_0^{t/\varepsilon^2} a_\varepsilon(x(s))ds$$

$$= \varepsilon \int_0^{t/\varepsilon^2} a(x(s))ds + \int_0^t a_1(x(s/\varepsilon^2))ds.$$

It is easy to see that the second term satisfies the average principle

$$\int_0^t a_1(x(s/\varepsilon^2))ds \Longrightarrow \widehat{a}_1 t, \quad \varepsilon \to 0.$$

The first term requires a more thorough explanation. The integral functional with time-scaling

$$\alpha^\varepsilon(t) = \varepsilon \int_0^{t/\varepsilon^2} a(x(s))ds,$$

under the balance condition

$$\int_E \pi(dx)a(x) = 0,$$

induces fluctuations comparable to the accelerated moving determined by the velocity

$$\sigma_0(x) = a(x)R_0 a(x).$$

Indeed, the *potential kernel* $R_0(x, dy)$ can be interpreted as an intensity of transition between state x and dy. Now, the variance of the Wiener process

$$\sigma^2 = \int_E \pi(dx)\sigma_0(x),$$

can be interpreted as a characteristic of the accelerated moving of Wiener process.

3.4.2 Stochastic Additive Functionals

The diffusion approximation is applied to the stochastic additive functionals (Section 3.3) in the series scheme with accelerated switching

$$\xi^\varepsilon(t) = \xi_0 + \int_0^t \eta^\varepsilon(ds; x(s/\varepsilon^2)), \quad t \geq 0. \tag{3.56}$$

The family of processes with locally independent increments $\eta^\varepsilon(t;x), t \geq 0, x \in E$, depends also on the series parameter ε and is determined by the generators

$$\Gamma_\varepsilon(x)\varphi(u) = a_\varepsilon(u;x)\varphi'(u) + \varepsilon^{-2}\int_{\mathbb{R}^d}[\varphi(u+\varepsilon v) - \varphi(u)]\Gamma_\varepsilon(u, dv; x). \tag{3.57}$$

The process $x(t/\varepsilon^2), t \geq 0$, is a semi-Markov process as described in the previous section.

3.4. DIFFUSION APPROXIMATION

The selection of the first two moments of the jump values in (3.57) transforms the generators into the following form

$$\mathbb{\Gamma}_\varepsilon(x)\varphi(u) = g_\varepsilon(u;x)\varphi'(u) + \frac{1}{2}C_\varepsilon(u;x)\varphi''(u) + \gamma_\varepsilon(x)\varphi(u),$$

where

$$\gamma_\varepsilon(x)\varphi(u) := \varepsilon^{-2}\int_{\mathbb{R}^d}[\varphi(u+\varepsilon v) - \varphi(u) - \varepsilon v\varphi'(u) - \frac{\varepsilon^2 v^2}{2}\varphi''(u)]\Gamma_\varepsilon(u,dv;x).$$

Here:

$$g_\varepsilon(u;x) := a_\varepsilon(u;x) + b_\varepsilon(u;x),$$

$$b_\varepsilon(u;x) := \int_{\mathbb{R}^d} v\Gamma_\varepsilon(u,dv;x),$$

$$C_\varepsilon(u;x) := \int_{\mathbb{R}^d} vv^*\Gamma_\varepsilon(u,dv;x).$$

The intensity kernel has the representation

$$\Gamma_\varepsilon(u,dv;x) = \Gamma(u,dv;x) + \varepsilon\Gamma_1(u,dv;x). \qquad (3.58)$$

The velocity of the deterministic drift has the representation

$$g_\varepsilon(u;x) = g(u;x) + \varepsilon g_1(u;x). \qquad (3.59)$$

The time-scaling of the increments in (3.57) is made for the same reasons as in the average scheme (Section 3.3.1). The time-scaling of the intensity kernel is connected with the finiteness of the second moments of increments. The balance condition (3.60) provides the compensation of the velocity $\varepsilon^{-1}g(u;x)$ in the average scheme and appears in the diffusion scheme.

Let us state here the following additional conditions.

D3': The velocity functions $g(u;x)$ and $g_1(u;x)$ belong to $C^1(\mathbb{R}^d \times E)$, and the balance condition is fulfilled

$$\int_E \pi(dx)g(u;x) \equiv 0. \qquad (3.60)$$

D4: The operators

$$\gamma_\varepsilon(x)\varphi(u) := \varepsilon^{-1}\int_{\mathbb{R}^d}[\varphi(u+\varepsilon v)-\varphi(u)-\varepsilon v\varphi'(u)-\frac{\varepsilon^2 v^2}{2}v^2\varphi''(u)]\Gamma_\varepsilon(u,dv;x),$$

are negligible for $\varphi \in C_0^3(\mathbb{R}^d)$, that is,

$$\|\gamma_\varepsilon(x)\varphi\| \longrightarrow 0, \quad \varepsilon \to 0.$$

Theorem 3.4 *Under Assumptions D1, D2, D3' and D4, the following weak convergence holds*

$$\xi^\varepsilon(t) \Longrightarrow \xi^0(t), \quad \varepsilon \to 0,$$

provided that the diffusion coefficient $\widehat{B}(u)$ is positive for $u \in \mathbb{R}^d$. The limit diffusion process $\xi^0(t), t \geq 0$, is defined by the generator

$$\mathbb{L}\varphi(u) = \widehat{g}(u)\varphi'(u) + \frac{1}{2}\widehat{B}(u)\varphi''(u).$$

The velocity of the drift is

$$\widehat{g}(u) = \widehat{g}_1(u) + \widehat{g}_2(u) + \widehat{g}_3(u),$$

where

$$\widehat{g}_k(u) := \int_E \pi(dx)g_k(u;x), \quad k = 1, 2, 3,$$

and:

$$g_2(u;x) := g(u;x)R_0 g'_u(u;x), \quad g_3(u;x) := \mu(x)g(u;x)g'_u(u;x).$$

The covariance function is

$$\widehat{B}(u) = \widehat{B}_0(u) + \widehat{B}_{00}(u) + \widehat{C}_0(u),$$

where:

$$\widehat{C}_0(u) = \int_E \pi(dx)C_0(u;x), \quad \widehat{B}_{00}(u) = \int_E \pi(dx)g_{00}(u;x),$$

$$\widehat{B}_0(u) = \int_E \pi(dx)g_0(u;x), \quad \widehat{C}_0(u;x) := \frac{1}{2}\int_{\mathbb{R}^d} vv^*\Gamma(u,dv;x),$$

$$g_0(u;x) := g(u;x)R_0 g(u;x), \quad g_{00}(u;x) := \mu(x)g(u;x)g^*(u;x).$$

Let us consider a stochastic additive functional $\xi^\varepsilon(t), t \geq 0$, represented by

$$\xi^\varepsilon(t) = \xi_0 + \int_0^t \zeta^\varepsilon(ds; x(s/\varepsilon^2)), \quad t \geq 0. \tag{3.61}$$

3.4. DIFFUSION APPROXIMATION

The family of *Markov processes with independent increments* $\zeta^\varepsilon(t;x), t \geq 0, x \in E$, is determined by the generators

$$\mathbb{\Gamma}_\varepsilon(x)\varphi(u) = a_\varepsilon(x)\varphi'(u) + \varepsilon^{-2} \int_{\mathbb{R}^d} [\varphi(u+\varepsilon v) - \varphi(u)] \Gamma_\varepsilon(dv;x). \quad (3.62)$$

Subtracting the first two moments of jump values, the transformed generators take the following form:

$$\mathbb{\Gamma}_\varepsilon(x)\varphi(u) = g_\varepsilon(x)\varphi'(u) + \frac{1}{2}C_\varepsilon(x)\varphi''(u) + \gamma_\varepsilon(x)\varphi(u).$$

$$\gamma_\varepsilon(x)\varphi(u) := \varepsilon^{-2} \int_{\mathbb{R}^d} [\varphi(u+\varepsilon v) - \varphi(u) - \varepsilon v\varphi'(u) - \frac{\varepsilon^2 v^2}{2}\varphi''(u)] \Gamma_\varepsilon(dv;x).$$

Here:

$$g_\varepsilon(x) := a_\varepsilon(x) + b_\varepsilon(x),$$

$$b_\varepsilon(x) := \int_{\mathbb{R}^d} v \Gamma_\varepsilon(dv;x),$$

$$C_\varepsilon(x) := \int_{\mathbb{R}^d} vv^* \Gamma_\varepsilon(dv;x).$$

The velocity of the deterministic drift has the representation

$$g^\varepsilon(x) = \varepsilon^{-1}g(x) + g_1(x). \quad (3.63)$$

The intensity kernel

$$\Gamma_\varepsilon(dv;x) = \Gamma(dv;x) + \varepsilon \Gamma_1(dv;x). \quad (3.64)$$

Then the following balance condition holds

$$\int_E \pi(dx)g(x) = 0. \quad (3.65)$$

The first two moments of the increments are bounded functions:

$$b_k(x) := \int_{\mathbb{R}^d} v\Gamma_k(dv;x), \quad C_k^2(x) := \int_{\mathbb{R}^d} vv^*\Gamma_k(dv;x), \quad k = 0, 1.$$

Corollary 3.7 *Under Assumptions D1-D2, the following weak convergence holds*

$$\int_0^t \zeta^\varepsilon(ds; x(s/\varepsilon^2)) \Longrightarrow \zeta^0(t), \quad \varepsilon \to 0.$$

The limit diffusion process $\zeta^0(t), t \geq 0$ is determined by the generator

$$\mathbb{L}^0 \varphi(u) = \widehat{g}_1 \varphi'(u) + \frac{1}{2}\widehat{B}\varphi''(u).$$

Here:

$$\widehat{g}_1 = \int_E \pi(dx) g_1(x), \quad \widehat{B} = \int_E \pi(dx)[2g_0(x) + C_0(x)],$$

$$g_0(x) := g(x) R_0 g(x), \quad C_0(x) := \int_{\mathbb{R}^d} vv^* \Gamma(dv; x).$$

3.4.3 Stochastic Evolutionary Systems

The evolutionary stochastic system in the diffusion approximation scheme is given by the evolutionary equation

$$\left| \begin{array}{l} \frac{d}{dt} U^\varepsilon(t) = g^\varepsilon(U^\varepsilon(t); x(t/\varepsilon^2)), \\ U^\varepsilon(0) = u. \end{array} \right.$$

The velocity g^ε has the following representation

$$g^\varepsilon(u; x) = \varepsilon^{-1} g(u; x) + g_1(u; x).$$

The balance condition (3.60) holds.

Corollary 3.8 *Under Assumptions D1, D2, and D3', the following weak convergence holds*

$$U^\varepsilon(t) \Longrightarrow \zeta^0(t), \quad \varepsilon \to 0,$$

provided that the diffusion coefficient $\widehat{B}$ is positive. The limit diffusion process $\zeta^0(t), t \geq 0$, is determined by the generator

$$\mathbb{L}^0 \varphi(u) = \widehat{g}(u) \varphi'(u) + \frac{1}{2}\widehat{B}(u)\varphi''(u).$$

The velocity of the drift is

$$\widehat{g}(u) = \widehat{g}_1(u) + \widehat{g}_2(u) + \widehat{g}_3(u),$$

where:

$$\widehat{g}_1(u) := \int_E \pi(dx) g_1(u; x), \quad \widehat{g}_2(u) := \int_E \pi(dx) g_2(u; x),$$

3.4. DIFFUSION APPROXIMATION

$$\widehat{g}_3(u) = \int_E \pi(dx)\mu(x)g(u;x)g'_u(u;x), \quad g_2(u;x) := g(u;x)R_0 g'_u(u;x).$$

The covariance function is

$$\widehat{B}(u) = 2\widehat{g}_0(u) + \widehat{g}_{00}(u),$$

where:

$$\widehat{g}_0(u) := \int_E \pi(dx)g_0(u;x), \quad g_0(u;x) := g(u;x)R_0 g(u;x),$$

$$\widehat{g}_{00}(u) := \int_E \pi(dx)\mu(x)g(u;x)g^*(u;x).$$

3.4.4 Increment Processes

The diffusion approximation for the increment processes in the series scheme is considered with the following time-scaling

$$\beta^\varepsilon(t) = \beta_0 + \varepsilon \sum_{n=1}^{\nu(t/\varepsilon^2)} a_\varepsilon(x_n), \quad t \geq 0. \tag{3.66}$$

The values of jumps are

$$a_\varepsilon(x) = a(x) + \varepsilon a_1(x).$$

The following balance condition holds

$$\int_E \rho(dx)a(x) = 0,$$

where $\rho(dx)$ is the stationary distribution of the embedded Markov chain $x_n, n \geq 0$.

Theorem 3.5 *Under Assumptions D1-D3, the following weak convergence holds*

$$\beta^\varepsilon(t) \Longrightarrow b + at + \sigma w(t), \quad \varepsilon \to 0,$$

provided that $\sigma^2 > 0$. The variance σ^2 is calculated by

$$\sigma^2 = \sigma_0^2 + \sigma_1^2,$$

where:

$$\sigma_0^2 = \int_E \rho(dx)a^2(x)/m, \quad \sigma_1^2 = 2\int_E \pi(dx)C_0(x),$$

$$C_0(x) := C(x)R_0C(x), \quad C(x) := b(x)/m(x),$$

$$b(x) := \int_E P(x,dy)a(y).$$

The drift velocity is

$$a = \int_E \rho(dx)a_1(x)/m, \quad m := \int_E \rho(dx)m(x).$$

Remark 3.4. As in the averaging scheme (Section 3.3.2), since the increment process (3.66) has its jumps at the renewal moments, the average effect is realized by using the stationary distribution of the embedded Markov chain $\rho(dx)$. The normalized factor $1/m$ transforms the discrete jumps of the increment process into the continuous characteristics of the limit process.

3.5 Diffusion Approximation with Equilibrium

The balance condition in the diffusion approximation for the stochastic additive functional in the series scheme, considered in Section 3.4.2, provides the homogeneous in time limit diffusion process. In applications there are situations in which the average approximation is not trivial, that is, the limit process must be considered as an equilibrium process, very often deterministic, determining the main behavior of the stochastic systems on the increasing time intervals. The problem of approximation of fluctuations of stochastic systems with respect to equilibrium is considered in this section.

3.5.1 Locally Independent Increment Processes

First we consider a stochastic system in series scheme with small series parameter $\varepsilon > 0, \varepsilon \to 0$, described by a Markov process with locally independent increments $\eta^\varepsilon(t), t \geq 0$, on the Euclidean space $\mathbb{R}^d, d \geq 1$, given by the generator

$$\mathbb{L}^\varepsilon \varphi(u) = \varepsilon^{-1} \int_{\mathbb{R}^d} [\varphi(u + \varepsilon v) - \varphi(u)] \Gamma_\varepsilon(u, dv). \tag{3.67}$$

3.5. DIFFUSION APPROXIMATION WITH EQUILIBRIUM

The main condition in the average scheme is the asymptotic representation of the first moment of jumps

$$b_\varepsilon(u) := \int_{\mathbb{R}^d} v \Gamma_\varepsilon(u, dv) = b(u) + \varepsilon b_1(u) + \varepsilon \theta^\varepsilon(u),$$

with bounded continuous functions $b(u)$, $b_1(u)$ and with the negligible term

$$\|\theta^\varepsilon\| \to 0, \quad \varepsilon \to 0.$$

Then the Markov process $\eta^\varepsilon(t), t \geq 0$, converges weakly

$$\eta^\varepsilon(t) \Longrightarrow \rho(t), \quad \varepsilon \to 0,$$

to the solution of the evolutionary equation

$$\left| \begin{array}{l} \frac{d}{dt}\rho(t) = b(\rho(t)), \\ \rho(0) = \eta^\varepsilon(0) = u. \end{array} \right.$$

If there exists an equilibrium point ρ for the velocity $b(u)$, that is,

$$b(\rho) = 0,$$

and the initial value of the process is close to the point ρ, (see (3.75)) then the weak convergence

$$\eta^\varepsilon(t) \Longrightarrow \rho, \quad \varepsilon \to 0, \quad t \to \infty, \tag{3.68}$$

holds.

Approximation of the fluctuation $\eta^\varepsilon(t) - \rho$ is considered in the following centered and normalized scheme

$$\zeta^\varepsilon(t) := \eta^\varepsilon(t/\varepsilon) - \varepsilon^{-1}\rho, \quad t \geq 0. \tag{3.69}$$

Such a normalization can be explained by noticing that

$$\zeta^\varepsilon(t) := [\varepsilon \eta^\varepsilon(t/\varepsilon) - \rho]/\varepsilon. \tag{3.70}$$

The convergence (3.68) provides the weak convergence

$$\varepsilon \eta^\varepsilon(t/\varepsilon) \Longrightarrow \rho, \quad \varepsilon \to 0, \quad t \to \infty.$$

Hence, the normalized scheme (3.70) is productive.

Theorem 3.6 *Let the intensity of the jump values of the Markov process $\eta^\varepsilon(t), t \geq 0$, given by the generator (3.67), have the asymptotic representations of the first two moments of jumps as $\varepsilon \to 0$:*

$$b^\varepsilon(z,u) := \int_{\mathbb{R}^d} v\Gamma_\varepsilon(z+\varepsilon u, dv) = b(z) + \varepsilon b(z,u) + \varepsilon \theta_1^\varepsilon(z,u), \quad (3.71)$$

$$B^\varepsilon(z,u) := \int_{\mathbb{R}^d} vv^*\Gamma_\varepsilon(z+\varepsilon u, dv) = B(z) + \theta_2^\varepsilon(z,u), \quad (3.72)$$

with the negligible residual terms

$$\|\theta_i^\varepsilon\| \to 0, \quad \varepsilon \to 0, \quad i = 1,2. \quad (3.73)$$

Then the normalized centered process (3.69) converges weakly, as $\varepsilon \to 0$, to the diffusion process $\zeta^0(t), t \geq 0$, given by the following generator

$$\mathbb{L}^0\varphi(u) = b(\rho,u)\varphi'(u) + \frac{1}{2}B(\rho)\varphi''(u). \quad (3.74)$$

The initial value of the limit diffusion is

$$\zeta^0(0) = \lim_{\varepsilon \to 0}[\varepsilon\eta^\varepsilon(0) - \rho]/\varepsilon, \quad (3.75)$$

that is $\varepsilon\eta^\varepsilon(0) \sim \rho + \varepsilon\zeta^0(0)$.

Remark 3.5. Let the intensity kernel be represented by

$$\Gamma_\varepsilon(u,dv) = \Gamma(u,dv) + \varepsilon\Gamma_1(u,dv), \quad (3.76)$$

and the kernel $\Gamma(u,dv)$ have continuous derivative in u. Then the asymptotic representation (3.71) has the following form

$$b^\varepsilon(z,u) = b(z) + \varepsilon[b_1(z) + ub'(u)] + \varepsilon\theta^\varepsilon(z,u),$$

where

$$b_1(z) := \int_{\mathbb{R}^d} v\Gamma_1(z,dv).$$

Corollary 3.9 *Under the conditions of Theorem 3.6 and the additional condition (3.76), the limit diffusion process $\zeta^0(t), t \geq 0$, is defined by the generator*

$$\mathbb{L}^0\varphi(u) = b_\rho(u)\varphi'(u) + \frac{1}{2}B(\rho)\varphi''(u),$$

where
$$b_\rho(u) := b_1(\rho) + ub'(\rho),$$
that is the Ornstein-Uhlenbeck diffusion process.

3.5.2 Stochastic Additive Functionals with Equilibrium

More complicated but some what similar is the diffusion approximation of the stochastic additive functional (Section 3.3.1) in the series scheme satisfying the average approximation conditions with non-zero average limit processes. That is, the stochastic additive functional with Markov switching in the average approximation scheme is represented as follows

$$\xi_\varepsilon(t) = \xi_0^\varepsilon + \int_0^t \eta^\varepsilon(ds; x(s/\varepsilon)), \quad t \geq 0.$$

The family of Markov processes with locally independent increments $\eta^\varepsilon(t; x), t \geq 0, x \in E$, with values in the Euclidean space $\mathbb{R}^d, d \geq 1$, is given by the generators (Section 3.3.1)

$$\Gamma_\varepsilon(x)\varphi(u) = g(u,x)\varphi'(u) + \varepsilon^{-1}\gamma_\varepsilon(x)\varphi(u), \qquad (3.77)$$

$$\gamma_\varepsilon(x)\varphi(u) = \int_{\mathbb{R}^d} [\varphi(u+\varepsilon v) - \varphi(u) - \varepsilon v\varphi'(u)]\Gamma(u,dv;x), \qquad (3.78)$$

defined on the Banach space $C^1(\mathbb{R}^d)$.

The switching Markov process $x(t), t \geq 0$, on the standard state space $(E, \mathcal{E})$ is given by the generator

$$Q\varphi(x) = q(x) \int_E P(x,dy)[\varphi(y) - \varphi(x)]. \qquad (3.79)$$

According to Theorem 3.1, the following weak convergence holds

$$\xi^\varepsilon(t) \Longrightarrow \widehat{\xi}(t), \quad \varepsilon \to 0.$$

The limit process $\widehat{\xi}(t), t \geq 0$, is a solution of the deterministic evolution equation

$$\left| \begin{array}{l} \frac{d}{dt}\widehat{\xi}(t) = \widehat{g}(\widehat{\xi}(t)), \\ \widehat{\xi}(0) = \widehat{\xi}_0. \end{array} \right. \qquad (3.80)$$

The average velocity $\widehat{g}(u), u \in \mathbb{R}^d$, is defined as follows

$$\widehat{g}(u) = \int_E \pi(dx) g(u; x).$$

Now we consider the centered stochastic additive functional

$$\zeta^\varepsilon(t) = \varepsilon^{-1}[\xi^\varepsilon(t) - \widehat{\xi}(t)], \quad t \geq 0, \tag{3.81}$$

with the re-scaled switching Markov process as follows

$$\xi^\varepsilon(t) = \xi_0^\varepsilon + \int_0^t \eta^\varepsilon(ds; x(s/\varepsilon^2)), \quad t \geq 0, \tag{3.82}$$

and with the more general of the stochastic additive functional $\eta^\varepsilon(t; x), t \geq 0, x \in E$,

$$\mathbb{\Gamma}^\varepsilon(x)\varphi(u) = g^\varepsilon(u; x)\varphi'(u) + \gamma_\varepsilon(x)\varphi(u), \tag{3.83}$$

where

$$\gamma_\varepsilon(x)\varphi(u) = \varepsilon^{-2} \int_{\mathbb{R}^d} [\varphi(u + \varepsilon v) - \varphi(u) - \varepsilon v \varphi'(u)] \Gamma_\varepsilon(u, dv; x). \tag{3.84}$$

This generalization means that the velocity of drift $g_\varepsilon(u; x)$ and the intensity kernel $\Gamma_\varepsilon(u, dv; x)$ now depend on the parameter series in the following way

$$g_\varepsilon(u; x) = g(u; x) + \varepsilon g_1(u; x), \tag{3.85}$$

and

$$\Gamma_\varepsilon(u, dv; x) = \Gamma(u, dv; x) + \varepsilon \Gamma_1(u, dv; x). \tag{3.86}$$

Subtracting the second moment of jump values in (3.83)-(3.84) gives the representation

$$\mathbb{\Gamma}^\varepsilon(x)\varphi(u) = g^\varepsilon(u; x)\varphi'(u) + \frac{1}{2} C_\varepsilon(u; x)\varphi''(u) + \gamma_\varepsilon^0(x)\varphi(u). \tag{3.87}$$

Here:

$$\gamma_\varepsilon^0(x) := \gamma_\varepsilon(x) - \frac{1}{2} C_\varepsilon(u; x)\varphi''(u)$$

$$= \varepsilon^{-2} \int_{\mathbb{R}^d} [\varphi(u + \varepsilon v) - \varphi(u) - \varepsilon v \varphi'(u) - \frac{\varepsilon^2 v^2}{2} \varphi''(u)] \Gamma_\varepsilon(u, dv; x),$$

3.5. DIFFUSION APPROXIMATION WITH EQUILIBRIUM

and
$$C_\varepsilon(u;x) := \int_{\mathbb{R}^d} vv^* \Gamma_\varepsilon(u, dv; x). \tag{3.88}$$

From (3.86), we get in (3.88)
$$C_\varepsilon(u;x) = C(u;x) + \varepsilon C_1(u;x),$$

where:
$$C(u;x) = \int_{\mathbb{R}^d} vv^* \Gamma(u, dv; x), \quad C_1(u;x) = \int_{\mathbb{R}^d} vv^* \Gamma_1(u, dv; x)$$

Theorem 3.7 *(Diffusion approximation without balance condition). Let the following conditions be fulfilled.*

D1': *The velocity and the intensity kernel are represented by (3.85) and (3.86).*

D2': *The velocity functions and the second moments of jumps have the following asymptotic expansion:*

$$g_\varepsilon(v + \varepsilon u; x) = g(v; x) + \varepsilon u g'_v(v; x) + \theta_1^\varepsilon(v, u; x),$$

$$g_1(v + \varepsilon u; x) = g_1(v; x) + \theta_2^\varepsilon(v, u; x),$$

$$C(v + \varepsilon u; x) = C(v; x) + \theta_3^\varepsilon(v, u; x),$$

with the negligible terms $\theta_k^\varepsilon(v, u; x), k = 1, 2, 3$ *satisfying the condition, for any* $R > 0$,

$$\sup_{\substack{x \in E \\ |u| \leq R}} |\theta_k^\varepsilon(v, u; x)| \to 0, \quad \varepsilon \to 0.$$

The negligible term $\gamma_\varepsilon^0(x)$ *satisfies the following condition*

$$\|\gamma_\varepsilon^0(x)\varphi\| \to 0, \quad \varepsilon \to 0, \quad \varphi \in C^3(\mathbb{R}^d).$$

D3": *The initial values satisfy*

$$\xi_0^\varepsilon = \widehat{\xi}_0 + \varepsilon \zeta_0^\varepsilon, \quad \zeta_0^\varepsilon \Rightarrow \widehat{\zeta}_0, \quad \sup_{\varepsilon > 0} \mathbb{E}|\zeta_0^\varepsilon| \leq C < +\infty.$$

Then the weak convergence holds

$$\zeta^\varepsilon(t) \Longrightarrow \widehat{\zeta}(t), \quad \varepsilon \to 0,$$

provided that $B(v) > 0$.

The limit diffusion process $\widehat{\zeta}(t), t \geq 0$, is determined by the generator of the coupled Markov process $\widehat{\zeta}(t), \widehat{\xi}(t), t \geq 0$,

$$\widehat{\mathbb{L}}\varphi(u,v) = b(v,u)\varphi'_u(u,v) + \frac{1}{2}B(v)\varphi''_{uu}(u,v) + \widehat{g}(v)\varphi'_v(u,v),$$

where:

$$b(v,u) = \widehat{g}_1(v) + u\widehat{g}'(v),$$

$$\widehat{g}(v) := \int_E \pi(dx)g(v;x), \quad \widehat{g}_1(v) := \int_E \pi(dx)g_1(v;x).$$

The covariance function is

$$B(v) = \widehat{B}_0(v) + \widehat{C}(v),$$

where:

$$\widehat{B}_0(v) = 2\int_E \pi(dx)\tilde{g}(u;x)R_0\tilde{g}(u;x),$$

$$\widehat{C}(v) = \int_E \pi(dx)C(v;x).$$

Here

$$\tilde{g}(v;x) := g(v;x) - \widehat{g}(v),$$

and R_0 is the potential operator of Q (Section 1.6).

This means that the coupled Markov process $\widehat{\zeta}(t), \widehat{\xi}(t), t \geq 0$, can be defined as a solution of the system of stochastic differential equations

$$d\widehat{\zeta}(t) = b(\widehat{\zeta}(t), \widehat{\xi}(t))dt + \sigma(\widehat{\xi}(t))dw(t),$$

$$d\widehat{\xi}(t) = \widehat{g}(\widehat{\xi}(t))dt.$$

The covariance function $\sigma(v)$ is determined from the representation

$$B(v) = \sigma(v)\sigma^*(v).$$

Remark 3.6. The limit diffusion process $\widehat{\zeta}(t), t \geq 0$, is not homogeneous in time and is determined by the generator

$$\widehat{\mathbb{L}}_t \varphi(u) = b(\widehat{\xi}(t), u)\varphi'(u) + \frac{1}{2} B(\widehat{\xi}(t))\varphi''(u).$$

The limit diffusion process is switched by the equilibrium process $\widehat{\xi}(t), t \geq 0$.

Remark 3.7. The stationary regime in the averaged process (3.80) is obtained when the velocity has an equilibrium point ρ, that is, $\widehat{g}(\rho) = 0$. Then the limit diffusion process $\widehat{\zeta}(t)$, $t \geq 0$, is of the Ornstein-Uhlenbeck type with generator

$$\widehat{\mathbb{L}}^0 \varphi(u) = b(u)\varphi'(u) + \frac{1}{2} B\varphi''(u),$$

where:

$$b(u) = b_0 + ub_1,$$

$$b_0 = \widehat{b}(\rho), \quad b_1 = \widehat{a}'(\rho), \quad B = B(\rho).$$

3.5.3 Stochastic Evolutionary Systems with Semi-Markov Switching

Now the stochastic evolutionary systems in diffusion approximation scheme considered in Section 3.4.3 is investigated without balance condition (3.60) but under assumption of average approximation conditions of Corollary 3.3, (Section 3.3).

The centered and normalized process is considered as follows

$$\zeta^\varepsilon(t) = \varepsilon^{-1}[U^\varepsilon(t) - \widehat{U}(t)], \tag{3.89}$$

The stochastic evolutionary system $U^\varepsilon(t)$ is described by a solution of the evolutionary equation in $\mathbb{R}^d$

$$\frac{d}{dt} U^\varepsilon(t) = a_\varepsilon(U^\varepsilon(t); x(t/\varepsilon^2)), \tag{3.90}$$

with

$$a_\varepsilon(u; x) = a(u; x) + \varepsilon a_1(u; x), \tag{3.91}$$

where $u \in \mathbb{R}^d$ and $x \in E$.

The switching semi-Markov process $x(t), t \geq 0$, on the standard state space $(E, \mathcal{E})$, is given by the semi-Markov kernel

$$Q(x, B, t) = P(x, B) F_x(t), \tag{3.92}$$

for $x \in E$, $B \in \mathcal{E}$, and $t \geq 0$, supposed to be uniformly ergodic with the stationary distribution $\pi(B), B \in \mathcal{E}$, satisfying the relation

$$\pi(dx) = \rho(dx) m(x) / m, \tag{3.93}$$

where $\rho(B), B \in \mathcal{E}$, is the stationary distribution of the embedded Markov chain $x_n, n \geq 0$, given by the stochastic kernel

$$P(x, B) := \mathbb{P}(x_{n+1} \in B \mid x_n = x). \tag{3.94}$$

As usual:

$$m(x) := \int_0^\infty \overline{F}_x(t) dt, \quad \overline{F}_x(t) := 1 - F_x(t), \quad m := \int_E \rho(dx) m(x). \tag{3.95}$$

The deterministic average process $\widehat{U}(t), t \geq 0$, is defined by a solution of the average evolutionary equation

$$\frac{d}{dt}\widehat{U}(t) = \widehat{a}(\widehat{U}(t)), \tag{3.96}$$

with the average velocity

$$\widehat{a}(u) = \int_E \pi(dx) a(u; x). \tag{3.97}$$

Theorem 3.8 *Let the stochastic evolutionary system (3.89) be defined by relations (3.89)-(3.97) and the following conditions be fulfilled.*

C1: *The switching semi-Markov process $x(t)$, $t \geq 0$, is uniformly ergodic with stationary distribution $\pi(dx)$ on the compact phase space E.*
C2: *The following asymptotic expansions take place:*

$$a(v + \varepsilon u; x) = a(v; x) + \varepsilon u a'_v(v; x) + \theta_0^\varepsilon(v, u; x).$$

$$a_1(v + \varepsilon u; x) = a_1(v; x) + \theta_1^\varepsilon(v, u; x),$$

where, for any $R > 0$,

$$\sup_{\substack{|v| \vee |u| \leq R \\ x \in E}} |\theta_i^\varepsilon(v, u; x)| \to 0, \quad \varepsilon \to 0, \quad i = 0, 1.$$

3.5. DIFFUSION APPROXIMATION WITH EQUILIBRIUM

Moreover, the velocity functions $a(u;x)$ and $a_1(u;x)$ satisfy the global solution of equations (3.90) and (3.91).
Then the weak convergence for $0 \le t \le T$,

$$\zeta^\varepsilon(t) \Longrightarrow \zeta^0(t), \quad \varepsilon \to 0,$$

takes place. The limit diffusion process $\zeta^0(t), t \ge 0$, is determined by the generator of the coupled process $\zeta^0(t), \widehat{U}(t), t \ge 0$,

$$\mathbb{L}\varphi(u,v) = b(u,v)\varphi'_u(u,v) + \frac{1}{2}B(v)\varphi''_{uu}(u,v) + \widehat{a}(v)\varphi'_v(u,v). \quad (3.98)$$

Here:

$$b(u,v) = \widehat{a}_1(v) + u\widehat{a}'(v), \quad (3.99)$$

$$\widehat{a}(v) = \int_E \pi(dx)a(v;x), \quad \widehat{a}_1(v) = \int_E \pi(dx)a_1(v;x).$$

The covariance matrix $B(v), v \in \mathbb{R}^d$, is determined by the relations:

$$B(v) = B_0(v) + B_1(v), \quad (3.100)$$

$$B_0(v) = 2\int_E \pi(dx)\widetilde{a}(v;x)R_0\widetilde{a}(v;x),$$

$$B_1(v) = \int_E \pi(dx)\mu(x)\widetilde{a}(v;x)\widetilde{a}^*(v;x) \quad (3.101)$$

$$\mu(x) = [m_2(x) - 2m^2(x)]/m(x)$$

$$\widetilde{a}(v;x) = a(v;x) - \widehat{a}(v).$$

In the particular case of Markov switching, we have $\mu(x) = 0$ (see Remark 3.3, page 83).

The limit diffusion process $\zeta^0(t), t \ge 0$, is nonhomogeneous in time and is solution of the following SDE

$$d\zeta^0(t) = [a_1(\widehat{U}(t)) + \widehat{a}'(\widehat{U}(t))\zeta^0(t)]dt + B^{1/2}(\widehat{U}(t))dw(t), \quad (3.102)$$

where $w(t), t \ge 0$ is the standard Wiener process in $\mathbb{R}^d$.

The stationary regime for the average process $\widehat{U}(t), t \ge 0$, is obtained when the average velocity $\widehat{a}(v)$ has an equilibrium point ρ, that is, $\widehat{a}(\rho) = 0$. Then the limit diffusion process $\widehat{\zeta}(t), t \ge 0$, is an Ornstein-Uhlenbeck process with the following generator

$$\widehat{\mathbb{L}}\varphi(u) = b(u)\varphi'(u) + \frac{1}{2}B\varphi''(u),$$

where
$$b(u) = b_1 + ub_0, \quad b_1 = \widehat{a}_1(\rho), \quad b_0 = \widehat{a}'(\rho), \quad B = B(\rho).$$

PROOF. The proof of Theorem 3.8 is divided into several steps. First, the extended Markov chain
$$\zeta_n^\varepsilon = \zeta^\varepsilon(\varepsilon^2 \tau_n), \quad \widehat{U}_n^\varepsilon = \widehat{U}(\varepsilon^2 \tau_n), \quad x_n = x(\tau_n), \quad n \geq 0, \quad (3.103)$$
is considered, where $\tau_n, n \geq 0$, is the sequence of the Markov renewal moments (moments of jumps of the semi-Markov process $x(t), t \geq 0$), that is:
$$\tau_{n+1} = \tau_n + \theta_{n+1}, \quad n \geq 0,$$
$$F_x(t) = \mathbb{P}(\theta_{n+1} \leq t \mid x_n = x).$$

Let us introduce the following families of semigroups:
$$\Gamma_t^\varepsilon(x)\varphi(u) = \varphi(U_x^\varepsilon(t)), \quad U_x^\varepsilon(0) = u \in \mathbb{R}^d, \quad (3.104)$$
where $U_x^\varepsilon(t), t \geq 0$, is a solution of the evolutionary system
$$\frac{d}{dt}U_x^\varepsilon(t) = a_\varepsilon(U_x^\varepsilon(t); x), \quad x \in E, \quad (3.105)$$
and, similarly,
$$\widehat{A}_t\varphi(v) = \varphi(\widehat{U}(t)), \quad \widehat{U}(0) = v \in \mathbb{R}^d, \quad (3.106)$$
where $\widehat{U}(t), t \geq 0$, is the solution of the average evolutionary system (3.96).

It is worth noticing that the generators of semigroups (3.104) and (3.106) are respectively:
$$\boldsymbol{\Gamma}_\varepsilon(x)\varphi(u) = a_\varepsilon(u; x)\varphi'(u),$$
$$\widehat{\mathbb{A}}\,\varphi(v) = \widehat{a}(v)\varphi'(v).$$

The following generators will be also used:
$$\boldsymbol{\Gamma}(x)\varphi(u) = a(u; x)\varphi'(u),$$
$$\widetilde{\boldsymbol{\Gamma}}(x)\varphi(u) = \widetilde{a}(u; x)\varphi'(u), \quad \widetilde{a}(u; x) := a(u; x) - \widehat{a}(u).$$

The main object in asymptotic analysis with semi-Markov processes is the compensating operator of the extended embedded Markov chain (3.103) given here in the next lemma.

3.5. DIFFUSION APPROXIMATION WITH EQUILIBRIUM

Lemma 3.1 *The compensating operator of the extended embedded Markov chain (3.103) is determined by the relation*

$$\mathbb{L}^\varepsilon \varphi(u,v,x) = \varepsilon^{-2} q(x) [\int_0^\infty F_x(dt) A^\varepsilon_{\varepsilon^2 t}(x,v) \widehat{A}_{\varepsilon^2 t}(v) P\varphi(u,v,x) - \varphi(u,v,x)], \quad (3.107)$$

where the semigroup $\Gamma^\varepsilon_t(x,v), t \geq 0$, *is defined by the generator:*

$$\mathbb{A}^\varepsilon(v;x)\varphi(u) = [a^\varepsilon(v+\varepsilon u;x) - \widetilde{a}(v)]\varphi'(u), \quad (3.108)$$

$$a^\varepsilon(u;x) := \varepsilon^{-1} a_\varepsilon(u;x) = \varepsilon^{-1} a(u;x) + a_1(u;x), \quad (3.109)$$

It is worth noticing that the generator $\mathbb{A}^\varepsilon(v;x)$ in (3.108) can be transformed by using condition C2 of Theorem 3.8, as follows

$$\mathbb{A}^\varepsilon(v;x) = \varepsilon^{-1} \mathbb{A}_\varepsilon(v;x), \quad (3.110)$$

$$\mathbb{A}_\varepsilon(v;x)\varphi(u) := [a_\varepsilon(v+\varepsilon u;x) - \widetilde{a}(v)]\varphi'(u)$$
$$= \widetilde{a}(v;x)\varphi'(u) + \varepsilon b(v,u;x)\varphi'(u) + \theta^\varepsilon(v,u;x)\varphi(u),$$

where by definition:

$$\widetilde{a}(v;x) = a(v;x) - \widehat{a}(v),$$

$$b(v,u;x) = a_1(v;x) + u a'_v(v;x).$$

PROOF OF LEMMA 3.1. The proof of this lemma is based on the conditional expectation of the extended embedded Markov chain (3.103) which is calculated by using (3.89)-(3.91) and (3.96):

$$\mathbb{E}[\varphi(\zeta^\varepsilon_{n+1}, \widehat{U}^\varepsilon_{n+1}, x_{n+1}) \mid \zeta^\varepsilon_n = u, \widehat{U}^\varepsilon_n = v, x_n = x]$$

$$= \int_0^\infty F_x(dt) \mathbb{E}[\varphi(u + \varepsilon^{-1}[\int_0^{\varepsilon^2 t} a_\varepsilon(U^\varepsilon_x(s);x)ds - \int_0^{\varepsilon^2 t} \widehat{a}(\widehat{U}(s))ds],$$

$$v + \int_0^{\varepsilon^2 t} \widehat{a}(\widehat{U}(s))ds, x_{n+1}) \mid U^\varepsilon_x(0) = v + \varepsilon u, \widehat{U}^\varepsilon_n = v, x_n = x]$$

$$= \int_0^\infty F_x(dt) \Gamma^\varepsilon_{\varepsilon^2 t}(x,v) \overline{\Gamma}^\varepsilon_{\varepsilon^2 t}(v) \widehat{A}_{\varepsilon^2 t} P\varphi(u,v,x) =: \mathbb{F}^\varepsilon(x) P\varphi(u,v,x).$$

□

The next step in the asymptotic analysis is to construct the asymptotic expansion of the compensating operator with respect to ε, (see Lemmas 5.3-5.4, Section 5.5.3).

Lemma 3.2 *The compensating operator (3.107)-(3.109) has the following asymptotic representation on test functions $\varphi \in C_0^{3,2}(\mathbb{R}^d \times \mathbb{R}^d)$*

$$\begin{aligned}\mathbb{L}^\varepsilon \varphi(u,v,x) = {}& \varepsilon^{-2} Q\varphi(\cdot,\cdot,x) + \varepsilon^{-1}\widetilde{\mathbb{A}}(v;v)P\varphi(u,\cdot,\cdot) \\ & + [\mathbb{L}_0(x,v)P\varphi(\cdot,v,\cdot) + \widehat{\mathbb{A}}(v)P\varphi(u,\cdot,\cdot)] \\ & + \theta_l^\varepsilon \varphi(u,v,x),\end{aligned} \quad (3.111)$$

with the negligible term

$$\left\| \sup_{x \in E} |\theta_l^\varepsilon \varphi(u,v,x)| \right\| \to 0, \quad \varepsilon \to 0.$$

Here, by definition,

$$Q\varphi(x) = q(x)[P - I]\varphi(x), \quad (3.112)$$

is the generator of the associated Markov process $x^0(t), t \geq 0$, with the intensity function

$$q(x) := 1/m(x), \quad m(x) := \int_0^\infty \overline{F}_x(t)dt.$$

The generator $\widetilde{\mathbb{A}}(v;x)$, and the operator $\mathbb{L}_0(v;x)$ are defined as follows:

$$\widetilde{\mathbb{A}}(v;x)\varphi(u) = \widetilde{a}(v;x)\varphi'(u), \quad (3.113)$$

and

$$\mathbb{L}_0(v;x)\varphi(u) = b(v,u;x)\varphi'(u) + \frac{1}{2}B_1(v;x)\varphi''(u), \quad (3.114)$$

$$b(v,u;x) := a_1(v;x) + u a_v'(v;x), \quad (3.115)$$
$$B_1(v;x) := \mu_2(x)\widetilde{a}(v;x)\widetilde{a}^*(v;x), \quad (3.116)$$
$$\mu_2(x) := m_2(x)/m(x),$$
$$m_2(x) := \int_0^\infty t^2 F_x(dt). \quad (3.117)$$

The proof of Lemma 3.2 is given in Section 5.5.3. □

Chapter 4

Stochastic Systems with Split and Merging

4.1 Introduction

In the study of real systems a special problem arises, connected to the generally high complexity of the state space.

Concerning this problem, in order to be able to give analytical or numerical tractable models, the state space must be simplified via a reduction of the number of states. This is possible when some subsets are connected between them by small transition probabilities and the states within such subsets are asymptotically connected. That is typically the case of reliability –and in most applications involving hitting time models, for which the state space is naturally cut into two subsets (the up states set and the down states set) [100,127]. In this case, transitions between the subsets are slow compared with those within the subsets. In the literature, the reduction of state space is also called *aggregation, lumping,* or *consolidation* of state space.

This chapter deals with *average* and *diffusion approximations* with *single* and *double asymptotic phase split and merging* of the switching process. The asymptotic merging provides a simpler process and for that reason is important for applications, as for example in reliability where in general two subsets of states are of interest: up and down states.

The main object studied here is the following stochastic additive functional (see Sections 2.6, 3.3.1, 3.4.2, and 3.5.2)

$$\xi^\varepsilon(t) = \xi^\varepsilon(0) + \int_0^t \eta^\varepsilon(ds; x^\varepsilon(s/\varepsilon)), \quad t \geq 0, \ \varepsilon > 0.$$

The switching semi-Markov process $x(t)$ is considered in two cases: er-

godic and *absorbing*.

Particular cases of the above additive functional that will be studied are the following three:

1. *Integral Functional*

$$\alpha^\varepsilon(t) = \int_0^t a^\varepsilon(x^\varepsilon(s/\varepsilon))ds, \quad t \geq 0, \ \varepsilon > 0. \tag{4.1}$$

2. *Dynamical System*

$$\frac{d}{dt}U^\varepsilon(t) = C^\varepsilon(U^\varepsilon(t); x^\varepsilon(s/\varepsilon)), \quad t \geq 0, \ \varepsilon > 0. \tag{4.2}$$

3. *Compound Poisson Process*

$$\zeta^\varepsilon(t) = \varepsilon \sum_{k=1}^{\nu(t/\varepsilon)} a^\varepsilon(x_k^\varepsilon), \quad t \geq 0, \ \varepsilon > 0, \tag{4.3}$$

where $\nu(t), t \geq 0$, is a Poisson process.

The above functional $\xi^\varepsilon(t), t \geq 0$, can also be written in the following form

$$\xi^\varepsilon(t) = \sum_{k=1}^{\nu^\varepsilon(t/\varepsilon)-1} \eta^\varepsilon(\varepsilon\theta_k; x_{k-1}^\varepsilon) + \eta^\varepsilon(\varepsilon\theta^\varepsilon(t); x^\varepsilon(t/\varepsilon)), \quad t \geq 0, \ \varepsilon > 0.$$

The generators $\mathbb{\Gamma}_\varepsilon(x), x \in E$, of the Markov processes with locally independent increments $\eta^\varepsilon(t; x)$, are given in Section 3.3, that is

$$\mathbb{\Gamma}_\varepsilon(x)\varphi(u) = a_\varepsilon(u; x)\varphi'(u)$$
$$+\varepsilon^{-1} \int_{\mathbb{R}^d} [\varphi(u + \varepsilon v) - \varphi(u) - \varepsilon v \varphi'(u)] \Gamma_\varepsilon(u, dv; x). \tag{4.4}$$

4.2 Phase Merging Scheme

4.2.1 Ergodic Merging

The general scheme of phase merging, described in the introduction, now will be realized for the semi-Markov processes $x^\varepsilon(t), t \geq 0$, with the standard phase (state) space $(E, \mathcal{E})$, in the series scheme with the small series parameter $\varepsilon \to 0, \varepsilon > 0$, on the split phase space (see Fig. 4.1)

$$E = \bigcup_{k=1}^N E_k, \quad E_k \bigcap E_{k'} = \emptyset, \quad k \neq k'. \tag{4.5}$$

4.2. PHASE MERGING SCHEME

Remark 4.1. More general split schemes can be used without essential changes in formulation, for example

$$E = \bigcup_{v \in V} E_v, \quad E_v \bigcap E_{v'} = \emptyset, \quad v \neq v',$$

where the factor space $(V, \mathcal{V})$ is a compact measurable space. The case where V is a finite set is of particular interest in applications.

The semi-Markov kernel is

$$Q^\varepsilon(x, B, t) = P^\varepsilon(x, B) F_x(t), \qquad (4.6)$$

where $x \in E, B \in \mathcal{E}, t \geq 0$.

Let us introduce the following assumptions:

ME1: The transition kernel of the embedded Markov chain x_n^ε, $n \geq 0$, has the following representation

$$P^\varepsilon(x, B) = P(x, B) + \varepsilon P_1(x, B). \qquad (4.7)$$

The *stochastic kernel* $P(x, B)$ is coordinated with the split phase space (4.5) as follows

$$P(x, E_k) = \mathbf{1}_k(x) := \begin{cases} 1, \, x \in E_k \\ 0, \, x \notin E_k. \end{cases} \qquad (4.8)$$

The stochastic kernel $P(x, B)$ determines the support Markov chain $x_n, n \geq 0$, on the separate classes $E_k, 1 \leq k \leq N$, (see Fig. 4.1 (b)). Moreover, the perturbing signed kernel $P_1(x, B)$ satisfies the *conservative condition*

$$P_1(x, E) = 0,$$

which is a direct consequence of (4.7) and $P^\varepsilon(x, E) = P(x, E) = 1$.

ME2: The associated Markov process $x^0(t), t \geq 0$, given by the generator

$$Q\varphi(x) = q(x) \int_E P(x, dy)[\varphi(y) - \varphi(x)], \qquad (4.9)$$

where $q(x) := 1/m(x)$, is uniformly ergodic in every class $E_k, 1 \leq k \leq N$, with the stationary distributions $\pi_k(dx), 1 \leq k \leq N$, satisfying the relations:

$$\pi_k(dx)q(x) = q_k\rho_k(dx), \quad q_k := \int_{E_k} \pi_k(dx)q(x) = 1/m_k, \quad (4.10)$$

$$m_k := \int_{E_k} \rho_k(dx)m(x).$$

As a consequence, the Markov chain $x_n, n \geq 0$, is uniformly ergodic with the stationary distributions $\rho_k(B), B \in \mathcal{E}_k = \mathcal{E} \cap E_k, 1 \leq k \leq N$, satisfying the integral equations

$$\rho_k(B) = \int_{E_k} \rho_k(dx)P(x,B), \quad B \in \mathcal{E}_k, \quad \rho_k(E_k) = 1. \quad (4.11)$$

ME3: The average exit probabilities

$$\widehat{p}_k := \int_{E_k} \rho_k(dx)P_1(x, E\backslash E_k) > 0, \quad 1 \leq k \leq N, \quad (4.12)$$

are positive, and the merged mean values

$$m_k := \int_{E_k} \rho_k(dx)m(x), \quad m(x) := \int_0^\infty \bar{F}_x(t)dt, \quad x \in E, \quad (4.13)$$

are positive and bounded.

The perturbing *signed kernel* $P_1(x, B)$ in (4.7) defines the transition probabilities between classes $E_k, 1 \leq k \leq N$. So, relation (4.7) means that the embedded Markov chain $x_n^\varepsilon, n \geq 0$, spends a long time in every class E_k and jumps from one class to another with the small probabilities $\varepsilon P_1(x, E\backslash E_k)$. It is worth noticing that under fast time-scaling the initial semi-Markov process can be approximated by some merged stochastic process on the merged phase space $\widehat{E} = \{1, ..., N\}$. The particularity of phase merging effect is that the approximating process will be Markovian.

Introduce the merging function (see Fig. 4.1. (c))

$$v(x) = k, \quad x \in E_k, \quad 1 \leq k \leq N, \quad (4.14)$$

and the merged process

$$\widehat{x}^\varepsilon(t) := v(x^\varepsilon(t/\varepsilon)), \quad t \geq 0, \quad (4.15)$$

on the merged phase space $\widehat{E} = \{1, ..., N\}$.

The phase merging principle establishes the weak convergence, as $\varepsilon \to 0$, of the merged process (4.15) to the limit Markov process.

4.2. PHASE MERGING SCHEME

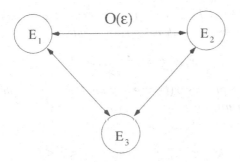

(a) Initial System S_ε

(b) Supporting System S

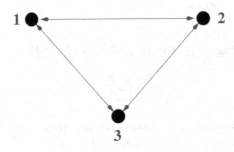

(c) Merged System $\widehat{S}$

Fig. 4.1   Asymptotic ergodic merging scheme

Theorem 4.1 *(Ergodic Phase merging principle).* Under Assumptions ME1-ME3, the following weak convergence holds

$$\widehat{x}^{\varepsilon}(t) \Longrightarrow \widehat{x}(t), \quad \varepsilon \to 0. \qquad (4.16)$$

The limit Markov process $\widehat{x}(t), t \geq 0$, on the merged phase space $\widehat{E} = \{1,...,N\}$ is determined by the generating matrix $\widehat{Q} = (\widehat{q}_{kr}, 1 \leq k, r \leq N)$, where:

$$\widehat{q}_{kr} = \widehat{q}_k \widehat{p}_{kr}, \quad k \neq r, \quad \widehat{q}_k = \widehat{p}_k / \widehat{m}_k, \quad 1 \leq k \leq N. \qquad (4.17)$$
$$\widehat{p}_{kr} = p_{kr}/\widehat{p}_k, \quad p_{kr} = \int_{E_k} \rho_k(dx) P_1(x, E_r), \quad 1 \leq k, r \leq N, k \neq r.$$

First let us precise the matrix $\widehat{Q}$ which is the generating matrix of some conservative Markov process. Indeed, from (4.7) and (4.8), we calculate:

$$\varepsilon \widehat{p}_{kr} = \int_{E_k} \rho_k(dx) \varepsilon P_1(x, E_r)$$
$$= \int_{E_k} \rho_k(dx) [P^{\varepsilon}(x, E_r) - P(x, E_r)]$$
$$= \int_{E_k} \rho_k(dx) P^{\varepsilon}(x, E_r) - \delta_{kr},$$

where δ_{kr} is the Kronecker symbol.

Hence, $\widehat{p}_{kr} \geq 0$, if $r \neq k$ and $\widehat{p}_{kk} \leq 0$. Using the conservative condition in ME1: $P_1(x, E) = 0$, and taking into account (4.12) we obtain the following:

$$\widehat{p}_k = -\int_{E_k} \rho_k(dx) P_1(x, E_k) = -p_{kk} = \sum_{r \neq k} p_{kr},$$

and after dividing by $\widehat{p}_k$, we get $\sum_{r \neq k} \widehat{p}_{kr} = 1$.

After multiplying by $\widehat{q}_k$, together with (4.17), it gives

$$\widehat{q}_k = \sum_{r \neq k} \widehat{q}_{kr}. \qquad (4.18)$$

That is the condition of conservation of the generating matrix $\widehat{Q}$. Condition (4.12) ensures that all states in $\widehat{E}$ are stable.

We will introduce the following additional assumption.

4.2. PHASE MERGING SCHEME

ME4: The merged Markov process $\widehat{x}(t), t \geq 0$, is ergodic, with the stationary distribution $\widehat{\pi} = (\pi_k, k \in \widehat{E})$.

In the particular case of the Markov initial process $x^\varepsilon(t)$, $t \geq 0$, with the semi-Markov kernel

$$Q^\varepsilon(x, B, t) = P^\varepsilon(x, B)[1 - e^{-q(x)t}],$$

the statement of the phase merging process theorem is valid with $m(x) = 1/q(x), x \in E$. Using the equation for the stationary distributions of the support Markov process

$$\pi_k(dx)q(x) = q_k \rho_k(dx),$$

we calculate:

$$\widehat{m}_k = \int_{E_k} \rho_k(dx)m(x) = \int_{E_k} \rho_k(dx)/q(x)$$
$$= q_k^{-1} \int_{E_k} \pi_k(dx) = q_k^{-1},$$

that is $q_k = 1/\widehat{m}_k$.

Hence, the intensity of the merged Markov process can be represented as in (4.17) with the merged intensity

$$\widehat{q}_k = q_k \widehat{p}_k, \quad 1 \leq k \leq N. \tag{4.19}$$

From the heuristic point of view the merging formulas (4.17) and (4.19) are natural. Indeed, in order to calculate an average exit probability by using the stationary distribution of the semi-Markov process defined by the semi-Markov kernel (4.6)-(4.7), we have to calculate:

$$p_k^\varepsilon = \int_{E_k} \rho_k(dx) P^\varepsilon(x, E \backslash E_k)$$
$$= \int_{E_k} \rho_k(dx)[P(x, E \backslash E_k) + \varepsilon P_1(x, E \backslash E_k)]$$
$$= \varepsilon \int_{E_k} \rho_k(dx) P_1(x, E \backslash E_k)$$
$$= \varepsilon \widehat{p}_k.$$

The relations $\widehat{q}_k = \widehat{p}_k/\widehat{m}_k, 1 \leq k \leq N$, also are natural, since the intensity of the limit Markov process has to be directly proportional to the average intensity $q_k = 1/\widehat{m}_k$, in the class E_k with factor $\widehat{p}_k$ which is the exit probability. Only one question remain, why are the stationary distributions of the support Markov process used? It is a natural consequence of the limit merging effect in the fast time-scaling scheme. Equalities (4.17) are obtained by using the phase merging principle based on a solution of singular perturbation problem (see Section 5.2).

4.2.2 Merging with Absorption

We will study here the merging scheme with an absorbing state. The semi-Markov process $x^\varepsilon(t), t \geq 0$, is considered on a split phase space

$$E^0 = E \bigcup \{0\}, \quad E = \bigcup_{k=1}^{N} E_k, \quad E_k \bigcap E_{k'} = \emptyset, \quad k \neq k', \quad (4.20)$$

with absorbing state 0. For example, in Fig. 4.2., we have: $E^0 = E_1 \cup E_2 \cup E_3 \cup \{0\}$; $E = E_1 \cup E_2 \cup E_3$; and $\widehat{E} = \{1, 2, 3\}$.

Let us introduce the following assumptions.

MA1: The semi-Markov kernel

$$Q^\varepsilon(x, B, t) = P^\varepsilon(x, B) F_x(t)$$

of which stochastic transition kernel is perturbed, that is

$$P^\varepsilon(x, B) = P(x, B) + \varepsilon P_1(x, B),$$

where $P(x, B)$ satisfies relation (4.8).

MA2: The perturbing kernel $P_1(x, B)$ satisfies the following *absorption condition*. There exists at least one $k \in \widehat{E}$, such that the absorption probability from k is positive, that is

$$p_{k0} := -\int_{E_k} \rho_k(dx) P_1(x, E) > 0. \quad (4.21)$$

MA3: The stochastic kernel $P(x, B)$ defines the support embedded Markov chain x_n, $n \geq 0$,

$$P(x, B) = \mathbb{P}(x_{n+1} \in B \mid x_n = x),$$

which is uniformly ergodic in every class E_k, $1 \leq k \leq N$, with the stationary distributions $\rho_k(B)$, $1 \leq k \leq N$, defined by a solution of the equations:

$$\rho_k(B) = \int_{E_k} \rho_k(dx) P(x, B), \quad \rho_k(E_k) = 1.$$

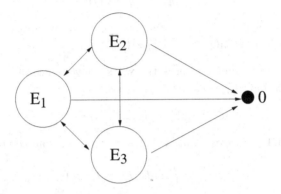

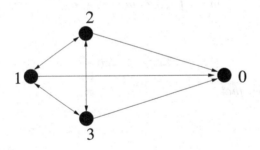

Fig. 4.2 Asymptotic merging scheme with absorption

Theorem 4.2 *(Absorbing phase merging scheme) Under split phase merging scheme (4.20) and Assumptions MA1-MA3 (Section 4.2.1) for the support Markov process, the following weak convergence takes place*

$$v(x^\varepsilon(t/\varepsilon)) \Longrightarrow \widehat{x}(t), \quad \varepsilon \to 0,$$

where the limit Markov process $\widehat{x}(t), 0 \le t \le \widehat{\zeta}$, is defined on the merged phase space $\widehat{E}^0 = \widehat{E} \cup \{0\}$, $\widehat{E} = \{1, ..., N\}$ by the generating matrix

$$\widehat{Q} = [\widehat{q}_{kr}; 0 \le k, r \le N],$$

where:

$$\widehat{q}_{kr} = \widehat{q}_k \widehat{p}_{kr}, \quad \widehat{p}_{kr} = p_{kr}/\widehat{p}_k \quad p_{kr} = \int_{E_k} \rho_k(dx) P_1(x, E_r), \quad k \neq r,$$

$$\widehat{q}_k = \widehat{p}_k q_k, \quad q_k = 1/m_k.$$

The random time $\widehat{\zeta}$, is the absorption (stoppage) time of the merged Markov process, that is, $\widehat{\zeta} := \inf\{t \geq 0 : \widehat{x}(t) = 0\}$.

The following result concerns the weak convergence of the absorption time $\widehat{\zeta}^\varepsilon$ of the initial process, that is,

$$\zeta^\varepsilon := \inf\{t \geq 0 : x^\varepsilon(t/\varepsilon) = 0\}.$$

Corollary 4.1 *Suppose that $N = 1$, and the following conditions hold:*

$$p := -\int_E \rho(dx) P_1(x, E) > 0,$$

and

$$m := \int_E \rho(dx) m(x) < +\infty.$$

Then

$$\zeta^\varepsilon \xrightarrow{d} \widehat{\zeta}, \quad \varepsilon \to 0,$$

where $\widehat{\zeta} \sim \mathcal{E}(p/m)$, that is,

$$\mathbb{P}(\widehat{\zeta} > t) = e^{-\Lambda t},$$

with $\Lambda = p/m$.

(See also Section 8.1).

4.2.3 Ergodic Double Merging

The stochastic systems can also be investigated in a double merging and averaging scheme. In fact, this kind of scheme is useful in practice when several orders of small transition probabilities arise. This also allows us to obtain several averaging and diffusion approximation results.

Consider a family of semi-Markov processes, $x^\varepsilon(t)$, $t \geq 0$, $\varepsilon > 0$, on a standard state space $(E, \mathcal{E})$, with semi-Markov kernel

$$Q^\varepsilon(x, B, t) = P^\varepsilon(x, B) F_x(t).$$

4.2. PHASE MERGING SCHEME

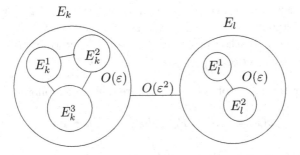

Fig. 4.3 Double merging

We consider the following finite split of the state space E (see Fig. 4.3):

$$E = \bigcup_{k=1}^{N} E_k, \quad E_k = \bigcup_{r=1}^{N_k} E_k^r, \ 1 \leq k \leq N,$$

$$E_k^r \bigcap E_{k'}^{r'} = \emptyset, \quad k \neq k' \text{ or } r \neq r'. \tag{4.22}$$

Let us introduce the following assumptions, specific to the double phase merging.

MD1: The stochastic kernel $P^\varepsilon(x, dy)$ has the following representation

$$P^\varepsilon(x, dy) = P(x, dy) + \varepsilon P_1(x, dy) + \varepsilon^2 P_2(x, dy), \tag{4.23}$$

where the stochastic kernel $P(x, dy)$ defines the associated Markov chain x_n, $n \geq 0$, and $P_1(x, dy)$ and $P_2(x, dy)$ are perturbing signed kernels. The first one concerns transitions between classes E_k^r and the second one between classes E_k. The perturbing kernels P_1 and P_2 satisfy the following conservative merging conditions:

$$P_1(x, E_k) = 0, \quad x \in E_k, \ 1 \leq k \leq N \tag{4.24}$$

$$P_2(x, E) = 0, \quad x \in E. \tag{4.25}$$

MD2: The associated Markov process $x^0(t)$, $t \geq 0$, is uniformly ergodic with generator Q, defined by

$$Q\varphi(x) = q(x) \int_E P(x, dy)[\varphi(y) - \varphi(x)], \tag{4.26}$$

with stationary distribution $\pi_k^r(dx), 1 \leq r \leq N_k, 1 \leq k \leq N$.
As a consequence, the associated Markov chain x_n, $n \geq 0$, is also uniformly ergodic in every class E_k^r, $1 \leq r \leq N_k, 1 \leq k \leq N$.

MD3: The merged Markov process $\widehat{x}(t), t \geq 0$ is uniformly ergodic, with stationary distribution $(\widehat{\pi}_k^r, 1 \leq r \leq N_k, 1 \leq k \leq N)$.

The perturbing operators $Q_k, k = 1,2$ are defined as follows

$$Q_k \varphi(x) = q(x) \int_E P_k(x, dy) \varphi(y), \quad k = 1, 2.$$

Let us define the following two merging functions:

$$\widehat{v}(x) = v_k^r, \quad \text{if} \quad x \in E_k^r,$$

and

$$\widehat{\widehat{v}}(x) = k, \quad \text{if} \quad x \in E_k.$$

Theorem 4.3 *(Ergodic double merging) Assume that the merging conditions MD1-MD3 hold, then the following weak convergences take place:*

$$\widehat{v}(x^\varepsilon(t/\varepsilon)) \Longrightarrow \widehat{x}(t), \quad \varepsilon \to 0, \tag{4.27}$$

$$\widehat{\widehat{v}}(x^\varepsilon(t/\varepsilon^2)) \Longrightarrow \widehat{\widehat{x}}(t) \quad \varepsilon \to 0. \tag{4.28}$$

The limit process $\widehat{x}(t)$ has the state space $\widehat{E} = \cup_{k=1}^N \widehat{E}_k$, $\widehat{E}_k = \{v_k^r : 1 \leq r \leq N_k\}$, and $\widehat{\widehat{x}}(t)$ the state space $\widehat{\widehat{E}} = \{1, 2, ..., N\}$. The generators of processes $\widehat{x}(t)$ and $\widehat{\widehat{x}}(t)$ are respectively $\widehat{Q}_1$ and $\widehat{\widehat{Q}}_2$, which are defined below.

The contracted operators $\widehat{Q}_1$ and $\widehat{\widehat{Q}}_2$ are defined as follows:

$$\Pi Q_1 \Pi = \widehat{Q}_1 \Pi$$

and

$$\Pi Q_2 \Pi = \widehat{Q}_2 \Pi$$

$$\widehat{\Pi} \widehat{Q}_2 \widehat{\Pi} = \widehat{\widehat{Q}}_2 \widehat{\Pi}.$$

The projectors Π and $\widehat{\Pi}$ are defined as follows:

$$\Pi \varphi(x) = \sum_{k=1}^N \sum_{r=1}^{N_k} \widehat{\varphi}_k^r \mathbf{1}_k^r(x)$$

where:
$$\widehat{\varphi}_k^r = \int_{E_k^r} \pi_k^r(dx)\varphi(x), \quad 1 \leq r \leq N_k, \ 1 \leq k \leq N$$

and
$$\begin{aligned}\widehat{\Pi}\varphi(x) &= \sum_{k=1}^{N}\sum_{r=1}^{N_k} \widehat{\pi}_k^r \varphi_k^r \mathbf{1}_k^r(x) \\ &= \sum_{k=1}^{N} \widehat{\widehat{\varphi}}_k \mathbf{1}_k(x)\end{aligned} \tag{4.29}$$

where:
$$\widehat{\widehat{\varphi}}_k = \sum_{r=1}^{N_k} \widehat{\pi}_k^r \widehat{\varphi}_k^r, \quad 1 \leq k \leq N,$$

$$\mathbf{1}_k^r(x) = \begin{cases} 1, x \in E_k^r \\ 0, x \notin E_k^r \end{cases}, \quad \mathbf{1}_k(x) = \begin{cases} 1, x \in E_k \\ 0, x \notin E_k \end{cases}.$$

Thus we have $\widehat{Q}_2 = (\widehat{q}_{\ell k}^{mr})$, where
$$\widehat{q}_{\ell k}^{mr} = \int_{E_\ell^m} \pi_\ell^m(dx) Q_2(x, E_k^r).$$

Moreover $\widehat{\widehat{Q}}_2 = \left(\widehat{\widehat{q}}_{\ell k}\right)$, where
$$\widehat{\widehat{q}}_{\ell k} = \sum_{m=1}^{N_\ell} \widehat{\pi}_\ell^m \sum_{r=1}^{N_k} \widehat{q}_{\ell k}^{mr} = \sum_{m=1}^{N_\ell} \widehat{\pi}_\ell^m \sum_{r=1}^{N_k} q_{\ell k}^{m\cdot}$$

and
$$q_{\ell k}^{m\cdot} = \int_{E_\ell^k} \pi_\ell^m(dx) Q_2(x, E_k).$$

Now, we have:
$$\begin{aligned}\sum_{k=1}^{N} \widehat{\widehat{q}}_{\ell k} &= \sum_{m=1}^{N_\ell} \widehat{\pi}_\ell^m \sum_{k=1}^{N} q_{\ell k}^{m\cdot} \\ &= \sum_{m=1}^{N_\ell} \widehat{\pi}_\ell^m \int_{E_\ell^k} \pi_\ell^m(dx) Q_2(x, E_k) \\ &= 0,\end{aligned}$$

CHAPTER 4. STOCHASTIC SYSTEMS WITH SPLIT AND MERGING

and

$$\widehat{q}_{\ell\ell} = \sum_{m=1}^{N_\ell} \widehat{\pi}_\ell^m q_{\ell\ell}^{m\cdot}$$
$$= \sum_{m=1}^{N_\ell} \widehat{\pi}_\ell^m \int_{E_\ell^m} \pi_\ell^m(dx) Q_2(x, E_\ell)$$
$$\leq 0.$$

Thus we have $\widehat{Q}_1 = \mathrm{diag}(Q_{ij}^1, ..., Q_{ij}^N)$, where $Q_{ij}^k = \left(q_{kk}^{ij}\right)$.

Let us introduce the following additional assumption, needed in the sequel.

MD4: The merged Markov process $\widehat{\widehat{x}}(t), t \geq 0$, is ergodic with stationary distribution $(\widehat{\widehat{\pi}}_k, 1 \leq k \leq N)$.

4.3 Average with Merging

In this section we consider switched stochastic systems with split and merging of the switching semi-Markov process. Let the processes $\eta^\varepsilon(t;x)$, $t \geq 0$, $x \in E$, $\varepsilon > 0$, be given by the generators (4.4),

$$\Gamma_\varepsilon(x)\varphi(u) = a_\varepsilon(u;x)\varphi'(u)$$
$$+\varepsilon^{-1} \int_{\mathbb{R}^d} [\varphi(u+\varepsilon v) - \varphi(u) - \varepsilon v \varphi'(u)] \Gamma_\varepsilon(u, dv; x). \quad (4.30)$$

Let the stochastic evolutionary system, $\xi^\varepsilon(t)$, be represented by

$$\xi^\varepsilon(t) = \xi^\varepsilon(0) + \int_0^t \eta^\varepsilon(ds; x^\varepsilon(s/\varepsilon)), \quad t \geq 0, \ \varepsilon > 0. \quad (4.31)$$

Let us introduce the following conditions.

A1: The drift velocity $a(u;x)$ belongs to the Banach space $\mathbf{B}^1$, with

$$a_\varepsilon(u;x) = a(u;x) + \theta^\varepsilon(u;x),$$

where $\theta^\varepsilon(u;x)$ goes to 0 as $\varepsilon \to 0$ uniformly on $(u;x)$.

And $\Gamma_\varepsilon(u, dv; x) \equiv \Gamma(u, dv; x)$ is independent of ε.

A2: The operator

$$\gamma_\varepsilon(x)\varphi(u) = \varepsilon^{-1} \int_{\mathbb{R}^d} [\varphi(u+\varepsilon v) - \varphi(u) - \varepsilon v \varphi'(u)] \Gamma(u, dv; x)$$

4.3. AVERAGE WITH MERGING

is negligible on $\mathbf{B}^1$, that is,

$$\sup_{\varphi \in \mathbf{B}^1} \|\gamma_\varepsilon(x)\varphi\| \to 0, \quad \varepsilon \to 0.$$

A3: Convergence in probability of the initial values of $\xi^\varepsilon(t), v(x^\varepsilon(t/\varepsilon)), t \geq 0$ hold, that is,

$$(\xi^\varepsilon(0), v(x^\varepsilon(0))) \xrightarrow{P} (\xi(0), \widehat{x}(0)),$$

and there exists a constant $c \in \mathbb{R}_+$, such that

$$\sup_{\varepsilon > 0} \mathbb{E}\, |\xi^\varepsilon(0)| \leq c < +\infty.$$

Remark 4.2. The operator $\gamma_\varepsilon(x)$ is the jump part after extraction of the drift part due to the jumps of the process $\eta^\varepsilon(t, x)$.

4.3.1 Ergodic Average

In this section we will give a theorem for the averaging of the evolutionary system $\xi^\varepsilon(t), t \geq 0$, in ergodic single split of the switching semi-Markov process $x^\varepsilon(t), t \geq 0$.

The switching semi-Markov process $x^\varepsilon(t)$, $t \geq 0$, is considered in a split phase space

$$E = \bigcup_{k=1}^{N} E_k, \quad E_k \bigcap E_{k'} = \emptyset, \quad k \neq k',$$

and supposed to satisfy the phase merging assumptions ME1-ME3 (Section 4.2.1).

Theorem 4.4 *(Ergodic Average) Let the switching semi-Markov process $x^\varepsilon(t)$, $t \geq 0$, satisfies the phase merging conditions ME1-ME3.*

Then, under Assumptions A1-A3, the stochastic evolutionary system $\xi^\varepsilon(t), t \geq 0$, (4.31), converges weakly to the averaged stochastic system $\widehat{U}(t)$:

$$\xi^\varepsilon(t) \Rightarrow \widehat{U}(t), \quad \varepsilon \to 0.$$

The limit process $\widehat{U}(t)$, $t \geq 0$, is defined by a solution of the evolutionary equation

$$\frac{d}{dt}\widehat{U}(t) = \widehat{a}(\widehat{U}(t); \widehat{x}(t)), \quad \widehat{U}(0) = \xi(0), \tag{4.32}$$

where the averaged velocity is determined by

$$\widehat{a}(u; k) = \int_{E_k} \pi_k(dx) a(u; x), \quad 1 \leq k \leq N.$$

The following corollary gives particular results of Theorem 4.4, in the three cases described in Section 4.1.

Corollary 4.2

1) The stochastic integral functional *(4.1)* converges weakly as follows

$$\int_0^t a(x^\varepsilon(s/\varepsilon))ds \Rightarrow \int_0^t \widehat{a}(\widehat{x}(s))ds \quad \varepsilon \to 0,$$

where

$$\widehat{a}(k) = \int_{E_k} \pi_k(dx) a(x).$$

2) The dynamical system defined by *(4.2)*, with

$$C^\varepsilon(u; x) = C(u; x) + \theta^\varepsilon(u; x),$$

where $\theta^\varepsilon(u; x)$ is the negligible term

$$\|\theta^\varepsilon(u; x)\| \to 0, \quad \varepsilon \to 0,$$

converges weakly to a dynamical system with switching process $\widehat{x}(t), t \geq 0$,

$$\frac{d}{dt}\widehat{U}(t) = \widehat{C}(\widehat{U}(t); \widehat{x}(t)),$$

where

$$\widehat{C}(u; k) = \int_{E_k} \pi_k(dx) C(u; x).$$

3) The compound Poisson process with Markov switching defined by *(4.3)* converges weakly as follows

$$\xi^\varepsilon(t) \Rightarrow \int_0^t \widehat{a}(\widehat{x}(s))ds, \quad \varepsilon \to 0.$$

4.3.2 Average with Absorption

The switching semi-Markov process $x^\varepsilon(t)$, $t \geq 0$, is considered in a split phase space (see relation (4.20), Section 4.2.2)

$$E^0 = E\bigcup\{0\}, \quad E = \bigcup_{k=1}^{N} E_k, \quad E_k \bigcap E_{k'} = \emptyset, \quad k \neq k', \quad (4.33)$$

with absorbing state 0, and supposed to satisfy the phase merging scheme, that is, Assumptions MA1-MA2 (Section 4.2.2).

Theorem 4.5 *(Average with Absorption) Let the switching semi-Markov process $x^\varepsilon(t)$, $t \geq 0$, satisfy the phase merging Assumptions MA1-MA2, (Section 4.2.2).*

Then, under Assumptions A1-A3, the stochastic evolutionary system $\xi^\varepsilon(t)$, $t \geq 0$, (4.31), converges weakly to the averaged stochastic system $\widehat{U}(t \wedge \widehat{\zeta})$:

$$\xi^\varepsilon(t) \Rightarrow \widehat{U}(t \wedge \widehat{\zeta}), \quad \varepsilon \to 0.$$

The limit process $\widehat{U}(t)$, $t \geq 0$, is defined by a solution of the evolutionary equation

$$\left| \begin{array}{l} \frac{d}{dt}\widehat{U}(t) = \widehat{a}(\widehat{U}(t), \widehat{x}(t)), \quad t \geq 0 \\ \widehat{U}(0) = \xi(0), \end{array} \right.$$

on the time interval $0 \leq t \leq \widehat{\zeta}$, ($\widehat{\zeta}$ is the absorption time of the merged Markov process $\widehat{x}(t)$, $t \geq 0$).

The averaged velocity is determined by

$$\widehat{a}(u; k) = \int_{E_k} \pi_k(dx) a(u; x), \quad 1 \leq k \leq N, \quad \widehat{a}(u; 0) = 0.$$

The following corollary gives particular results of Theorem 4.5, in the three cases described in Section 4.1.

Corollary 4.3

1) The stochastic integral functional (4.1) converges weakly as follows

$$\int_0^t a(x^\varepsilon(s/\varepsilon))ds \Rightarrow \int_0^{t \wedge \widehat{\zeta}} \widehat{a}(\widehat{x}(s))ds, \quad \varepsilon \to 0,$$

where

$$\widehat{a}(k) = \int_{E_k} \pi_k(dx)a(x).$$

In the particular case where $N = 1$, the stochastic integral functional converges weakly as follows:

$$\int_0^t a(x^\varepsilon(s/\varepsilon))ds \Rightarrow \widehat{a} \cdot (t \wedge \widehat{\zeta}), \quad \varepsilon \to 0, \quad \widehat{a} = \int_E \pi(dx)a(x).$$

2) The dynamical system defined by (4.2) converges weakly to a dynamical system with a simpler switching process $\widehat{x}(t), 0 \leq t \leq \widehat{\zeta}$, than the initial one $x^\varepsilon(t), t \geq 0$.

3) The compound Poisson process with Markov switching defined by (4.3) converges weakly as follows

$$\xi^\varepsilon(t) \Rightarrow \int_0^{t \wedge \widehat{\zeta}} \widehat{a}(\widehat{x}(s))ds, \quad \varepsilon \to 0.$$

4.3.3 Ergodic Average with Double Merging

The following theorem concerns averaging results for the evolutionary system $\xi^\varepsilon(t)$ in the double merging scheme (4.22), (Section 4.2.3), satisfying Assumptions MD1-MD3.

Theorem 4.6 *(Double average) Let the switching semi-Markov process $x^\varepsilon(t)$, $t \geq 0$, satisfy the conditions of double merging scheme MD1-MD4. Let the stochastic system be represented as follows*

$$\xi^\varepsilon(t) = \xi^\varepsilon(0) + \int_0^t \eta^\varepsilon(ds; x^\varepsilon(s/\varepsilon^2)), \tag{4.34}$$

where the processes $\eta^\varepsilon(t;x)$, $t \geq 0$, $x \in E$, are given by the generator of (4.4). Let Assumptions A1-A3 hold.

Then the weak convergence

$$\xi^\varepsilon(t) \Rightarrow \widehat{\widehat{U}}(t), \quad \varepsilon \to 0$$

takes place. The limit double averaged system $\widehat{\widehat{U}}(t), \widehat{\widehat{x}}(t), t \geq 0$, is defined also equivalently by a solution of the equation

$$\frac{d}{dt}\widehat{\widehat{U}}(t) = \widehat{\widehat{a}}(\widehat{\widehat{U}}(t), \widehat{\widehat{x}}(t)), \quad \widehat{\widehat{U}}(0) = \xi(0), \tag{4.35}$$

where

$$\widehat{\widehat{a}}(u;k) = \sum_{r=1}^{N_k} \widehat{\pi}_k^r \widehat{a}_k^r(u), \quad \widehat{a}_k^r(u) := \int_{E_k^r} \pi_k^r(dx) a(u;x).$$

Remark 4.3. The stochastic ergodic system (4.35) can be considered in ergodic average scheme (see Section 4.3.1). The same ergodic average result can be obtained for the initial stochastic system (4.34) with time-scaling ε^3 instead of ε^2.

Remark 4.4. A result analogous to Corollary 4.1 can be obtained for the double merged process $\widehat{\widehat{x}}(t)$, $t \geq 0$, in the cases of stochastic integral functional (4.1), of dynamical system (4.2), and of the compound Poisson process with Markov switching (4.3) [100].

4.3.4 Double Average with Absorption

The following result is an averaging result for the evolutionary system $\xi^\varepsilon(t)$ in the double merging scheme (4.22).

Define $\widehat{\widehat{\zeta}}$ the absorption time of the process $\widehat{\widehat{x}}(t)$, by

$$\widehat{\widehat{\zeta}} = \min\{t \geq 0 : \widehat{\widehat{x}}(t) = 0\}.$$

Corollary 4.4 *(Double average) Let the switching Markov process $x^\varepsilon(t)$, $t \geq 0$, satisfy the conditions of double merging scheme (4.22). Let the stochastic system be represented as follows*

$$\xi^\varepsilon(t) = \xi^\varepsilon(0) + \int_0^t \eta^\varepsilon(ds; x^\varepsilon(s/\varepsilon^2)),$$

where the processes $\eta^\varepsilon(t;x)$, $t \geq 0$, $x \in E$, are given by the generator of (4.4). Let Conditions A1-A3 (Section 4.3) hold.

Then the weak convergence

$$\xi^\varepsilon(t) \Rightarrow \widehat{\widehat{U}}(t \wedge \widehat{\widehat{\zeta}}) \quad \varepsilon \to 0,$$

takes place. The limit double averaged system $\widehat{\widehat{U}}(t)$, $t \geq 0$, is defined by a solution of the equation

$$\left| \begin{array}{l} \frac{d}{dt}\widehat{\widehat{U}}(t) = \widehat{\widehat{a}}(\widehat{\widehat{U}}(t), \widehat{\widehat{x}}(t)), \\ \widehat{\widehat{U}}(0) = \xi(0), \end{array} \right.$$

where

$$\widehat{\widehat{a}}(u;k) = \sum_{r=1}^{N_k} \widehat{\pi}_k^r \widehat{a}_k^r(u).$$

The stopping time, for $N = 1$, $\widehat{\widehat{\zeta}}$, has an exponential distribution with the parameter

$$\widehat{\widehat{\Lambda}} = qp,$$

where:

$$q = \int_E \pi(dx)q(x), \quad p = \int_E \rho(dx)p(x),$$

and $p(x)$ is defined as follows

$$P^\varepsilon(x,\{0\}) = -\varepsilon^2 P_2(x,E) = \varepsilon^2 p(x).$$

4.4 Diffusion Approximation with Split and Merging

In this section we consider the additive functional $\xi^\varepsilon(t), t \geq 0$, under the following time-scaling of the switching process

$$\xi^\varepsilon(t) = \xi^\varepsilon(0) + \int_0^t \eta^\varepsilon(ds; x^\varepsilon(s/\varepsilon^2)). \tag{4.36}$$

The processes $\eta^\varepsilon(t;x), t \geq 0, x \in E, \varepsilon > 0$, are given by the generator (compare with (4.30))

$$\Gamma_\varepsilon(x)\varphi(u) = a^\varepsilon(u;x)\varphi'(u)$$
$$+ \varepsilon^{-2} \int_{\mathbb{R}^d} [\varphi(u+\varepsilon v) - \varphi(u) - \varepsilon v \varphi'(u)] \Gamma_\varepsilon(u, dv; x). \tag{4.37}$$

Let us consider the following conditions.

4.4. DIFFUSION APPROXIMATION WITH SPLIT AND MERGING

D1: The drift velocity function has the following representation

$$a^\varepsilon(u;x) = \varepsilon^{-1}a(u;x) + a_1(u;x),$$

where $a(u;x)$ and $a_1(u;x)$ belong to the Banach space $\mathbf{B}^2$.

BC1: Balance condition

$$\widehat{a}(u) - \int_E \pi(dx)a(u;x) \equiv 0. \tag{4.38}$$

D2: The operators

$$\gamma_\varepsilon(x)\varphi(u) = \varepsilon^{-1}\int_{\mathbb{R}^d}[\varphi(u+\varepsilon v) - \varphi(u) - \varepsilon v\varphi'(u) - \frac{\varepsilon^2}{2}v^2\varphi''(u)]\Gamma_\varepsilon(u,dv;x)$$

are negligible on $\mathbf{B}^2$, that is,

$$\sup_{\varphi\in\mathbf{B}^2}\|\gamma_\varepsilon(x)\varphi\| \to 0, \quad \varepsilon \to 0.$$

4.4.1 Ergodic Split and Merging

The switching Markov processes $x^\varepsilon(t), t \geq 0$, are considered on the split phase space (4.5) and the support Markov process $x(t), t \geq 0$, defined by the generator (4.9) is uniformly ergodic in every class $E_k, 1 \leq k \leq N$, with the stationary distributions $\pi_k(dx), 1 \leq k \leq N$, satisfying relation (4.10). The stochastic additive functional $\xi^\varepsilon(t), t \geq 0$, given by relation (4.36) satisfies conditions D1-D2 and the balance condition BC1 (4.38).

Let the following assumption holds.

MS1: *Merging and split condition.* The transition kernel of the embedded Markov chain x_n^ε of the switching Markov process $x^\varepsilon(t)$, $t \geq 0$, has the following representation

$$P^\varepsilon(x,B) = P(x,B) + \varepsilon^2 P_1(x,B),$$

where the kernels $P(x,B)$ and $P_1(x,B)$ satisfy conditions ME1-ME3 of the phase merging scheme.

Theorem 4.7 *(Ergodic split and merging) Let the switching Markov processes $x^\varepsilon(t), t \geq 0$, satisfy Condition MS1.*
Then the weak convergence

$$\xi^\varepsilon(t) \Rightarrow \widehat{\xi}(t), \quad \varepsilon \to 0,$$

CHAPTER 4. STOCHASTIC SYSTEMS WITH SPLIT AND MERGING

takes place.

The limit diffusion process $\widehat{\xi}(t), t \geq 0$, switched by the merged Markov process $\widehat{x}(t), t \geq 0$, is defined by the generator of the coupled Markov process $\widehat{\xi}(t), \widehat{x}(t), t \geq 0$,

$$\mathbb{L}\varphi(u,k) = \widehat{b}(u;k)\varphi'_u(u,k) + \frac{1}{2}\widehat{B}(u;k)\varphi''_{uu}(u,k) + \widehat{Q}\varphi(\cdot,k), \quad (4.39)$$

where the generator $\widehat{Q}$ defines the merged process $\widehat{x}(t), t \geq 0$, on the merged phase space $\widehat{E} = \{1, 2, ..., N\}$ (see Theorem 4.1). The drift coefficient is defined by

$$\widehat{b}(u;k) = \widehat{a}_1(u;k) + \widehat{b}_1(u;k) + \widehat{b}_2(u;k),$$

with:

$$\widehat{a}_1(u;k) = \int_{E_k} \pi_k(dx) a_1(u;x),$$

$$\widehat{b}_1(u;k) = \int_{E_k} \pi_k(dx) b_1(u;x),$$

$$\widehat{b}_2(u;x) = \int_{E_k} \pi_k(dx) b_2(u;x),$$

$$b_1(u;x) := a(u;x) R_0 a'_u(u;x),$$

$$b_2(u;x) := \mu(x) a(u;x) a'_u(u;x).$$

The covariance function is defined by

$$\widehat{B}(u;k) = \widehat{B}_0(u;k) + \widehat{B}_{00}(u;k) + \widehat{C}_0(u;k)$$

with:

$$\widehat{B}_0(u;k) = 2\int_{E_k} \pi_k(dx) a_0(u;x),$$

$$\widehat{B}_{00}(u;k) = \int_{E_k} \pi_k(dx) B_{00}(u;x),$$

$$\widehat{C}_0(u;k) = \int_{E_k} \pi_k(dx) C_0(u;x),$$

$$B_{00}(u;x) = \mu(x) a(u;x) a^*(u;x),$$

$$a_0(u;x) = a(u;x) R_0 a(u;x),$$

$$C_0(u;x) = \frac{1}{2} \int_{\mathbb{R}^d} vv^* \Gamma(u, dv; x),$$

$$\mu(x) := [m_2(x) - 2m^2(x)]/m(x).$$

4.4. DIFFUSION APPROXIMATION WITH SPLIT AND MERGING

Operator R_0 is the potential of generator Q (Section 1.6).

Remark 4.5. The limit diffusion process $\widehat{\xi}(t), t \geq 0$, can be defined by a solution of the following stochastic differential equation

$$d\widehat{\xi}(t) = \widehat{b}(\widehat{\xi}(t); \widehat{x}(t)) + \widehat{\sigma}(\widehat{\xi}(t); \widehat{x}(t)) dw(t),$$

where the variance function is

$$\widehat{\sigma}(u; k)\widehat{\sigma}^*(u; k) = \widehat{B}(u; k),$$

and $w(t), t \geq 0$, is the standard Wiener process.

The following corollary concerns particular cases of the above theorem.

Corollary 4.5

1) The stochastic integral functional (4.1) converges weakly

$$\int_0^t a^\varepsilon(x^\varepsilon(s/\varepsilon^2))ds \Rightarrow \widehat{\xi}(t), \quad \varepsilon \to 0.$$

The limit process $\widehat{\xi}(t)$, $t \geq 0$, is a diffusion process with generator (4.41), where:

$$\widehat{b}(u; k) \equiv \widehat{a}_1(k) = \int_{E_k} \pi_k(dx) a_1(x), \text{ and } \widehat{B}(u; k) \equiv \int_{E_k} \pi_k(dx) a(x) R_0 a(x).$$

2) The dynamical system

$$\frac{d}{dt} U^\varepsilon(t) = a^\varepsilon(U^\varepsilon(t); x^\varepsilon(t/\varepsilon^2)), \quad t \geq 0, \varepsilon > 0, \quad (4.40)$$

converges weakly to a diffusion as in the above theorem.

3) The compound Poisson process with Markov switching defined by generators (4.3) converges weakly

$$\xi^\varepsilon(t/\varepsilon^2) \Rightarrow \widehat{\xi}(t), \quad \varepsilon \to 0,$$

where the limit process $\widehat{\xi}(t), t \geq 0$, is a diffusion process with generator (4.41), with drift

$$\widehat{a}_1 = \int_E \pi(dx) a_1(x), \quad a_1(x) = \int_{\mathbb{R}^d} v \Gamma_1(dv; x),$$

and covariance coefficient

$$\widehat{B} = \int_E \pi(dx)[a_0(x) + C_0(x)],$$

with:

$$a_0(x) = a(x)R_0 a(x), \quad C_0(x) = \frac{1}{2}\int_{\mathbb{R}^d} vv^*\Gamma(dv;x).$$

4.4.2 Split and Merging with Absorption

Here, the switching Markov process $x^\varepsilon(t), t \geq 0, \varepsilon > 0$, is supposed to be as in the previous Section 4.4.1 but with $N = 1$, for simplification and without the conservative condition ME1, Section 4.2.1.

The stochastic additive functionals $\xi^\varepsilon(t), t \geq 0$, are given in (4.36)-(4.37) and satisfy Conditions D1-D2, and the balance condition BC1.

Theorem 4.8 *(Split and merging with absorption) Let the switching Markov processes $x^\varepsilon(t), t \geq 0, \varepsilon > 0$, satisfy Condition MS1.*

Then the weak convergence

$$\xi^\varepsilon(t) \Rightarrow \widehat{\xi}(t), \quad 0 \leq t \leq \widehat{\zeta}, \quad \varepsilon \to 0,$$

takes place.

The limit diffusion process $\widehat{\xi}(t)$, $0 \leq t \leq \widehat{\zeta}$, is defined by the generator

$$\widehat{\mathbb{L}}\varphi(u) = \widehat{b}(u)\varphi'(u) + \frac{1}{2}\widehat{B}(u)\varphi''(u) - \widehat{\Lambda}\varphi(u). \qquad (4.41)$$

The drift coefficient is defined by

$$\widehat{b}(u) = \widehat{a}_1(u) + \widehat{b}_1(u),$$

where

$$\widehat{a}_1(u) = \int_E \pi(dx)a_1(u;x) \quad \text{and} \quad \widehat{b}_1(u) = \int_E \pi(dx)a(u;x)R_0 a'_u(u;x).$$

The covariance function is defined by

$$\widehat{B}(u) = 2\int_E \pi(dx)[a(u;x)R_0 a(u;x) + C_0(u;x)],$$

where

$$C_0(u;x) = \frac{1}{2}\int_{\mathbb{R}^d} vv^*\Gamma(u,dv;x)$$

4.4. DIFFUSION APPROXIMATION WITH SPLIT AND MERGING

where v^* is the transpose of the vector v.

The absorption time $\widehat{\zeta}$ is exponentially distributed with intensity

$$\widehat{\Lambda} = q\widehat{p}.$$

The following corollary concerns particular cases of Theorem 4.8 in the three cases given in Section 4.1.

Corollary 4.6

1) The stochastic integral functional (4.1) converges weakly

$$\int_0^t a^\varepsilon(x^\varepsilon(s/\varepsilon^2))ds \Rightarrow \widehat{\xi}(t \wedge \widehat{\zeta}), \quad \varepsilon \to 0.$$

The limit process $\widehat{\xi}(t)$, $t \geq 0$, is a diffusion process with generator (4.41), where

$$b(u) \equiv \widehat{a}_1 = \int_E \pi(dx)a_1(x), \text{ and } \widehat{B}(u) \equiv \int_E \pi(dx)a(x)R_0a(x).$$

2) The dynamical system (4.40) converges weakly to a diffusion as in the above theorem.

3) The compound Poisson process with Markov switching defined by generators (4.3) converges weakly

$$\xi^\varepsilon(t/\varepsilon^2) \Rightarrow \widehat{\xi}(t \wedge \widehat{\zeta}), \quad \varepsilon \to 0,$$

where the limit process $\widehat{\xi}(t)$, $t \geq 0$, is a diffusion process with generator (4.41), with drift

$$\widehat{a}_1 = \int_E \pi(dx)a_1(x), \quad a_1(x) = \int_{\mathbb{R}^d} v\Gamma_1(dv;x),$$

and covariance coefficient

$$\widehat{B} = 2\int_E \pi(dx)[a_0(x) + C_0(x)],$$

with:

$$a_0(x) = a(x)R_0a(x), \quad C_0(x) = \frac{1}{2}\int_{\mathbb{R}^d} vv^*\Gamma(dv;x).$$

4.4.3 Ergodic Split and Double Merging

The switching Markov processes $x^\varepsilon(t), t \geq 0, \varepsilon > 0$, are considered on the split phase space (4.5) and satisfy conditions ME1-ME4, Section 4.2.1.

The stochastic additive functionals $\xi^\varepsilon(t), t \geq 0$, are given with accelerated switching

$$\xi^\varepsilon(t) = \xi^\varepsilon(0) + \int_0^t \eta^\varepsilon(ds; x^\varepsilon(t/\varepsilon^3)), \quad t \geq 0, \varepsilon > 0. \tag{4.42}$$

The following condition will be used next.

BC2: Balance condition $\sum_{k=1}^N \widehat{\pi}_k \widehat{a}_k(u) \equiv 0$, where, by definition,

$$\widehat{a}_k(u) := \int_{E_k} \pi_k(dx) a(u; x), \quad 1 \leq k \leq N. \tag{4.43}$$

Theorem 4.9 *(Diffusion approximation in ergodic split and double merging) Under Conditions MD1-MD2 and balance condition BC2, the following weak convergence holds*

$$\xi^\varepsilon(t) \Rightarrow \widehat{\widehat{\xi}}(t), \quad \varepsilon \to 0.$$

The limit diffusion process $\widehat{\widehat{\xi}}(t), t \geq 0$, *is defined by the generator*

$$\mathbb{L}\varphi(u) = \widehat{\widehat{b}}(u)\varphi'(u) + \frac{1}{2}\widehat{\widehat{B}}(u)\varphi''(u). \tag{4.44}$$

The drift coefficient is defined by

$$\widehat{\widehat{b}}(u) = \widehat{\widehat{a}}_1(u) + \widehat{\widehat{b}}_1(u),$$

where:

$$\widehat{\widehat{a}}_1(u) = \sum_{k=1}^N \widehat{\pi}_k \int_{E_k} \pi_k(dx) a_1(u; x),$$

$$\widehat{\widehat{b}}_1(u) = \sum_{k=1}^N \widehat{\pi}_k \int_{E_k} \pi_k(dx) a(u; x) R_0 a'_u(u; x).$$

The covariance matrix is defined by

$$\widehat{\widehat{B}}(u) = 2 \sum_{k=1}^N \widehat{\pi}_k \int_{E_k} \pi_k(dx)[a_0(u; x) + C_0(u; x)],$$

where:
$$a_0(u;k) = a(u;k)R_0 a(u;k),$$
$$C_0(u;x) = \frac{1}{2}\int_{\mathbb{R}^d} vv^*\Gamma(u,dv;x).$$

Corollary 4.7

1) The stochastic integral functional (4.1) converges weakly

$$\int_0^t a^\varepsilon(x^\varepsilon(s/\varepsilon^3))ds \Rightarrow \widehat{\widehat{\xi}}(t), \quad \varepsilon \to 0.$$

The limit process $\widehat{\widehat{\xi}}(t)$, $t \geq 0$, is a diffusion process with generator (4.44), where the drift coefficient is defined by

$$\widehat{\widehat{b}}(u) = \widehat{\widehat{a}}_1(u),$$

with

$$\widehat{\widehat{a}}_1(u) = \sum_{k=1}^N \widehat{\pi}_k \int_{E_k} \pi_k(dx) a_1(u;x).$$

The covariance matrix is defined by

$$\widehat{\widehat{B}}(u) = 2\sum_{k=1}^N \widehat{\pi}_k \int_{E_k} \pi_k(dx) a_0(u;x),$$

where

$$a_0(u;k) = a(u;k)R_0 a(u;k).$$

2) The compound Poisson processes with Markov switching defined by generators (4.3) converge weakly

$$\xi^\varepsilon(t/\varepsilon^3) \Rightarrow \widehat{\widehat{\xi}}(t), \quad \varepsilon \to 0,$$

where the limit process $\widehat{\widehat{\xi}}(t)$, $t \geq 0$, is a diffusion process having generator (4.44), with:

$$\widehat{\widehat{b}} \equiv 0, \quad \widehat{\widehat{B}} = \sum_{k=1}^N \widehat{\pi}_k \widehat{C}(k), \quad \widehat{C}(k) = \int_{E_k} \pi_k(dx) C(x).$$

4.4.4 Double Split and Merging

Here, the switching Markov processes $x^\varepsilon(t), t \geq 0$, are considered in the double split phase space (4.22) and satisfy conditions MD1-MD4, Section 4.2.3.

The stochastic additive functionals $\xi^\varepsilon(t), t \geq 0$, are considered with the accelerated switching

$$\xi^\varepsilon(t) = \xi^\varepsilon(0) + \int_0^t \eta^\varepsilon(ds; x^\varepsilon(t/\varepsilon^3)), \quad t \geq 0, \varepsilon > 0. \qquad (4.45)$$

The following condition will be used next.

BC3: Balance condition

$$\sum_{r=1}^{N_k} \widehat{\pi}_k^r \widehat{a}_k^r(u) \equiv 0, \quad 1 \leq k \leq N, \qquad (4.46)$$

where

$$\widehat{a}_k^r(u) := \int_{E_k} \pi_k^r(dx) a(u; x).$$

Theorem 4.10 *(Double merging) Under conditions D1-D3, and the balance condition BC3, the following weak convergence holds*

$$\xi^\varepsilon(t) \Rightarrow \widehat{\widehat{\xi}}(t), \quad \varepsilon \to 0.$$

The limit diffusion process $\widehat{\widehat{\xi}}(t)$, $t \geq 0$, switched by the twice merged Markov process $\widehat{\widehat{x}}(t)$, $t \geq 0$, is defined by the generator of the coupled Markov process: $\widehat{\widehat{\xi}}(t), \widehat{\widehat{x}}(t)$, $t \geq 0$,

$$\mathbb{L}\varphi(u, k) = \widehat{\widehat{b}}(u; k)\varphi'_u(u, k) + \frac{1}{2}\widehat{\widehat{B}}(u; k)\varphi''_{uu}(u, k) + \widehat{\widehat{Q}}\varphi(\cdot, k), \qquad (4.47)$$

where the generating matrix $\widehat{\widehat{Q}}$ of the double merged Markov process $\widehat{\widehat{x}}(t)$, $t \geq 0$, is defined by the relations in Section 4.2.3.

The drift function is defined by

$$\widehat{\widehat{b}}(u; k) = \widehat{\widehat{a}}_1(u; k) + \widehat{\widehat{b}}_1(u; k),$$

4.4. DIFFUSION APPROXIMATION WITH SPLIT AND MERGING

where:

$$\widehat{\widehat{a}}_1(u;k) = \sum_{r=1}^{N_k} \widehat{\pi}_k^r \int_{E_k^r} \pi_k^r(dx) a_1(u;x),$$

$$\widehat{\widehat{b}}_1(u;k) = \sum_{r=1}^{N_k} \widehat{\pi}_k^r \widehat{b}_k^r(u),$$

$$\widehat{b}_k^r(u) = \widehat{a}_k^r(u) \widehat{R}_0 \widehat{a}_k^r{'}(u),$$

$$\widehat{a}_k^r(u) = \int_{E_k^r} \pi_k^r(dx) a(u;x).$$

The covariance function is defined by

$$\widehat{\widehat{B}}(u;k) = \sum_{r=1}^{N_k} \widehat{\pi}_k^r \widehat{B}_k^r(u),$$

where:

$$\widehat{B}_k^r(u) = \widehat{a}_k^r(u) \widehat{R}_0 \widehat{a}_k^r(u) + \widehat{C}_k^r(u),$$

$$\widehat{C}_k^r(u) = \int_{E_k^r} \pi_k^r(dx) C_0(u;x).$$

Here, the operator $\widehat{R}_0$ is the potential operator of the merged Markov process $\widehat{x}(t), t \geq 0$, defined by the generating matrix $\widehat{Q}$.

Corollary 4.8
1) The stochastic integral functional (4.1) converges weakly

$$\int_0^t a^\varepsilon(x^\varepsilon(s/\varepsilon^3))ds \Rightarrow \widehat{\widehat{\xi}}(t), \quad \varepsilon \to 0.$$

The limit process $\widehat{\widehat{\xi}}(t)$, $t \geq 0$, is a diffusion process with generator (4.47), where:

$$\widehat{\widehat{b}}(u;k) \equiv \widehat{\widehat{a}}_1(k), \quad \widehat{\widehat{a}}_1(k) = \sum_{r=1}^{N_k} \widehat{\pi}_k^r \int_{E_k^r} \pi_k^r(dx) a_1(x),$$

and

$$\widehat{\widehat{B}}(u;k) \equiv \sum_{r=1}^{N_k} \widehat{\pi}_k^r \widehat{B}_k^r, \quad \widehat{B}_k^r = \widehat{a}_k^r \widehat{R}_0 \widehat{a}_k^r, \quad \widehat{a}_k^r := \int_{E_k^r} \pi_k^r(dx) a(x).$$

2) The compound Poisson processes with Markov switching defined by generators (4.3) converge weakly:

$$\xi^\varepsilon(t/\varepsilon^3) \Rightarrow \widehat{\widehat{\xi}}(t), \quad \varepsilon \to 0,$$

where the limit process $\widehat{\widehat{\xi}}(t)$, $t \geq 0$, is a diffusion process with generator (4.47), where

$$\widehat{\widehat{b}} \equiv 0, \quad \widehat{\widehat{B}}(k) = \sum_{r=1}^{N_k} \widehat{\pi}_k^r \widehat{C}_k^r, \quad \widehat{C}_k^r = \int_{E_k^r} \widehat{\pi}_k^r(dx) C(x).$$

4.4.5 Double Split and Double Merging

Here, the switching Markov processes $x^\varepsilon(t), t \geq 0, \varepsilon > 0$, are considered as in the previous Section 4.4.4, and satisfy conditions MD1-MD3, Section 4.2.3.

The stochastic additive functionals $\xi^\varepsilon(t), t \geq 0$, are considered with the more accelerated switching

$$\xi^\varepsilon(t) = \xi^\varepsilon(0) + \int_0^t \eta^\varepsilon(ds; x^\varepsilon(t/\varepsilon^4)), \quad t \geq 0, \varepsilon > 0.$$

Theorem 4.11 *(Double split and double merging) Under Conditions D1-D3, and the balance condition BC2 (Section 4.4.4), the following weak convergence takes place*

$$\xi^\varepsilon(t) \Rightarrow \widehat{\widehat{\xi}}(t), \quad \varepsilon \to 0.$$

The limit diffusion process $\widehat{\widehat{\xi}}(t), t \geq 0$, *is defined by the generator*

$$\mathbb{L}\varphi(u) = \widehat{\widehat{b}}(u)\varphi'(u) + \frac{1}{2}\widehat{\widehat{B}}(u)\varphi''(u). \tag{4.48}$$

The drift coefficient is defined by

$$\widehat{\widehat{b}}(u) = \widehat{\widehat{a}}_1(u) + \widehat{\widehat{a}}_0(u),$$

where:

$$\widehat{\widehat{a}}_1(u) = \sum_{k=1}^N \widehat{\widehat{\pi}}_k \widehat{\widehat{a}}_1(u;k), \quad \widehat{\widehat{a}}_1(u;k) = \sum_{r=1}^{N_k} \widehat{\pi}_k^r \widehat{a}_1(u;k,r),$$

4.4. DIFFUSION APPROXIMATION WITH SPLIT AND MERGING

$$\widehat{a}_1(u;k,r) = \int_{E_k^r} \widehat{\pi}_k^r(dx) a_1(u;x),$$

$$\widehat{\widehat{a}}_0(u) = \sum_{k=1}^N \widehat{\pi}_k \widehat{\widehat{a}}_0(u;k), \quad \widehat{\widehat{a}}_0(u;k) = \widehat{a}(u;k)\widehat{\widehat{R}}_0 \widehat{\widehat{a}}'_u(u;k),$$

$$\widehat{a}(u;k) = \sum_{r=1}^{N_k} \widehat{\pi}_k^r \widehat{a}(u;k,r), \quad \widehat{a}(u;k,r) = \int_{E_k^r} \widehat{\pi}_k^r(dx) a(u;x).$$

The covariance matrix is defined by

$$\widehat{\widehat{B}}(u) = 2 \sum_{k=1}^N \widehat{\pi}_k \widehat{\widehat{B}}(u;k),$$

where:

$$\widehat{\widehat{B}}(u;k) = \widehat{\widehat{B}}_0(u;k) + \widehat{\widehat{C}}_0(u;k),$$

$$\widehat{\widehat{B}}_0(u;k) = \widehat{a}(u;k)\widehat{\widehat{R}}_0\widehat{a}(u;k),$$

$$\widehat{\widehat{C}}_0(u;k) = \sum_{r=1}^{N_k} \widehat{\pi}_k^r \widehat{C}_0(u;k,r),$$

$$\widehat{C}_0(u;k,r) = \int_{E_k^r} \widehat{\pi}_k^r(dx) C_0(u;x).$$

Here, the potential operator $\widehat{\widehat{R}}_0$ corresponds to the twice contracted operator $\widehat{\widehat{Q}}_2$, that is

$$\widehat{\widehat{Q}}_2 \widehat{\widehat{R}}_0 = \widehat{\widehat{R}}_0 \widehat{\widehat{Q}}_2 = \widehat{\widehat{\Pi}} - I.$$

Corollary 4.9

1) The stochastic integral functional (4.1) converges weakly

$$\int_0^t a^\varepsilon(x^\varepsilon(s/\varepsilon^4))ds \Rightarrow \widehat{\widehat{\xi}}(t), \quad \varepsilon \to 0.$$

The limit process $\widehat{\widehat{\xi}}(t)$, $t \geq 0$, is a diffusion process with generator (4.48), where:

$$\widehat{\widehat{b}}(u) \equiv \widehat{\widehat{a}}_1 = \sum_{k=1}^{N} \widehat{\pi}_k \widehat{a}_1(x), \quad \widehat{a}_1(k) = \sum_{r=1}^{N_k} \widehat{\pi}_k^r \int_{E_k^r} \pi_k^r(dx) a_1(x),$$

$$\widehat{\widehat{B}}(u) \equiv \sum_{k=1}^{N} \widehat{\pi}_k \widehat{B}_k, \quad \widehat{B}_k = \widehat{a}(k) \widehat{R}_0 \widehat{a}(k), \quad \widehat{a}(k) = \sum_{r=1}^{N_k} \widehat{\pi}_k^r \int_{E_k^r} \pi_k^r(dx) a(x).$$

2) The compound Poisson processes with Markov switching defined by generators (4.3) converge weakly

$$\xi^\varepsilon(t/\varepsilon^4) \Rightarrow \widehat{\widehat{\xi}}(t), \quad \varepsilon \to 0,$$

where the limit process $\widehat{\widehat{\xi}}(t)$, $t \geq 0$, is a diffusion process with generator (4.48), where:

$$\widehat{\widehat{b}} \equiv 0, \quad \widehat{\widehat{B}} = \sum_{k=1}^{N} \widehat{\pi}_k \widehat{C}(k), \quad \widehat{C}(k) = \sum_{r=1}^{N_k} \widehat{\pi}_k^r \int_{E_k^r} \pi_k^r(dx) C(x).$$

4.5 Integral Functionals in Split Phase Space

In this section we will consider the integral functional

$$\alpha^\varepsilon(t) = \alpha_0 + \int_0^t a(x^\varepsilon(s)) ds, \quad (4.49)$$

with Markov switching $x^\varepsilon(t), t \geq 0$, in single and double ergodic split.

4.5.1 Ergodic Split

Let us define the potential matrix

$$\widehat{R}_0 = [r_{kl}; 1 \leq k, l \leq N],$$

by the following relations :

$$\widehat{Q} \widehat{R}_0 = \widehat{R}_0 \widehat{Q} = \widehat{\Pi} - I = [\pi_{lk} = \pi_k - \delta_{lk}; 1 \leq l, k \leq N],$$

where the generator $\widehat{Q}$ is defined in Theorem 4.1.

4.5. INTEGRAL FUNCTIONALS IN SPLIT PHASE SPACE

The centering shift-coefficient $\widehat{\widehat{a}}$ is defined by the relation

$$\widehat{\widehat{a}} = \sum_{k=1}^{N} \pi_k \widehat{a}_k, \quad \widehat{a}_k := \int_{E_k} \pi_k(dx)a(x), \quad 1 \leq k \leq N;$$

or, in an equivalent form,

$$\widehat{\widehat{a}} = q \sum_{k=1}^{N} \rho_k \widehat{a}_k, \quad \widehat{a}_k := \int_{E_k} \rho_k(dx)a(x), \quad 1 \leq k \leq N, \quad (4.50)$$

where

$$q = \sum_{k=1}^{N} \pi_k q_k,$$

and

$$\rho_k = \pi_k q_k / q, \quad 1 \leq k \leq N,$$

is the stationary distribution of the embedded Markov chain $\widehat{x}_n, n \geq 0$. The vector $\tilde{a}$ is defined as $\tilde{a} = (\tilde{a}_k := a_k - \widehat{\widehat{a}}, 1 \leq k \leq N)$.

Proposition 4.1 *Let the merged condition of Theorem 4.3 be fulfilled, and the limit merged Markov process $\widehat{x}(t)$, $t \geq 0$, be ergodic with the stationary distribution $\pi = (\pi_k, 1 \leq k \leq N)$. Then the normalized centered integral functional*

$$\xi^\varepsilon(t) = \varepsilon^2 \alpha^\varepsilon(t/\varepsilon^3) - \varepsilon^{-1} t \widehat{\widehat{a}}$$

converges weakly, as $\varepsilon \to 0$, to a diffusion process $\xi(t)$, $t \geq 0$, with zero mean and variance

$$\sigma^2 = 2 \sum_{k,l=0}^{N} \pi_k \tilde{a}_k r_{kl} \tilde{a}_l. \quad (4.51)$$

The variance (4.51) can be represented by the *Liptser's formula* [132] in the following way

$$\sigma^2 = -2 \sum_{i=1}^{N} \pi_i \tilde{a}_i b_i, \quad (4.52)$$

where $(b_1, \ldots, b_N)$ is the solution of the equations

$$\sum_{j=1}^{N} \tilde{q}_{ij} b_j = a(i) - a(0), \quad i = 1, \ldots N \quad (4.53)$$

and $\tilde{Q} = [\tilde{q}_{ij}; 1 \leq i, j \leq N]$ is nonsingular matrix defined by

$$\tilde{q}_{ij} = q_{ij} - q_{0j}. \quad (4.54)$$

As was shown in [101], the variance (4.51) can be represented in the following form ($q_i := -q_{ii}$)

$$\sigma^2 = 2 \sum_{i=1}^{N} \pi_i q_i b_i^2 - \sum_{i \neq j \geq 1}^{N} b_i b_j [\pi_i q_{ij} + \pi_j q_{ji}]. \quad (4.55)$$

From (4.53)-(4.54) we obtain:

$$\sum_{i=0}^{N} \pi_i \sum_{j=1}^{N} \tilde{q}_{ij} b_j = \sum_{j=1}^{N} b_j \sum_{i=0}^{N} \pi_i (q_{ij} - q_{0j}) = -\sum_{j=1}^{N} q_{0j} b_j.$$

By using the balance condition

$$\sum_{i=0}^{N} \pi_i \tilde{a}_i = 0, \quad (4.56)$$

in the right hand side of (4.53), we get

$$\sum_{i=0}^{N} \pi_i (\tilde{a}_i - \tilde{a}_0) = -\tilde{a}_0,$$

and hence

$$\sum_{j=1}^{N} q_{0j} b_j = \tilde{a}_0. \quad (4.57)$$

Now from (4.53) and (4.57), we have:

$$\sum_{j=1}^{N} \tilde{q}_{ij} b_j = \sum_{j=1}^{N} q_{ij} b_j - \sum_{j=1}^{N} q_{0j} b_j = \tilde{a}_i - \tilde{a}_0,$$

and hence

$$\sum_{j=1}^{N} q_{ij} b_j = \tilde{a}_i, \quad 0 \leq i \leq N. \quad (4.58)$$

4.5. INTEGRAL FUNCTIONALS IN SPLIT PHASE SPACE

Let us consider the negative term in (4.55):

$$\sum_{\substack{i=1\\i\neq j}}^{N} \pi_i q_{ij} b_i b_j = \sum_{i=1}^{N} \pi_i b_i (\sum_{j\neq i}^{N} q_{ij} b_j + q_i b_i)$$

$$= \sum_{i=1}^{N} \pi_i b_i \sum_{j=1}^{N} q_{ij} b_j + \sum_{i=1}^{N} \pi_i b_i^2 q_i \quad \text{(from (4.58))}$$

$$= \sum_{i=1}^{N} \pi_i b_i \tilde{a}_i + \sum_{i=1}^{N} \pi_i b_i^2 q_i.$$

Thus the variance (4.52) is transformed into the form (4.55).

4.5.2 Double Split and Merging

Let $\widehat{\widehat{R}}_0 = (\widehat{\widehat{r}}_{k\ell},\ 1 \leq k, \ell \leq N)$ be the potential matrix of operator $\widehat{\widehat{Q}}_2$ (see Theorem 4.3) defined by relations:

$$\widehat{\widehat{Q}}_2 \widehat{\widehat{R}}_0 = \widehat{\widehat{R}}_0 \widehat{\widehat{Q}}_2 = \widehat{\widehat{\Pi}} - I = (\pi_{k\ell} = \pi_k - \delta_{k\ell},\ 1 \leq k, \ell \leq N).$$

In the same way, $\widehat{\widehat{R}}_0$ is the potential matrix of operator $\widehat{Q}_1$.

Let $w(t)$, $t \geq 0$, be the standard Wiener process. Then the following result takes place.

Proposition 4.2 *If the merging condition of Theorem 4.3 holds true and the limit merged Markov process $\widehat{x}(t)$, $t \geq 0$, has a stationary distribution, $\widehat{\pi} = (\pi_k^r,\ 1 \leq r \leq N_k,\ 1 \leq k \leq N)$, then, under the balance condition $\widehat{\widehat{A}} = 0$, the following weak convergence takes place,*

$$\varepsilon^{-1} \int_0^t a(x^\varepsilon(s/\varepsilon^4))ds \Longrightarrow \sigma w(t), \quad \varepsilon \to 0,$$

with variance

$$\sigma^2 = -2 \sum_{k=1}^{N} \sum_{\ell=1}^{N} \widehat{\pi}_k \widehat{\widehat{a}}_k \widehat{\widehat{r}}_{k\ell} \widehat{\widehat{a}}_\ell, \tag{4.59}$$

where $\widehat{\widehat{a}}_k = \sum_{r=1}^{N_k} \widehat{\pi}_k^r \int_{E_k^r} \pi_k^r(dx) a(x),\ 1 \leq k \leq N.$

4.5.3 Triple Split and Merging

Proposition 4.3 *If the merging condition of Theorem 4.3 holds true and the limit merged Markov process $\widehat{\widehat{x}}(t)$, $t \geq 0$, has a stationary distribution, $\widehat{\widehat{\pi}} = (\widehat{\widehat{\pi}}_k, 1 \leq k \leq N)$, then, under the balance condition $\widehat{\widehat{A}} = 0$, the following weak convergence takes place,*

$$\varepsilon^{-1} \int_0^t \widehat{\widehat{a}}(\widehat{\widehat{x}}(s/\varepsilon^2))ds \Longrightarrow \sigma w(t), \quad \varepsilon \to 0,$$

with variance σ^2 given by relation (4.59).

Proposition 4.4 *If the merging condition of Theorem 4.3 holds true and the support Markov process $x(t)$, $t \geq 0$, has a stationary distribution, $\pi(dx) = (\pi_k^r(dx), 1 \leq r \leq N_k, 1 \leq k \leq N)$, then the following weak convergence takes place,*

$$\eta^\varepsilon(t) := \varepsilon^{-1} \int_0^t \widetilde{a}(x^\varepsilon(s/\varepsilon^4))ds \Longrightarrow \sigma w(t), \quad \varepsilon \to 0,$$

provided that the balance condition holds

$$\widehat{\widehat{\Pi}}\widehat{\Pi}\Pi \widetilde{a}(x) = 0,$$

where $\widetilde{a} := a(x) - \widehat{\widehat{a}}$, and with variance σ^2 given by

$$\sigma^2 = -2 \sum_{k=1}^N \sum_{\ell=1}^N \widetilde{\widehat{a}}_k \widetilde{\widehat{r}}_{k\ell} \widetilde{\widehat{a}}_\ell,$$

where $\widetilde{\widehat{a}}_k = \widehat{\widehat{a}}_k - \widehat{\widehat{a}}$, $1 \leq k \leq N$.

Concerning positiveness of the variances σ^2 defined by the above formulas see Appendix C.

Chapter 5

Phase Merging Principles

5.1 Introduction

The phase merging principles for switching processes constructed in this chapter are based on a solution of a *singular perturbation problem* for an asymptotic representation of singular perturbed operators.

Solving singular perturbation problems will also be basic in constructing and verifying phase merging principles, averaging, and diffusion approximation schemes.

Therefore, first, the singular perturbation problems for the *reducible-invertible operators* of stochastic systems presented in Chapters 3 and 4 are solved. The solution of these problems will yield the first part of the proof of weak convergence of stochastic processes, corresponding to the convergence of finite-dimensional distributions of the laws of the stochastic processes. The main assumption is that the switching processes are strongly ergodic, which implies that the generators are reducible-invertible. This means that the singular perturbation problem has a solution provided that some additional non restrictive conditions hold.

The basic singular perturbation problems used here are given in Propositions 5.1-5.5. Particular stochastic systems switched by a semi-Markov process are especially considered. The average approximation is given in Propositions 5.6-5.7. The diffusion approximation is given in Propositions 5.8-5.17.

5.2 Perturbation of Reducible-Invertible Operators

5.2.1 *Preliminaries*

Here, we will give the main steps of the solution of the *singular perturbation problem*. Let $x(t), t \geq 0$, be a uniformly ergodic Markov process with state space $(E, \mathcal{E})$, generator Q, and stationary distribution π. We suppose that E is split into N finite ergodic classes, say $E_1, ..., E_N$, with

$$E = \cup_{k=1}^{N} E_k, \quad E_k \cap E_l = \emptyset, \quad k \neq l,$$

with stationary distributions π_k, $1 \leq k \leq N$, on each class.

Let Π be the projector onto the null space $\mathcal{N}_Q$ of the generator Q (Section 1.6) acting as follows on the test functions φ

$$\Pi\varphi(x) = \sum_{k=1}^{N} \widehat{\varphi}_k \mathbf{1}_k(x), \quad \widehat{\varphi}_k := \int_{E_k} \pi_k(dx)\varphi(x), \quad (5.1)$$

and

$$\mathbf{1}_k(x) := \begin{cases} 1, & x \in E_k \\ 0, & x \notin E_k. \end{cases}$$

As a consequence, the contracted space $\widehat{\mathcal{N}}_Q$ is an N-dimensional Euclidean space $\mathbb{R}^N$. So, the contracted vector $\widehat{\varphi} := \Pi\varphi$ is $\widehat{\varphi} := (\widehat{\varphi}_k, 1 \leq k \leq N)$.

The problems of singular perturbation for reducible-invertible operators are the main tools for achieving phase merging principles. Let us recall that a reducible-invertible operator is normally solvable (see Section 1.6).

Let Q be a bounded reducible-invertible operator on a Banach space $\mathbf{B}$. Then we have the following representation

$$\mathbf{B} = \mathcal{N}_Q \oplus \mathcal{R}_Q. \quad (5.2)$$

The null-space is not empty, $\dim \mathcal{N}_Q \geq 1$.

The decomposition (5.2), generates the projection Π on the subspace $\mathcal{N}_Q$

$$\Pi\varphi := \begin{cases} \varphi, & \varphi \in \mathcal{N}_Q, \\ 0, & \varphi \in \mathcal{R}_Q. \end{cases}$$

5.2. PERTURBATION OF REDUCIBLE-INVERTIBLE OPERATORS

The operator $I - \Pi$ is the projector on the space $\mathcal{R}_Q$

$$(I - \Pi)\varphi := \begin{cases} 0, & \varphi \in \mathcal{N}_Q \\ \varphi, & \varphi \in \mathcal{R}_Q, \end{cases}$$

where I is the identity operator in **B**.

Let Q be an reducible-invertible operator, and its potential (operator) R_0 (see Section 1.6).

The solution of the following equation

$$Q\varphi = \psi, \qquad (5.3)$$

in the space $\mathcal{R}_Q$, is represented by

$$\varphi = -R_0\psi, \qquad (5.4)$$

where:

$$QR_0 = R_0Q = \Pi - I.$$

For a uniformly ergodic Markov process with generator Q, and semigroup $P_t, t \geq 0$, the potential R_0 is a bounded operator defined by

$$R_0 = \int_0^\infty (P_t - \Pi)dt,$$

where the projector operator Π is defined as follows

$$\Pi\varphi(u) = \int_E \pi(dx)\varphi(x)\mathbf{1}(x) = \widehat{\varphi}\mathbf{1}(x),$$

$$\widehat{\varphi} := \int_E \pi(dx)\varphi(x), \quad \mathbf{1}(x) = 1, \quad x \in E,$$

with $\pi(B), B \in \mathcal{E}$ the stationary distribution of the Markov process.

5.2.2 Solution of Singular Perturbation Problems

A solution of the asymptotic singular perturbation problem for the reducible-invertible operator Q in the series scheme with small parameter series $\varepsilon > 0, \varepsilon \to 0$, and perturbing operator Q_1 is formulated in the following way. We have to construct the vector $\varphi^\varepsilon = \varphi + \varepsilon\varphi_1$ and the vector ψ which satisfy the asymptotic representation

$$[\varepsilon^{-1}Q + Q_1]\varphi^\varepsilon = \psi + \varepsilon\theta^\varepsilon, \qquad (5.5)$$

with uniformly bounded in norm vector θ^ε, that is,

$$\|\theta^\varepsilon\| \leq C, \quad \varepsilon \to 0.$$

It is worth noticing that in such a problem the operator Q corresponds to the generator of a uniformly ergodic Markov process. Usually the operator Q_1 is a generator too, but may be just a perturbing jump kernel operator.

Such a problem amounts to the asymptotic solution of the following equation for a given vector ψ^ε

$$[Q + \varepsilon Q_1]\varphi^\varepsilon = \psi^\varepsilon.$$

A similar equation appears when the inverse operator of a singular operator is constructed, that is,

$$[Q + \varepsilon Q_1]^{-1} = \varepsilon^{-1} Q^0 + Q^1 + \dots$$

While there exist many situations which cannot be classified, it is possible to find out some logically complete variants of these problems.

Equation (5.5) can be represented as follows

$$[\varepsilon^{-1} Q + Q_1](\varphi + \varepsilon \varphi_1) = \varepsilon^{-1} Q\varphi + [Q\varphi_1 + Q_1 \varphi] + \varepsilon Q_1 \varphi_1. \qquad (5.6)$$

In order to obtain the right-hand side of (5.5), we set:

$$Q\varphi = 0, \quad Q\varphi_1 + Q_1 \varphi = \psi, \quad Q_1 \varphi_1 = \theta^\varepsilon. \qquad (5.7)$$

From (5.7) we get $\varphi \in \mathcal{N}_Q$. The third equality in (5.7) means that the vector θ^ε is independent of ε. Hence, the boundedness of the remaining vector in (5.5) provides the boundedness of the function $Q_1 \varphi_1$. Now, the main problem is to solve the second equation of (5.7), that is,

$$Q\varphi_1 = \psi - Q_1 \varphi. \qquad (5.8)$$

The *solvability condition* for (5.8) with the reducible-invertible operator Q has the following form

$$\Pi(\psi - Q_1 \varphi) = 0, \qquad (5.9)$$

where Π is the projector onto $\mathcal{N}_Q$. Taking into account that $\varphi \in \mathcal{N}_Q$, that is $\Pi \varphi = \varphi$, (5.9) leads to

$$\Pi Q_1 \Pi \varphi = \Pi \psi. \qquad (5.10)$$

5.2. PERTURBATION OF REDUCIBLE-INVERTIBLE OPERATORS

The decisive step of the singular perturbation problem (5.5) comes now. The operator $\Pi Q_1 \Pi$ acts in the subspace $\mathcal{N}_Q$ and $\Pi Q_1 \Pi \varphi = 0$ if $\varphi \in \mathcal{R}_Q$.

Let us introduce the contracted operator $\widehat{Q}_1$ on the contracted space $\widehat{\mathcal{N}}_Q$ by the following relation

$$\Pi Q_1 \Pi = \widehat{Q}_1 \Pi, \qquad (5.11)$$

and set also $\widehat{\psi} := \widehat{\Pi \psi} \in \widehat{\mathcal{N}}_Q$. So, equality (5.10) becomes

$$\widehat{\psi} = \widehat{Q}_1 \widehat{\varphi}. \qquad (5.12)$$

Relation (5.12) establishes a connection between two vectors $\widehat{\varphi}$ and $\widehat{\psi}$ in $\widehat{\mathcal{N}}_Q$. Now, by formula (5.4), we get from Equation (5.8):

$$\varphi_1 = R_0(Q_1 \varphi - \psi), \quad \Pi \varphi_1 = 0. \qquad (5.13)$$

Substituting (5.12) into (5.13), we get

$$\varphi_1 = R_0 \widetilde{Q}_1 \varphi, \quad \widetilde{Q}_1 := Q_1 - \widehat{Q}_1. \qquad (5.14)$$

Finally, the vector θ^ε has the following representation:

$$\theta^\varepsilon = Q_1 \varphi_1 = Q_1 R_0 \widetilde{Q}_1 \varphi. \qquad (5.15)$$

Equations (5.12)-(5.15) give the solution of the singular perturbation problem (5.5).

Let us now calculate the contracted operator $\widehat{Q}_1$, associated to the kernel $Q_1(x, dy)$ on E, which acts on **B** as follows

$$Q_1 \varphi(x) := \int_E Q_1(x, dy) \varphi(y), \qquad (5.16)$$

where we suppose that $Q_1(x, E) = q_1(x)$, satisfies

$$\|q_1\| := \sup_{x \in E} |q_1(x)| < \infty.$$

The *contracted operator* $\widehat{Q}_1$ is defined by the relation

$$\widehat{Q}_1 \Pi = \Pi Q_1 \Pi. \qquad (5.17)$$

Thus from (5.1) and (5.16), we have:

$$\Pi Q_1 \Pi \varphi(x) = \Pi Q_1 \sum_{k=1}^{N} \widehat{\varphi}_k \mathbf{1}_k(x) = \sum_{k=1}^{N} \widehat{\varphi}_k \Pi \int_{E_k} Q_1(x, dy)$$

$$= \sum_{k=1}^{N} \widehat{\varphi}_k \Pi Q_1(x, E_k) = \sum_{k=1}^{N} \widehat{\varphi}_k \sum_{r=1}^{N} \widehat{q}_{rk} \mathbf{1}_r(x), \quad (5.18)$$

where

$$\widehat{q}_{rk} := \int_{E_r} \pi_r(dx) Q_1(x, E_k), \quad 1 \le r, k \le N.$$

Hence, we get

$$\Pi Q_1 \Pi \varphi(x) = \sum_{r=1}^{N} \mathbf{1}_r(x) \sum_{k=1}^{N} \widehat{q}_{rk} \widehat{\varphi}_k.$$

Now, we conclude that the contracted operator $\widehat{Q}_1$ is determined by the matrix

$$\widehat{Q}_1 := (\widehat{q}_{rk}; 1 \le r, k \le N),$$

and acts on the Euclidean space $\mathbb{R}^N$ of vectors $\widehat{\varphi} = (\varphi_k, 1 \le k \le N)$ as follows

$$\widehat{Q}_1 \widehat{\varphi} = \widehat{\psi}, \quad \widehat{\psi}_r := \sum_{k=1}^{N} \widehat{q}_{rk} \widehat{\varphi}_k, \quad 1 \le r \le N.$$

For future reference, we formulate the solution of singular perturbation problem (5.5) as follows.

Proposition 5.1 *Let the bounded operator Q, on the Banach space $\mathbf{B}$, be reducible-invertible with projector Π on the null-space $\mathcal{N}_Q$, $\dim \mathcal{N}_Q \ge 1$ and potential operator R_0.*

Let the perturbing operator Q_1 on $\mathbf{B}$ be closed with a dense domain $\mathbf{B}_0 \subseteq \mathbf{B}$, $\overline{\mathbf{B}}_0 = \mathbf{B}$, and a non-zero contracted operator $\widehat{Q}_1$.

Then the asymptotic representation

$$[\varepsilon^{-1} Q + Q_1](\varphi + \varepsilon \varphi_1) = \widehat{Q}_1 \varphi + \varepsilon \theta^\varepsilon, \quad (5.19)$$

is realized by the vectors $\varphi \in \mathcal{N}_Q$ and

$$\varphi_1 = R_0 \widetilde{Q}_1 \varphi, \quad \theta^\varepsilon = Q_1 R_0 \widetilde{Q}_1 \varphi.$$

Here R_0 is the potential of Q, and $\widetilde{Q}_1$ is given by (5.14).

5.2. PERTURBATION OF REDUCIBLE-INVERTIBLE OPERATORS

PROOF. The proof was given above in the case of the bounded operator Q_1. For the case of the densely closed defined on **B** operator Q_1, see[116]. □

Corollary 5.1 *Under the assumptions of Proposition 5.1, the asymptotic representation*

$$[\varepsilon^{-1}Q + Q_1 + \varepsilon\theta_2^\varepsilon](\varphi + \varepsilon\varphi_1) = \widehat{Q}_1\varphi + \varepsilon\theta^\varepsilon\varphi, \qquad (5.20)$$

is realized by the vectors $\varphi \in \mathcal{N}_Q$ and $\varphi_1 = R_0\ddot{Q}_1\varphi$. The remaining term θ^ε is represented as follows

$$\theta^\varepsilon\varphi = [\theta_2^\varepsilon + \theta_1^\varepsilon R_0\widetilde{Q}_1]\varphi. \qquad (5.21)$$

Here by definition: $\theta_1^\varepsilon := Q_1 + \varepsilon\theta_2^\varepsilon$.

PROOF. By taking into account the following equalities, the proof becomes straightforward:

$$[\varepsilon^{-1}Q + Q_1 + \varepsilon\theta_2^\varepsilon](\varphi + \varepsilon\varphi_1) = [\varepsilon^{-1}Q + Q_1 + \varepsilon\theta_2^\varepsilon]\varphi + \varepsilon[\varepsilon^{-1}Q + \theta_1^\varepsilon]\varphi_1$$
$$= \widehat{Q}_1\varphi + \varepsilon[\theta_2^\varepsilon\varphi + \theta_1^\varepsilon\varphi_1].$$

□

Solving the singular perturbation problem (5.19) is trivial if the following *balance condition* holds

$$\Pi Q_1 \Pi = 0. \qquad (5.22)$$

Nevertheless, there are cases of non trivial solution of singular perturbation problem under the balance condition.

Proposition 5.2 *Let the operator Q on **B** be a bounded reducible-invertible with the projection operator Π and the potential R_0. Assume that the operator Q_1 satisfies the balance condition (5.22), and that Q_1 and Q_2 are closed with common domain $\mathbf{B}_0$ dense in **B** and operator $Q_0 := Q_2 + Q_1 R_0 Q_1$ whose contraction on the null-space $\widehat{\mathcal{N}}_Q$ is the non-zero operator $\widehat{Q}_0$.*

Then the asymptotic representation

$$[\varepsilon^{-2}Q + \varepsilon^{-1}Q_1 + Q_2](\varphi + \varepsilon\varphi_1 + \varepsilon^2\varphi_2) = \widehat{Q}_0\varphi + \varepsilon\theta^\varepsilon, \qquad (5.23)$$

is realized by the vectors determined by the equations:

$$\varphi_1 = R_0 Q_1 \varphi \qquad (5.24)$$

$$\varphi_2 = R_0 \widetilde{Q}_0 \varphi, \quad \widetilde{Q}_0 := Q_0 - \widehat{Q}_0, \qquad (5.25)$$

$$\theta^\varepsilon = [Q_1 + \varepsilon Q_2]\varphi_2 + Q_2\varphi_1 \tag{5.26}$$

PROOF. From (5.23), we get:

$$\begin{aligned} Q\varphi &= 0, \\ Q\varphi_1 + Q_1\varphi &= 0, \\ Q\varphi_2 + Q_1\varphi_1 + Q_2\varphi &= \psi. \end{aligned} \tag{5.27}$$

The first equation in (5.27) gives $\varphi \in \mathcal{N}_Q$. The second equation together with balance condition (5.22) gives (5.24).

Rewriting the third equation using (5.24) we get

$$Q\varphi_2 + Q_0\varphi = \psi.$$

The solvability condition for this equation gives

$$\widehat{\psi} = \widehat{Q}_0\widehat{\varphi},$$

that is the main part in (5.23). □

Remark 5.1. Proposition 5.2 is valid for the closed operator Q with the common dense domain B_0. In this case, the potential operator R_0 is a closed densely defined operator [116].

In various situations the operator $\widehat{Q}_1$ is reducible-invertible[116].

The phase merging principles constructed in the next sections are based on a solution of the singular perturbation problem for an asymptotic representation singular perturbation operator with a remaining term.

Proposition 5.3 *Under the conditions of Proposition 5.1, assume that the contracted operator $\widehat{Q}_1$ is reducible-invertible with null-space $\widehat{\mathcal{N}}_{\widehat{Q}_1} \subset \widehat{\mathcal{N}}_Q$ and projection operator $\widehat{\Pi}$ on $\widehat{\mathcal{N}}_Q$.*

The operator Q_2, under the conditions of Proposition 5.2 has a twice contracted non-zero operator $\widehat{\widehat{Q}}_2$ on $\widehat{\mathcal{N}}_{\widehat{Q}_1}$, determined by the relations:

$$\widehat{\Pi}\widehat{Q}_2\widehat{\Pi} = \widehat{\widehat{Q}}_2\widehat{\Pi}, \quad \widehat{Q}_2\Pi = \Pi Q_2\Pi. \tag{5.28}$$

Then the asymptotic representation

$$[\varepsilon^{-2}Q + \varepsilon^{-1}Q_1 + Q_2](\varphi + \varepsilon\varphi_1 + \varepsilon^2\varphi_2) = \widehat{\widehat{Q}}_2\varphi + \varepsilon\theta^\varepsilon, \tag{5.29}$$

5.2. PERTURBATION OF REDUCIBLE-INVERTIBLE OPERATORS

is realized by the vectors given by the relations: $\varphi \in \mathcal{N}_Q$,

$$\varphi_1 = \widehat{R}_0 \widetilde{\widehat{Q}}_2 \widehat{\varphi} \tag{5.30}$$

$$\varphi_2 = R_0[Q_2 - \widehat{\widehat{Q}}_2]\varphi + R_0 Q_1 \varphi_1 \tag{5.31}$$

$$\theta^\varepsilon = [Q_1 + \varepsilon Q_2]\varphi_2 + Q_2 \varphi_1 \tag{5.32}$$

where $\widehat{R}_0$ is the potential of the reducible-invertible operator $\widehat{Q}_1$, and

$$\widetilde{\widehat{Q}}_2 := \widehat{Q}_2 - \widehat{\widehat{Q}}_2.$$

PROOF. From (5.29), we obtain:

$$Q\varphi = 0 \tag{5.33}$$
$$Q\varphi_1 + Q_1\varphi = 0 \tag{5.34}$$
$$Q\varphi_2 + Q_1\varphi_1 + Q_2\varphi = \widehat{\widehat{Q}}_2 \varphi \tag{5.35}$$
$$(Q_1 + \varepsilon Q_2)\varphi_2 + Q_2\varphi_1 = \theta^\varepsilon. \tag{5.36}$$

By (5.33), we get that $\varphi \in \mathcal{N}_Q$. The solvability condition for (5.34) gives

$$\Pi Q_1 \Pi \varphi = 0,$$

or, after contraction on the subspace $\widehat{\mathcal{N}}_Q$,

$$\widehat{Q}_1 \widehat{\varphi} = 0, \tag{5.37}$$

that is $\widehat{\varphi} \in \widehat{\mathcal{N}}_{\widehat{Q}_1}$.

Writing the solvability condition for φ_2 by using projector Π, we obtain from (5.35)

$$\widehat{Q}_1 \widehat{\varphi}_1 + \widehat{Q}_2 \widehat{\varphi} = \widehat{\widehat{Q}}_2 \varphi,$$

or in another form

$$\widehat{Q}_1 \widehat{\varphi}_1 = -\widetilde{\widehat{Q}}_2 \varphi.$$

Now, the solvability condition for $\widehat{\varphi}_1$ holds because of relation (5.28).

The solvability condition for the operator $\widetilde{\widehat{Q}}_2$ is verified. Hence, the vector $\widehat{\varphi}_1$ is defined by (5.30). Now, (5.35), for the vector φ_2, can be written

$$Q\varphi_2 = -[Q_2 - \widehat{\widehat{Q}}_2]\varphi - Q_1\varphi_1.$$

So that the solution φ_2 is represented by (5.31), and (5.32) is obvious. $\square$

A solution of the singular perturbation problem (5.29) is trivial if the twice contracted operator $\widehat{\widehat{Q}}_2$ is null. However, there are cases of non trivial solution of singular perturbation problem under this condition.

Proposition 5.4 *Let the bounded operator Q on the Banach space $\mathbf{B}$ be reducible-invertible with projector Π and potential R_0. Assume that:*

(1) *the operators Q_1, Q_2 and Q_3 are closed with common domain $\mathbf{B}_0$ dense in $\mathbf{B}$;*

(2) *the contracted operator $\widehat{Q}_1$ on the null-space $\mathcal{N}_Q$ is reducible-invertible with projector $\widehat{\Pi}$ and potential $\widehat{R}_0$;*

$$\widehat{Q}_1 \widehat{R}_0 = \widehat{R}_0 \widehat{Q}_1 = \widehat{\Pi} - I,$$

(3) *the twice contracted operator $\widehat{\widehat{Q}}_2$ is null.*

Then the asymptotic representation

$$[\varepsilon^{-3}Q + \varepsilon^{-2}Q_1 + \varepsilon^{-1}Q_2 + Q_3](\varphi + \varepsilon\varphi_1 + \varepsilon^2\varphi_2 + \varepsilon^3\varphi_3) = \widehat{\widehat{Q}}_0\varphi + \theta^\varepsilon\varphi, \tag{5.38}$$

is realized by the contracted operator:

$$\widehat{\widehat{Q}}_0 := \widehat{Q}_3 + \widehat{Q}_2 \widehat{R}_0 \widehat{Q}_2, \quad \widehat{\widehat{Q}}_0 \widehat{\Pi} = \widehat{\Pi} \widehat{\widehat{Q}}_0 \widehat{\Pi}, \tag{5.39}$$

on the null-space $\widehat{\mathcal{N}}_{\widehat{Q}_1}$, and by the following vectors:

$$\varphi_1 = \widehat{R}_0 \widehat{Q}_2 \varphi, \tag{5.40}$$

$$\varphi_2 = \widehat{R}_0 \widetilde{\widehat{Q}}_2 \varphi, \quad \widetilde{\widehat{Q}}_0 := \widehat{\widehat{Q}}_0 - \widehat{\widehat{Q}}_0, \tag{5.41}$$

$$\varphi_3 = R_0[Q_0 - \widetilde{\widehat{Q}}_0] + Q_1 \widehat{R}_0 \widetilde{\widehat{Q}}_0]\varphi, \tag{5.42}$$

with the negligible term

$$\theta^\varepsilon = Q_3 \varphi_1 + (Q_2 + \varepsilon Q_3)\varphi_2 + (Q_1 + \varepsilon Q_2 + \varepsilon^2 Q_3)\varphi_3. \tag{5.43}$$

5.2. PERTURBATION OF REDUCIBLE-INVERTIBLE OPERATORS

PROOF. Comparing the coefficients, with respect to the degrees of the parameter ε, of the expansion of the left-hand side of (5.38) with those of the right-hand side, we get the following relations:

$$Q\varphi = 0 \tag{5.44}$$

$$Q\varphi_1 + Q_1\varphi = 0 \tag{5.45}$$

$$Q\varphi_2 + Q_1\varphi_1 + Q_2\varphi = 0 \tag{5.46}$$

$$Q\varphi_3 + Q_1\varphi_2 + Q_2\varphi_1 + Q_3\varphi = \psi \tag{5.47}$$

From (5.44) we get $\varphi \in \mathcal{N}_Q$. Equation (5.45) gives $\varphi \in \mathcal{N}_{Q_1}$. Hence $Q_1\varphi = 0$, $Q\varphi_1 = 0$, that is $\varphi_1 \in \mathcal{N}_Q$. Let us now investigate Equation (5.46). The solvability condition is

$$\Pi[Q_1\varphi_1 + Q_2\varphi] = 0,$$

or, in another form

$$\widehat{Q}_1\varphi_1 + \widehat{Q}_2\varphi = 0, \tag{5.48}$$

from which we get

$$\varphi_1 = \widehat{R}_0\widehat{Q}_2\varphi. \tag{5.49}$$

The solvability condition for (5.48) is by Assumption (3),

$$\widehat{\Pi}\widehat{Q}_2\varphi = \widehat{\widehat{Q}}_2\varphi = 0.$$

Hence, the vector φ_1 is defined by Equation (5.40).

Finally, the solvability condition for Equation (5.47) is

$$\widehat{Q}_1\varphi_2 + \widehat{Q}_2\varphi_1 + \widehat{Q}_3\varphi = \widehat{\psi}, \quad \psi \in \mathcal{N}_Q.$$

This equation for the vector φ_2 has to satisfy the solvability condition

$$\widehat{\Pi}[\widehat{Q}_2\varphi_1 + \widehat{Q}_3\varphi] = \widehat{\widehat{\psi}}, \quad \widehat{\psi} \in \widehat{\mathcal{N}}_{\widehat{Q}_1}$$

or, in another form, using the representation of the vector φ_1,

$$\widehat{\Pi}[\widehat{Q}_2\widehat{R}_0\widehat{Q}_2 + \widehat{Q}_3]\varphi = \widehat{\widehat{\psi}}.$$

That is the representation of the main part of the asymptotic relation (5.38), with the contracted operator $\widehat{Q}_0$ in (5.39). $\square$

Now the singular perturbation problem can be solved under some additional assumption on the perturbing operators $Q_1, Q_2, \ldots$. We propose only the next result.

Proposition 5.5 *Let the bounded operator Q on the Banach space $\mathbf{B}$ be a reducible-invertible with projector Π and potential R_0. Assume that:*

(1) *the operators Q_k, $k = 1, 2, 3, 4$, are closed with the common domain $\mathbf{B}_0$ dense in $\mathbf{B}$;*

(2) *the contracted operator $\widehat{Q}_1$ on the null-space $\mathcal{N}_Q$ is reducible invertible with projector $\widehat{\Pi}$ and potential $\widehat{R}_0$;*

(3) *the twice contracted operator $\widehat{\widehat{Q}}_2$ on the null-space $\widehat{\mathcal{N}}_{\widehat{Q}_1}$ is reducible-invertible with projector $\widehat{\widehat{\Pi}}$ and potential $\widehat{\widehat{R}}_0$.*

Then the asymptotic representation

$$[\varepsilon^{-4}Q + \varepsilon^{-3}Q_1 + \varepsilon^{-2}Q_2 + \varepsilon^{-1}Q_3 + Q_4](\varphi + \varepsilon\varphi_1 + \varepsilon^2\varphi_2 + \varepsilon^3\varphi_3 + \varepsilon^4\varphi_4)$$
$$= \widehat{\widehat{Q}}_0\varphi + \theta^\varepsilon\varphi, \qquad (5.50)$$

is realized with the contracted operator for

$$\widehat{\widehat{Q}}_0 := \widehat{\widehat{Q}}_4 + \widehat{\widehat{Q}}_3\widehat{\widehat{R}}_0\widehat{\widehat{Q}}_3,$$

on the null space $\widehat{\mathcal{N}}_{\widehat{Q}_2}$, and with the negligible term $\|\theta^\varepsilon\varphi\| \to 0$, as $\varepsilon \to 0$.

PROOF. Similarly to the proof of Proposition 5.4, the last step can be obtained by the following equation

$$\widehat{\widehat{Q}}_2\varphi_2 + [\widehat{\widehat{Q}}_3\widehat{\widehat{R}}_0\widehat{\widehat{Q}}_3 + \widehat{\widehat{Q}}_4]\varphi = \widehat{\widehat{Q}}_0\varphi.$$

The solvability condition for this equation gives the statement of Proposition 5.5. □

The formulated Propositions 5.1-5.5 have an adequate interpretation as phase merging principles for the stochastic systems considered in the next section.

5.3 Average Merging Principle

Averaging is an important step in stochastic approximation of systems. We present in this section averaging results for switched stochastic systems: stochastic evolutionary systems, additive functionals, random evolu-

5.3. AVERAGE MERGING PRINCIPLE

tions, where the switching semi-Markov or Markov process is time-scaled by "ε^{-1}".

5.3.1 Stochastic Evolutionary Systems

First, the stochastic evolutionary system with the switching ergodic Markov process in the series scheme is considered (see Section 3.3.1)

$$\left| \begin{array}{l} \frac{d}{dt} U^\varepsilon(t) = g(U^\varepsilon(t); x(t/\varepsilon)), \\ U^\varepsilon(0) = u. \end{array} \right. \tag{5.51}$$

The switching Markov process $x(t), t \geq 0$, is defined by the generator

$$Q\varphi(x) = q(x) \int_E P(x, dy)[\varphi(y) - \varphi(x)]. \tag{5.52}$$

The stationary distribution $\pi(dx)$ of the ergodic process with generator (5.52) defines the projector

$$\Pi\varphi(x) = \widehat{\varphi}\mathbf{1}(x), \quad \widehat{\varphi} := \int_E \pi(dx)\varphi(x), \quad \mathbf{1}(x) \equiv 1. \tag{5.53}$$

The velocity function $g(u; x), u \in \mathbb{R}^d, x \in E$, is supposed to provide the global solution of the deterministic evolutionary equations:

$$\left| \begin{array}{l} \frac{d}{dt} U_x(t) = g(U_x(t); x) \\ U_x(0) = u \end{array} \right., \quad x \in E. \tag{5.54}$$

The coupled Markov process $U^\varepsilon(t), x^\varepsilon(t) := x(t/\varepsilon), t \geq 0$, can be characterized by the generator (see Proposition 3.3)

$$\mathbb{L}^\varepsilon \varphi(u, x) = [\varepsilon^{-1} Q + \mathbb{\Gamma}(x)]\varphi(u, x), \tag{5.55}$$

where the generator $\mathbb{\Gamma}(x), x \in E$, is defined by the relation

$$\mathbb{\Gamma}(x)\varphi(u) = g(u; x)\varphi'(u). \tag{5.56}$$

Note that the generator (5.56) is induced by the family of semigroups

$$\mathbb{\Gamma}_t(x)\varphi(u) := \varphi(U_x(t)), \quad U_x(0) = u. \tag{5.57}$$

The average merging principle in ergodic merging scheme is realized by a solution of the singular perturbation problem for the generator (5.55).

Proposition 5.6 *The solution of the singular perturbation problem for generator (5.55) is given by the relation*

$$\mathbb{L}^\varepsilon \varphi^\varepsilon(u,x) = \widehat{\mathbb{\Gamma}}\varphi(u) + \varepsilon \theta^\varepsilon(x)\varphi(u), \qquad (5.58)$$

on the perturbed functions $\varphi^\varepsilon(u,x) = \varphi(u) + \varepsilon \varphi_1(u,x)$, *with* $\varphi \in C_0^2(\mathbb{R}^d)$. *The average operator* $\widehat{\mathbb{\Gamma}}$ *is determined by the relations*

$$\widehat{\mathbb{\Gamma}}\varphi(u) = \widehat{g}(u)\varphi'(u), \quad \widehat{g}(u) := \int_E \pi(dx)g(u;x). \qquad (5.59)$$

The remaining term $\theta^\varepsilon(x)$ *is defined by*

$$\theta^\varepsilon(x) = \mathbb{\Gamma}(x)R_0\widetilde{\mathbb{\Gamma}}(x), \quad \widetilde{\mathbb{\Gamma}}(x) := \mathbb{\Gamma}(x) - \widehat{\mathbb{\Gamma}}. \qquad (5.60)$$

It is worth noticing that the average operator (5.59) provides the average evolutionary system

$$\frac{d}{dt}\widehat{U}(t) = \widehat{g}(\widehat{U}(t)),$$

that is exactly as in Corollary 3.3.

PROOF. According to the solution of the singular perturbation problem, given in Proposition 5.1, we obtain (5.58) and (5.60) with the average operator $\widehat{\mathbb{\Gamma}}$, given by the relation

$$\Pi\mathbb{\Gamma}(x)\Pi\varphi(u) = \Pi\mathbb{\Gamma}(x)\varphi(u). \qquad (5.61)$$

Now calculate the average operator in (5.61):

$$\Pi\mathbb{\Gamma}(x)\Pi\varphi(u) = \Pi\mathbb{\Gamma}(x)\varphi(u)$$
$$= \Pi g(u;x)\varphi'(u)$$
$$= \widehat{g}(u)\varphi'(u)$$
$$= \widehat{\mathbb{\Gamma}}\varphi(u).$$

The average velocity is defined in (5.59). □

5.3.2 Stochastic Additive Functionals

The average phase merging principle for the stochastic additive functionals (see Section 3.3.1)

$$\xi^\varepsilon(t) = \xi^\varepsilon(0) + \int_0^t \eta^\varepsilon(ds; x(s/\varepsilon)), \quad t \geq 0, \qquad (5.62)$$

5.3. AVERAGE MERGING PRINCIPLE

with switching ergodic semi-Markov process $x(t), t \geq 0$, on the standard phase space $(E, \mathcal{E})$, given by the semi-Markov kernel

$$Q(x, B, t) = P(x, B) F_x(t), \quad x \in E, B \in \mathcal{E}, t \geq 0, \quad (5.63)$$

is realized by a solution of the singular perturbation problem for the compensating operator of the extended Markov renewal process represented in the following asymptotic forms (see Proposition 3.1):

$$\mathbb{L}^\varepsilon \varphi(u, x) = [\varepsilon^{-1} Q + \mathbf{\Gamma}(x) P + \varepsilon \theta_2^\varepsilon(x)] \varphi$$
$$= [\varepsilon^{-1} Q + \theta_1^\varepsilon(x)] \varphi. \quad (5.64)$$

The generator Q of the associated Markov process is defined by (5.52) with the intensity functions:

$$q(x) = 1/m(x), \quad m(x) := \int_0^\infty \overline{F}_x(t) dt, \quad x \in E,$$

$$q = 1/m, \quad m := \int_E \rho(dx) m(x).$$

The generators $\mathbf{\Gamma}(x), x \in E$, are defined by relation (5.56). The remaining terms $\theta_k^\varepsilon(x), k = 1, 2$, are given by the following relations (see Proposition 3.1)

$$\theta_k^\varepsilon(x) := \mathbf{\Gamma}_\varepsilon^k(x) \mathbb{F}_\varepsilon^{(k)}(x) Q_0, \quad k = 1, 2. \quad (5.65)$$

Here $\mathbf{\Gamma}_\varepsilon(x) := \mathbf{\Gamma}(x) + \varepsilon \gamma_\varepsilon(x)$. The remaining term $\gamma_\varepsilon(x)$ is represented in (3.36).

Proposition 5.7 *The solution of the singular perturbation problem for the operator (5.64) is given by the relation*

$$\mathbb{L}^\varepsilon \varphi^\varepsilon(u, x) = \widehat{\mathbf{\Gamma}} \varphi(u) + \varepsilon \theta_l^\varepsilon(x) \varphi(u), \quad (5.66)$$

on the perturbed test functions $\varphi^\varepsilon(u, x) = \varphi(u) + \varepsilon \varphi_1(u, x)$, *with* $\varphi \in C_0^2(\mathbb{R}^d)$. *The average operator* $\widehat{\mathbf{\Gamma}}$ *is determined by relation (5.59). The remaining term* $\theta_l^\varepsilon(x)$ *is defined by*

$$\theta_l^\varepsilon(x) = \theta_2^\varepsilon(x) + \theta_1^\varepsilon(x) R_0 \widetilde{\mathbf{\Gamma}}(x), \quad (5.67)$$

where $\widetilde{\mathbf{\Gamma}}(x) := \mathbf{\Gamma}(x) - \widehat{\mathbf{\Gamma}}$.

It is worth noticing that the average merging principle gives the same result (5.59) for Markov and semi-Markov switching. In the semi-Markov case, the difference lies only in the definition of the intensity function $q(x) = 1/m(x)$, and the remaining terms.

PROOF. According to the solution of singular perturbation problem, given in Proposition 5.1, we obtain (5.66) with the average operator $\widehat{\mathbb{\Gamma}}$, given by the relation:

$$\widehat{\mathbb{\Gamma}}\Pi = \Pi\mathbb{\Gamma}(x)P\Pi = \Pi\mathbb{\Gamma}(x)\Pi, \qquad \text{(since } P\Pi = \Pi\text{).} \tag{5.68}$$

The remaining operator $\theta_1^\varepsilon(x)$ is calculated as follows:

$$\mathbb{L}^\varepsilon \varphi^\varepsilon = [\varepsilon^{-1}Q + \mathbb{\Gamma}(x)P + \varepsilon\theta_2^\varepsilon(x)]\varphi(u) + \varepsilon[\varepsilon^{-1}Q + \theta_1^\varepsilon(x)]\varphi_1(u,x)$$
$$= \widehat{\mathbb{\Gamma}}\varphi(u) + \theta_2^\varepsilon(x)\varphi(u) + \theta_1^\varepsilon(x)\varphi_1(u,x).$$

Hence, due to the relation $\varphi_1 = R_0\widetilde{\mathbb{\Gamma}}(x)\varphi$, we get (5.67).

Now, the calculation of the average operator in (5.68) gives us the same result (5.59) as in the case of the Markov switching, for the stochastic evolutionary system (5.51). □

Remark 5.2. The average merging principle for stochastic evolutionary systems with semi-Markov switching can be represented as a corollary of Proposition 5.7, formulated for the stochastic additive functionals.

5.3.3 Increment Processes

The increment process in the series scheme with the semi-Markov switching, associated to the jump random evolution, considered in Section 3.2.2, is defined by the family of bounded operators $\mathbb{D}^\varepsilon(x), x \in E$,

$$\mathbb{D}^\varepsilon(x)\varphi(u) := \varphi(u + \varepsilon a(x)), \quad x \in E, \tag{5.69}$$

which has the asymptotic expansion (3.21) on the test functions $\varphi \in C_0^2(\mathbb{R}^d)$, that is

$$\mathbb{D}^\varepsilon(x) = I + \varepsilon \mathbb{D}(x) + \mathbb{D}_1^\varepsilon(x). \tag{5.70}$$

The definition (5.69) provides that

$$\mathbb{D}(x)\varphi(u) = a(x)\varphi'(u).$$

5.3. AVERAGE MERGING PRINCIPLE

The compensating operator of the extended Markov renewal process

$$\xi_n^\varepsilon, \quad x_n^\varepsilon, \quad \tau_n^\varepsilon, \quad n \geq 0,$$

is represented on the test functions $\varphi(u, x)$ as follows, (see (3.23)-(3.25)):

$$\mathbb{L}^\varepsilon \varphi(u, x) = [\varepsilon^{-1} Q + Q_0 \mathbb{D}(x) + Q_0 \mathbb{D}_1^\varepsilon(x)] \varphi(u, x),$$
$$= [\varepsilon^{-1} Q + Q_0 \mathbb{D}_0^\varepsilon(x)] \varphi(u, x), \tag{5.71}$$

where

$$\mathbb{D}_0^\varepsilon(x) = \mathbb{D}(x) + \mathbb{D}_1^\varepsilon(x), \tag{5.72}$$

with the negligible term

$$\|\mathbb{D}_1^\varepsilon(x)\varphi\| \to 0, \quad \varepsilon \to 0.$$

The average merging principle in the ergodic merged scheme can be obtained by using Proposition 5.1 for the truncated compensating operator

$$\overline{\mathbb{L}}^\varepsilon \varphi(u, x) = [\varepsilon^{-1} Q + Q_0 \mathbb{D}(x)] \varphi(u, x). \tag{5.73}$$

Considering the truncated operator (5.73) on the perturbed test functions $\varphi^\varepsilon(u, x) = \varphi(u) + \varepsilon \varphi_1(u, x)$, with $\varphi \in C_0^2(\mathbb{R}^d)$, we get, by Proposition 5.1,

$$\overline{\mathbb{L}}^\varepsilon \varphi^\varepsilon(u, x) = \widehat{\mathbb{D}}_0 \varphi(u) + \varepsilon \theta_0^\varepsilon(x) \varphi(u), \tag{5.74}$$

where the negligible term is $\theta_0^\varepsilon(x) \varphi(u) = Q_0 \mathbb{D}(x) \varphi_1(u)$, or, in explicit form:

$$\theta_0^\varepsilon(x) = a_0(x) R_0 \widetilde{a}(x) \varphi''(\widetilde{u}),$$
$$a_0(x) := Q_0 a(x), \quad \widetilde{a}_0(x) := a_0(x) - \widehat{a}.$$

The average operator $\widehat{\mathbb{D}}_0$ is determined by the relation

$$\widehat{\mathbb{D}}_0 \Pi = \Pi Q_0 \mathbb{D}(x) \Pi. \tag{5.75}$$

Let us calculate the average operator $\widehat{Q}_0$:

$$\begin{aligned}
\Pi Q_0 \mathbb{D}(x)\Pi\varphi(u) &= \Pi Q_0 \mathbb{D}(x)\varphi(u) \\
&= \Pi Q_0 a(x)\varphi'(u) \\
&= \int_E \pi(dx)q(x)Pa(x)\varphi'(u) \\
&= q\int_E \rho(dx)Pa(x)\varphi'(u) \\
&= q\int_E \rho(dx)a(x)\varphi'(u) \\
&= \widehat{a}\varphi'(u),
\end{aligned}$$

where:

$$\widehat{a} = q\int_E \rho(dx)a(x) = \int_E \rho(dx)a(x)/m.$$

Hence, the average operator $\widehat{\mathbb{D}}_0$ is represented as follows

$$\widehat{\mathbb{D}}_0\varphi(u) = \widehat{a}\varphi'(u), \quad \widehat{a} = \int_E \rho(dx)a(x)/m.$$

that is exactly the result of Theorem 3.2.

5.3.4 Continuous Random Evolutions

The continuous random evolution with semi-Markov switching in the average approximation scheme is given by a solution of the evolutionary equation in the Banach space $C(\mathbb{R}^d)$ (see Section 3.2.1),

$$\left| \begin{array}{l} \frac{d}{dt}\Phi^\varepsilon(t) = \mathbb{\Gamma}(x(t/\varepsilon))\Phi^\varepsilon(t), \\ \Phi^\varepsilon(0) = I, \end{array} \right. \tag{5.76}$$

with a given family of generators $\mathbb{\Gamma}(x), x \in E$, generating the semigroups $\Gamma_t(x), t \geq 0, x \in E$.

The coupled random evolution (see Section 2.7)

$$\Phi^\varepsilon(t, x^\varepsilon(t)) = \Phi^\varepsilon(t)\varphi(u, x(t/\varepsilon)) \tag{5.77}$$

on the Banach space $C(\mathbb{R}^d \times E)$ can be characterized by the compensating

5.3. AVERAGE MERGING PRINCIPLE

operator of the extended Markov renewal process (Section 3.2.1)

$$\mathbb{L}^\varepsilon \varphi(u,x) = \varepsilon^{-1} q(x) \Big[\int_0^\infty F_x(ds) \Gamma_{\varepsilon s}^\varepsilon(x) \int_E P(x,dy)\varphi(u,y) - \varphi(u,x) \Big]. \tag{5.78}$$

The factors "ε^{-1}" and "εs" concern the fast time-scaling in (5.76).

The average merging principle in ergodic merging scheme for the continuous random evolutions with semi-Markov switching in the series scheme can be obtained by using Proposition 5.1 for solving the singular perturbation problem for the truncated compensating operator

$$\overline{\mathbb{L}}^\varepsilon = \varepsilon^{-1} Q + \mathbb{\Gamma}(x) P. \tag{5.79}$$

According to Proposition 5.1, the operator (5.79) on the perturbed test functions $\varphi^\varepsilon(u,x) = \varphi(u) + \varepsilon\varphi_1(u,x)$, with $\varphi \in \mathbf{B}_0$, dense in $C(\mathbb{R}^d)$, has the representation

$$\overline{\mathbb{L}}^\varepsilon \varphi^\varepsilon(u,x) = \widehat{\mathbb{\Gamma}}\varphi(u) + \varepsilon \theta^\varepsilon(x)\varphi(u), \tag{5.80}$$

where $\theta^\varepsilon(x)\varphi(u) = \mathbb{\Gamma}(x) R_0 P \widetilde{\mathbb{\Gamma}}(x)\varphi(u)$, and $\widetilde{\mathbb{\Gamma}}(x) := \mathbb{\Gamma}(x) - \widehat{\mathbb{\Gamma}}$.
The limit average generator $\widehat{\mathbb{\Gamma}}$ is defined by

$$\widehat{\mathbb{\Gamma}}\Pi = \Pi\mathbb{\Gamma}(x) P\Pi = \Pi\mathbb{\Gamma}(x)\Pi. \tag{5.81}$$

Hence

$$\widehat{\mathbb{\Gamma}} = \int_E \pi(dx)\mathbb{\Gamma}(x). \tag{5.82}$$

The remaining term $\theta^\varepsilon(x)$ is supposed to be bounded on $\mathbf{B}_0$, that is,

$$\|\theta^\varepsilon(x)\varphi\| \leq b < +\infty, \quad \varphi \in \mathbf{B}_0.$$

5.3.5 Jump Random Evolutions

The jump random evolution in the average merging scheme with semi-Markov switching is determined by a solution of the difference equation (see Section 2.7.2) on the Banach space $C(\mathbb{R}^d)$

$$\Phi^\varepsilon(\tau_n^\varepsilon) = [\mathbb{D}^\varepsilon(x_n^\varepsilon) - I]\Phi^\varepsilon(\tau_n^\varepsilon), \quad n \geq 0, \quad \Phi^\varepsilon(\tau_0^\varepsilon) = \Phi^\varepsilon(0) = I, \tag{5.83}$$

and

$$\Phi^\varepsilon(t) := \Phi^\varepsilon(\tau^\varepsilon(t)), \quad \tau^\varepsilon(t) = \tau_{\nu^\varepsilon(t/\varepsilon)}^\varepsilon, \tag{5.84}$$

given by the family of the bounded operators $\mathbb{D}^\varepsilon(x), x \in E$.
The coupled jump random evolution

$$\Phi^\varepsilon(t, x^\varepsilon(t)) = \Phi^\varepsilon(t)\varphi(u, x(t/\varepsilon)) \tag{5.85}$$

on the Banach space $C(\mathbb{R}^d \times E)$ functions $\varphi(u, x), u \in \mathbb{R}^d, x \in E$, can be characterized by the compensating operator (see Section 3.2.2)

$$\mathbb{L}^\varepsilon \varphi(u, x) = \varepsilon^{-1} q(x) \left[\int_0^\infty F_x(ds) \int_E P(x, dy) D^\varepsilon(y) \varphi(u, y) - \varphi(u, x) \right]. \tag{5.86}$$

The main assumption in what follows is that the family of bounded operators $\mathbb{D}^\varepsilon(x)$, $x \in E$, has the following asymptotic expansion on the test functions $\varphi \in \mathbf{B}_0$, dense in $C(\mathbb{R}^d)$,

$$\mathbb{D}^\varepsilon(x) = I + \varepsilon \mathbb{D}(x) + \mathbb{D}_1^\varepsilon(x), \tag{5.87}$$

with the family of generators $\mathbb{D}(x), x \in E$, having common dense domain of definition on $\mathbf{B}_0$. The negligible term is supposed to satisfy

$$\sup_{x \in E} \|\mathbb{D}_1^\varepsilon(x)\varphi\| \to 0, \quad \varepsilon \to 0, \quad \varphi \in \mathbf{B}_0. \tag{5.88}$$

The average merging principle in ergodic merging scheme for the jump random evolution with semi-Markov switching in series scheme can be obtained by using Proposition 5.1, applied to the truncated operator (see (5.73)).

$$\mathbb{L}_0^\varepsilon = \varepsilon^{-1} Q + Q_0 \mathbb{D}(x). \tag{5.89}$$

According to Proposition 5.1, the generator (5.89) on the perturbed test functions $\varphi^\varepsilon(u, x) = \varphi(u) + \varepsilon \varphi_1(u, x)$, has the representation

$$\mathbb{L}^\varepsilon \varphi^\varepsilon(u, x) = \widehat{\mathbb{D}} \varphi(u) + \theta_d^\varepsilon(u, x). \tag{5.90}$$

The limit average generator $\widehat{\mathbb{D}}$ is determined by the relation

$$\widehat{\mathbb{D}} \Pi = \Pi Q_0 \mathbb{D}(x) P \Pi. \tag{5.91}$$

Taking into account that $Q_0\varphi(x) = q(x)P\varphi(x)$, we calculate:

$$\Pi Q_0 \mathbb{D}(x) = \int_E \pi(dx)q(x) \int_E P(x,dy)\mathbb{D}(y)$$
$$= q \int_E \rho(dx) \int_E P(x,dy)D(y)$$
$$- q \int_E \rho(dy)D(y)$$
$$= \int_E \rho(dx)\mathbb{D}(x)/m, \quad m := \int_E \rho(dx)m(x).$$

Hence

$$\widehat{\mathbb{D}} = \int_E \rho(dx)\mathbb{D}(x)/m. \tag{5.92}$$

5.3.6 Random Evolutions with Markov Switching

The average merging principle for the continuous random evolution with Markov switching can be obtained by using Proposition 5.1 applied to the generator

$$\mathbb{L}^\varepsilon = \varepsilon^{-1}Q + \Gamma(x), \tag{5.93}$$

characterizing the coupled random evolution (see Section 3.2.1, Proposition 3.3)

$$\Phi^\varepsilon(t, x^\varepsilon(t)) = \Phi^\varepsilon(t)\varphi(u, x(t/\varepsilon)). \tag{5.94}$$

According to Proposition 5.1, the average generator is defined by the relation

$$\widehat{\Gamma}\Pi = \Pi\Gamma(x)\Pi, \tag{5.95}$$

that is

$$\widehat{\Gamma} = \int_E \pi(dx)\Gamma(x). \tag{5.96}$$

The jump random evolution with Markov switching in average scheme is characterized by the generator (see Section 3.2.2, Proposition 3.4)

$$\mathbb{L}_d^\varepsilon = \varepsilon^{-1}[Q + Q_0(\mathbb{D}^\varepsilon(x) - I)], \tag{5.97}$$

where, as usual, $Q_0\varphi(x) = q(x)P\varphi(x)$.

The main assumption in the asymptotic expansion, as $\varepsilon \to 0$, is

$$\mathbb{D}^\varepsilon(x) - I = \varepsilon \mathbb{D}(x) + \mathbb{D}_1^\varepsilon(x). \tag{5.98}$$

The negligible term is supposed to satisfy

$$\|\mathbb{D}_1^\varepsilon(x)\varphi\| \to 0, \quad \varepsilon \to 0,$$

on $\varphi \in \mathbf{B}_0$.

The average phase merging principle for the jump random evolution with Markov switching can be obtained by using Proposition 5.1, applied to the truncated generator

$$\mathbb{L}_0^\varepsilon = \varepsilon^{-1} Q + Q_0 \mathbb{D}(x). \tag{5.99}$$

The average generator is determined from the relation

$$\widehat{\mathbb{D}} \Pi = \Pi Q_0 \mathbb{D}(x) \Pi. \tag{5.100}$$

Let us calculate:

$$\Pi Q_0 \mathbb{D}(x) = \int_E \pi(dx) q(x) \int_E P(x, dy) D(y)$$
$$= q \int_E \rho(dx) \int_E P(x, dy) D - y)$$
$$= q \int_E \rho(dx) \mathbb{D}(x).$$

Hence

$$\widehat{\mathbb{D}} = \int_E \rho(dx) \mathbb{D}(x)/m, \quad m := \int_E \rho(dx) m(x). \tag{5.101}$$

5.4 Diffusion Approximation Principle

In this section we verify the algorithms of diffusion approximation for the stochastic systems in ergodic merging scheme formulated in Section 3.4 (Theorems 3.3-3.5, and Corollaries 3.5-3.7). This is obtained by using solution of the singular perturbation problem given in Proposition 5.2, Section 5.2.

5.4.1 Stochastic Integral Functionals

Let us show that the diffusion approximation principle is satisfied by the stochastic integral functionals in series scheme with accelerated ergodic Markov switching (see Section 3.4.1)

$$\alpha^\varepsilon(t) = \alpha_0 + \int_0^t a_\varepsilon(x(s/\varepsilon^2))ds, \quad t \geq 0. \tag{5.102}$$

The velocity function is supposed to depend on the series parameter ε as follows

$$a^\varepsilon(x) = \varepsilon^{-1}a(x) + a_1(x), \quad x \in E. \tag{5.103}$$

The first term of the right hand side in (5.103) satisfies the *balance condition*

$$\int_E \pi(dx)a(x) = 0. \tag{5.104}$$

The switching process $x(t), t \geq 0$, is supposed to be Markovian and given by the generator

$$Q\varphi(x) = q(x)\int_E P(x,dy)[\varphi(y) - \varphi(x)]. \tag{5.105}$$

The coupled Markov process $\alpha^\varepsilon(t), x^\varepsilon(t) := x(t/\varepsilon^2), t \geq 0$, can be determined by the generator (see Section 3.4.1)

$$\mathbb{L}^\varepsilon\varphi(u,x) = [\varepsilon^{-2}Q + \mathbb{A}_\varepsilon(x)]\varphi(u,x), \tag{5.106}$$

where the family of generators $\mathbb{A}_\varepsilon(x), x \in E$, is defined by the velocity (5.103) (see Section 3.2.1, Proposition 3.3):

$$\begin{aligned}\mathbb{A}_\varepsilon(x)\varphi(u) &= a^\varepsilon(x)\varphi'(u) \\ &= \varepsilon^{-1}a(x)\varphi'(u) + a_1(x)\varphi'(u) \\ &= [\varepsilon^{-1}\mathbb{A}(x) + \mathbb{A}_1(x)]\varphi(u), \end{aligned} \tag{5.107}$$

where:

$$\mathbb{A}(x)\varphi(u) = a(x)\varphi'(u), \quad \text{and} \quad \mathbb{A}_1(x)\varphi(u) = a_1(x)\varphi'(u). \tag{5.108}$$

The balance condition (5.104) can be expressed in terms of the projector Π of the generator Q of the switching uniformly ergodic Markov process $x(t), t \geq 0$, as in Theorem 3.3.

The limit generator $\mathbb{L}$ in Corollary 3.5, obtained by a solution of the singular perturbation problem for the generator (5.106):

$$\mathbb{L}^\varepsilon \varphi^\varepsilon(u,x) = [\varepsilon^{-2}Q + \varepsilon^{-1}\mathbb{A}(x) + \mathbb{A}_1(x)][\varphi(u) + \varepsilon\varphi_1(u,x) + \varepsilon^2\varphi_2(u,x)]$$
$$= \mathbb{L}\varphi(u) + \theta^\varepsilon(u,x), \qquad (5.109)$$

has the following form (see Proposition 5.2, Section 5.2)

$$\mathbb{L} = \Pi\mathbb{A}_1(x)\Pi + \Pi\mathbb{A}(x)R_0\mathbb{A}(x)\Pi, \qquad (5.110)$$

with negligible term $\theta^\varepsilon(u,x)$

$$\|\theta^\varepsilon\| \longrightarrow 0, \quad \varepsilon \to 0,$$

and where R_0 is the potential of Q (Section 1.6).

Now, we calculate using representation (5.107):

$$\widehat{\mathbb{A}}_1\Pi\varphi(u) = \Pi\mathbb{A}_1(x)\Pi\varphi(u)$$
$$= \Pi\mathbb{A}_1(x)\varphi(u)$$
$$= \Pi a_1(x)\varphi'(u)$$
$$= a_1\varphi'(u),$$

where $a_1 := \int_E \pi(dx)a_1(x)$.

By a similar calculus, we have:

$$\Pi\mathbb{A}(x)R_0\mathbb{A}(x)\Pi\varphi(u) = \Pi\mathbb{A}(x)R_0\mathbb{A}(x)\Pi\varphi(u)$$
$$= \Pi\mathbb{A}(x)R_0\mathbb{A}(x)\varphi(u)$$
$$= \Pi\mathbb{A}(x)R_0 a(x)\varphi'(u)$$
$$= \Pi a(x)R_0 a(x)\varphi''(u)$$
$$= \frac{1}{2}\sigma_0^2\varphi''(u),$$

where

$$\sigma_0^2 := 2\int_E \pi(dx)a_0(x), \quad a_0(x) := a(x)R_0 a(x).$$

Note that actually $\sigma_0^2 \geq 0$ (see Appendix C).

Therefore, the limit generator $\mathbb{L}$ in Corollary 3.5 is represented by

$$\mathbb{L}\varphi(u) = a_1\varphi'(u) + \frac{1}{2}\sigma_0^2\varphi''(u), \qquad (5.111)$$

5.4. DIFFUSION APPROXIMATION PRINCIPLE

that is the limit diffusion process is

$$\alpha^0(t) = \alpha_0 + a_1 t + \sigma_0 w(t), \quad t \geq 0, \tag{5.112}$$

exactly as in Corollary 3.5.

The proof of Theorem 3.3 is based on the representation of the stochastic integral functional (5.102) by the associated semi-Markov random evolution given by a solution of the evolutional equation (3.6). The corresponding family of generators $\mathbb{\Gamma}^\varepsilon(x), x \in E$, in (3.7), is given in (5.107), that is $\mathbb{\Gamma}^\varepsilon(x) = \mathbb{A}_\varepsilon(x)$.

Here $A_s^\varepsilon(x), s \geq 0, x \in E$, is the family of semigroups, determined by the evolutions

$$\alpha_x^\varepsilon(s) = u + s a_\varepsilon(x), \quad s \geq 0, x \in E, \tag{5.113}$$

that is

$$A_s^\varepsilon(x)\varphi(u) = \varphi(\alpha_x^\varepsilon(s)). \tag{5.114}$$

The compensating operator of the extended Markov renewal process

$$\alpha_n^\varepsilon = \alpha^\varepsilon(\tau_n^\varepsilon), \quad x_n^\varepsilon, \quad \tau_n^\varepsilon = \varepsilon^2 \tau_n, \quad n \geq 0. \tag{5.115}$$

is given by relations (3.8)-(3.9), that is

$$\mathbb{L}^\varepsilon \varphi(u, x) := \varepsilon^{-2} q(x) \left[\int_0^\infty F_x(ds) A_{\varepsilon^2 s}^\varepsilon(x) P \varphi(u, x) - \varphi(u, x) \right]. \tag{5.116}$$

Now we can use the asymptotic representation (3.10) given in Proposition 3.2 with obvious changes $\mathbb{\Gamma}(x) = \mathbb{A}(x)$, and $\mathbb{\Gamma}_1(x) = \mathbb{A}_1(x)$. $\square$

Proposition 5.8 *The generator $\mathbb{L}$ of the limit diffusion process in Theorem 3.3 is calculated by a solution of the singular perturbation problem for the truncated generator*

$$\mathbb{L}_0^\varepsilon = \varepsilon^{-2} Q + \varepsilon^{-1} \mathbb{A}(x) P + Q_2(x) P,$$

as follows

$$\mathbb{L} = \Pi Q_2(x) \Pi + \Pi \mathbb{A}(x) P R_0 \mathbb{A}(x) P \Pi, \tag{5.117}$$

where

$$Q_2(x) = \mathbb{A}_1(x) + \mu_2(x) \mathbb{A}^2(x), \quad \mu_2(x) = m_2(x)/2m(x).$$

The operator R_0 is the potential of Q (see Section 1.6)

$$QR_0 = R_0Q = \Pi - I, \tag{5.118}$$

or, equivalently

$$q[P - I]R_0 = \Pi - I.$$

Hence

$$PR_0 = R_0 + m(\Pi - I), \tag{5.119}$$

where $q(x) = 1/m(x)$.

PROOF. We calculate:

$$\begin{aligned}\Pi Q_2(x)P\Pi\varphi(u) &= \Pi Q_2(x)\Pi\varphi(u) = \Pi Q_2(x)\varphi(u) \\ &= \Pi[\mathbb{A}_1(x) + \mu_2(x)\mathbb{A}^2(x)]\varphi(u) \\ &= \Pi[a_1(x)\varphi'(u) + \mu_2(x)a^2(x)\varphi''(u)] \\ &= a_1\varphi'(u) + \frac{1}{2}B_1\varphi''(u),\end{aligned}$$

where:

$$a_1 := \int_E \pi(dx)a_1(x), \quad \text{and } B_1 := 2\int_E \pi(dx)\mu_2(x)a^2(x). \tag{5.120}$$

Next:

$$\begin{aligned}\Pi\mathbb{A}(x)PR_0\mathbb{A}(x)P\Pi\varphi(u) &= \Pi\mathbb{A}(x)PR_0\mathbb{A}(x)\varphi(u) \\ &= \Pi\mathbb{A}(x)PR_0a(x)\varphi'(u) \\ &= \Pi a(x)PR_0a(x)\varphi'(u) \\ &= \Pi a(x)R_0a(x)\varphi''(u) - \Pi m(x)a^2(x)\varphi''(u), \\ &\quad \text{(by using (5.119))} \\ &= \frac{1}{2}B_0\varphi''(u) - \frac{1}{2}B_{01}\varphi''(u),\end{aligned}$$

where, by definition:

$$B_0 := 2\int_E \pi(dx)a_0(x)/m, \quad a_0(x) := a(x)R_0a(x),$$

$$B_{01} := 2\int_E \pi(dx)m(x)a^2(x).$$

5.4. DIFFUSION APPROXIMATION PRINCIPLE

Hence, we get the following representation of the limit generator

$$\mathbb{L}\varphi(u) = a_1\varphi'(u) + \frac{1}{2}B\varphi''(u),$$

where:

$$a_1 := \int_E \pi(dx)a_1(x), \quad B := B_0 + B_{00},$$

with:

$$B_0 := 2\int_E \pi(dx)a_0(x), \quad B_{00} := \int_E \pi(dx)\mu(x)a^2(x),$$

$$\mu(x) := [m_2(x) - 2m^2(x)]/m(x).$$

Note that $\mu(x)$ can be seen as a distance from exponential distributions of $F_x(t)$ of sojourn times (see Remark 3.3, page 83).

5.4.2 Continuous Random Evolutions

The diffusion approximation principle in ergodic merging scheme can be verified for a continuous random evolution in the series scheme with accelerated semi-Markov switching given by a solution of the evolutionary equation (Section 3.2.1)

$$\begin{vmatrix} \frac{d}{dt}\Phi^\varepsilon(t) = \mathbb{\Gamma}^\varepsilon(x(t/\varepsilon^2))\Phi^\varepsilon(t), & t \geq 0, \\ \Phi^\varepsilon(0) = I, \end{vmatrix} \quad (5.121)$$

on the Banach space **B**, with the given family of generators $\mathbb{\Gamma}^\varepsilon(x), x \in E$, of the semigroup $\Gamma_t^\varepsilon(x), t \geq 0, x \in E$.

The generators $\mathbb{\Gamma}^\varepsilon(x), x \in E$, have the following form

$$\mathbb{\Gamma}_\varepsilon(x) = \varepsilon^{-1}\mathbb{\Gamma}(x) + \mathbb{\Gamma}_1(x), \quad x \in E. \quad (5.122)$$

In what follows, the generators $\mathbb{\Gamma}(x)$ and $\mathbb{\Gamma}_1(x), x \in E$, are supposed to have a common domain of definition $\mathbf{B}_0$, dense in **B**.

The *coupled random evolution* (Section 2.7.3, Definition 2.11)

$$\Phi^\varepsilon(t, x(t/\varepsilon^2)) := \Phi^\varepsilon(t)\varphi(u, x(t/\varepsilon^2)), \quad x(0) = x,$$

can be characterized by the compensating operator (Section 3.2.1)

$$\mathbb{L}^\varepsilon\varphi(u,x) = \varepsilon^{-2}q(x)\left[\int_0^t F_x(ds)\Gamma^\varepsilon_{\varepsilon^2 s}(x)\int_E P(x,dy)\varphi(u,y) - \varphi(u,x)\right]. \tag{5.123}$$

The factor "ε^{-2}" corresponds to the accelerating time-scaling of the switched semi-Markov process $x(t), t \geq 0$, in (5.121), given by the semi-Markov kernel

$$Q(x, dy, ds) = P(x, dy)F_x(ds).$$

The fast time-scaling "ε^2" of the semigroup in (5.123) provides a diffusion approximation of the increments under the balance condition for the first term in (3.10),

$$\Pi\Gamma(x)\Pi = 0, \tag{5.124}$$

where Π is the projector of the associated ergodic Markov process $x^0(t), t \geq 0$, defined by the generator

$$Q\varphi(u) = q(x)\int_E P(x, dy)[\varphi(y) - \varphi(x)], \tag{5.125}$$

with

$$q(x) = 1/m(x), \quad m(x) = \int_0^\infty \overline{F}_x(t)dt.$$

The key problem in the diffusion approximation is to construct an asymptotic representation for the compensating operator (5.123), by using Proposition 5.2 and Assumptions (5.122) and (5.124).

The diffusion approximation principle for the semi-Markov continuous random evolution in the series scheme (5.121) with the switching ergodic semi-Markov process $x(t), t \geq 0$, is realized by a solution of the singular perturbation problem for the truncated operator (see Proposition 5.2)

$$\mathbb{L}^\varepsilon_0 = \varepsilon^{-2}Q + \varepsilon^{-1}Q_1(x)P + Q_2(x)P. \tag{5.126}$$

The limit generator is given by

$$\mathbb{L}\Pi = \Pi Q_2(x)P\Pi + \Pi Q_1(x)PR_0Q_1(x)P\Pi,$$

or, in our case

$$\mathbb{L}\Pi = \Pi Q_2(x)P\Pi + \Pi\Gamma(x)PR_0\Gamma(x)P\Pi, \tag{5.127}$$

5.4. DIFFUSION APPROXIMATION PRINCIPLE

where:
$$Q_1(x)\varphi(u,x) := \mathbf{\Gamma}(x)\varphi(u,x),$$
$$Q_2(x)\varphi(u,x) = [\mathbf{\Gamma}_1(x) + \mu_2(x)\mathbf{\Gamma}^2(x)]\varphi(u,x).$$

Let us now compute the limit generator in explicit form. The first term in (5.127) gives:
$$\Pi Q_2(x)\Pi = \Pi\mathbf{\Gamma}_1(x) + \mu_2(x)\mathbf{\Gamma}^2(x)\Pi$$
$$= (\widehat{\mathbf{\Gamma}}_1 + \widehat{\mathbf{\Gamma}}_{01})\Pi, \tag{5.128}$$

where, by definition:
$$\widehat{\mathbf{\Gamma}}_1 := \int_E \pi(dx)\mathbf{\Gamma}_1(x), \quad \widehat{\mathbf{\Gamma}}_{01} := \int_E \pi(dx)\mu_2(x)\mathbf{\Gamma}^2(x).$$

Recall that the potential operator R_0 satisfies the equation (see Section 1.6):
$$QR_0 = R_0Q = \Pi - I, \quad Q = q[P - I].$$

Hence,
$$PR_0 = R_0 + m[\Pi - I].$$

Next, we calculate the second term in (5.127):
$$\Pi\mathbf{\Gamma}(x)PR_0\mathbf{\Gamma}(x)\Pi = \Pi\mathbf{\Gamma}(x)R_0\mathbf{\Gamma}(x)\Pi - \Pi m(x)\mathbf{\Gamma}^2(x)\Pi$$
$$= (\widehat{\mathbf{\Gamma}}_0 - \widehat{\mathbf{\Gamma}}_{02})\Pi,$$

where, by definition,
$$\widehat{\mathbf{\Gamma}}_0 := \int_E \pi(dx)\mathbf{\Gamma}(x)R_0\mathbf{\Gamma}(x), \tag{5.129}$$

Gathering all the above calculations we get
$$\mathbf{L} = \widehat{\mathbf{\Gamma}}_0 + \widehat{\mathbf{\Gamma}}_{00} + \widehat{\mathbf{\Gamma}}_1, \tag{5.130}$$

where, by definition:
$$\widehat{\mathbf{\Gamma}}_{00} := \widehat{\mathbf{\Gamma}}_{01} - \widehat{\mathbf{\Gamma}}_{02} = \int_E \pi(dx)\mu(x)\mathbf{\Gamma}^2(x), \tag{5.131}$$

$$\mu(x) := [m_2(x) - 2m^2(x)]/m(x). \tag{5.132}$$

It is worth noticing that the formulas (5.127)-(5.132) give the preliminary "blank-cheque" for constructing the limit generator in the diffusion approximation scheme for stochastic systems with ergodic semi-Markov switching considered in Section 3.4.

The generators of the limit diffusion processes are constructed by formulas (5.129)-(5.132) with the following generators of the corresponding continuous random evolutions (Section 3.4).

1. In Theorem 3.4:

$$\mathbb{\Gamma}(x)\varphi(u) = g(u;x)\varphi'(u),$$
$$\mathbb{\Gamma}_1(x)\varphi(u) = g_1(u;x)\varphi'(u) + C_0(u;x)\varphi''(u). \qquad (5.133)$$

2. In Corollary 3.6:

$$\mathbb{\Gamma}(x)\varphi(u) = g(x)\varphi'(u),$$
$$\mathbb{\Gamma}_1(x)\varphi(u) = g_1(x)\varphi'(u) + C_0(x)\varphi''(u). \qquad (5.134)$$

3. In Corollary 3.7:

$$\mathbb{\Gamma}(x)\varphi(u) = g(u;x)\varphi'(u),$$
$$\mathbb{\Gamma}_1(x)\varphi(u) = g_1(u;x)\varphi'(u). \qquad (5.135)$$

Now, the generators of the limit diffusion processes for stochastic systems in Section 3.4 can be calculated in explicit form by formulas (5.129)-(5.132) and (5.133)-(5.135).

First, we calculate the generator for stochastic evolutionary system (Section 3.4.3, Corollary 3.7):

$$\begin{aligned}\mathbb{\Gamma}_0(x)\varphi(u) &:= \mathbb{\Gamma}(x)R_0\mathbb{\Gamma}(x)\varphi(u) \\ &= \mathbb{\Gamma}(x)R_0 g(u;x)\varphi'(u) \\ &= g(u;x)R_0[g(u;x)\varphi'(u)]'_u \\ &= g(u;x)R_0 g(u;x)\varphi''(u) + g(u;x)R_0 g'_u(u;x)\varphi'(u) \\ &= g_0(u;x)\varphi''(u) + g_2(u;x)\varphi'(u),\end{aligned}$$

where, as in Theorem 3.4:

$$g_0(u;x) := g(u;x)R_0 g(u;x),$$

$$g_2(u;x) := g(u;x)R_0 g'_u(u;x).$$

5.4. DIFFUSION APPROXIMATION PRINCIPLE

Next, we calculate:

$$\begin{aligned}
\mathbb{\Gamma}_{00}(x)\varphi(u) &:= \mu(x)\mathbb{\Gamma}^2(x)\varphi(u) \\
&= \mu(x)[g(u;x)g'_u(u;x)\varphi'(u) \\
&\quad + g(u;x)g^*(u;x)\varphi''(u)] \\
&=: g_3(u;x)\varphi'(u) + g_{00}(u;x)\varphi''(u).
\end{aligned}$$

Now, we get, by using (5.129)-(5.132):

$$\begin{aligned}
\widehat{\mathbb{\Gamma}}_1\varphi(u) &= \widehat{g}_1(u)\varphi'(u), \\
\widehat{\mathbb{\Gamma}}_0\varphi(u) &= \widehat{g}_0(u)\varphi''(u) + \widehat{g}_2(u)\varphi'(u), \\
\widehat{\mathbb{\Gamma}}_{00}\varphi(u) &= \widehat{g}_{00}(u)\varphi''(u) + \widehat{g}_3(u)\varphi'(u),
\end{aligned} \qquad (5.136)$$

where, by definition:

$$\widehat{g}_{00}(u) = \int_E \pi(dx)\mu(x)g(u;x)g^*(u;x), \qquad (5.137)$$

and

$$\widehat{g}_3(u) = \int_E \pi(dx)\mu(x)g(u;x)g'_u(u;x). \qquad (5.138)$$

Gathering (5.136)-(5.138), we obtain the generator of the limit diffusion process in Corollary 3.7.

Analogous calculation can be done for the stochastic additive functionals considered in Sections 3.4.2 (Theorem 3.4).

5.4.3 *Jump Random Evolutions*

The increment process in the series scheme with the semi-Markov switching, considered in Section 3.4.4, in the diffusion approximation scheme, is here considered with the following accelerated scaling

$$\beta^\varepsilon(t) = \beta_0 + \varepsilon \sum_{k=1}^{\nu(t/\varepsilon^2)} a_\varepsilon(x_k), \quad t \geq 0. \qquad (5.139)$$

The values of jumps are defined by the bounded deterministic function $a_\varepsilon(x), x \in E$, which takes values in the Euclidean space $\mathbb{R}^d$, and has the following representation

$$a_\varepsilon(x) = a(x) + \varepsilon a_1(x). \qquad (5.140)$$

The first term satisfies the balance condition

$$\int_E \rho(dx)a(x) = 0, \qquad (5.141)$$

where $\rho(dx)$ is the stationary distribution of the embedded Markov chain $x_n, n \geq 0$.

To verify the algorithm of diffusion approximation formulated in Theorem 3.5, the associated jump random evolution is considered (Section 3.2.2), defined by the family of bounded operators

$$\mathbb{D}^\varepsilon(x)\varphi(u) := \varphi(u + \varepsilon a_\varepsilon(x)), \quad x \in E, \qquad (5.142)$$

given on the test functions $\varphi \in C_0^3(\mathbb{R}^d)$.

By using representation (5.140), the following asymptotic expansion is valid

$$\mathbb{D}^\varepsilon(x) := I + \varepsilon \mathbb{D}(x) + \varepsilon^2 \mathbb{D}_1(x) + \varepsilon^2 \theta^\varepsilon(x), \qquad (5.143)$$

where, by definition:

$$\mathbb{D}(x)\varphi(u) = a(x)\varphi'(u), \qquad (5.144)$$

$$\mathbb{D}_1\varphi(u) = a_1(x)\varphi'(u) + \frac{1}{2}a^2(x)\varphi''(u), \qquad (5.145)$$

and

$$\|\theta^\varepsilon(x)\varphi\| \to 0, \quad \varepsilon \to 0, \quad \varphi \in C_0^3(\mathbb{R}^d). \qquad (5.146)$$

Proposition 5.9 *The diffusion approximation principle for the semi-Markov jump random evolution in the series scheme with the switching ergodic semi-Markov process, satisfying Conditions D1-D3 of Theorem 3.3 (Section 3.4.1), and the family of jump operators $\mathbb{D}^\varepsilon(x), x \in E$, is realized by a solution of the singular perturbation problem for the truncated operator*

$$\mathbb{L}_0^\varepsilon = \varepsilon^{-2} Q + \varepsilon^{-1} Q_0 \mathbb{D}(x) + Q_0 \mathbb{D}_1(x). \qquad (5.147)$$

PROOF. The proof is obtained by using the asymptotic representation (3.30) of the compensating operator (3.29) for the jump random evolution (3.26)-(3.28).

5.4. DIFFUSION APPROXIMATION PRINCIPLE

Considering the operator (5.147) on the perturbed test functions $\varphi^\varepsilon(u,x) = \varphi(u) + \varepsilon\varphi_1(u,x) + \varepsilon^2\varphi(u,x)$, with $\varphi \in C_0^3(\mathbb{R}^d)$ we get, by Proposition 5.2,

$$\mathbb{L}_0^\varepsilon \varphi^\varepsilon(u,x) = \mathbb{L}\varphi(u) + \theta_l^\varepsilon(x)\varphi(u),$$

with the negligible term

$$\|\theta_l^\varepsilon(x)\varphi\| \to 0, \quad \varepsilon \to 0, \quad \varphi \in C_0^3(\mathbb{R}^d).$$

The limit operator $\mathbb{L}$ is calculated by the formula (see Proposition 5.2)

$$\mathbb{L}\Pi = \Pi Q_0 \mathbb{D}_1(x)\Pi + \Pi Q_0 \mathbb{D}(x) R_0 Q_0 \mathbb{D}(x)\Pi,$$

where R_0 is the potential of operator Q (see Section 5.2).

Now we calculate using representation (5.144)-(5.145) and $Q_0\varphi(x) = q(x)P\varphi(x)$:

$$\begin{aligned}
\mathbb{L}_1 \Pi \varphi &= \Pi Q_0 \mathbb{D}_1(x) \Pi \varphi(u) \\
&= \Pi Q_0 \mathbb{D}_1(x) \varphi(u) \\
&= \Pi Q_0 a_1(x)\varphi'(u) + \frac{1}{2}\Pi Q_0 a^2(x)\varphi''(u) \\
&= \int_E \pi(dx) q(x) P a_1(x)\varphi'(x) + \frac{1}{2}\int_E \pi(dx) q(x) P a^2(x)\varphi''(u) \\
&= q \int_E \rho(dx) a_1(x)\varphi'(x) + \frac{1}{2} q \int_E \rho(dx) a^2(x)\varphi''(u).
\end{aligned}$$

Hence

$$\mathbb{L}_1 \varphi(u) = a\varphi'(u) + \frac{1}{2}\sigma_0^2 \varphi''(u).$$

Here (compare with Theorem 3.5)

$$a := q \int_E \rho(dx) a_1(x), \quad \sigma_0^2 := q \int_E \rho(dx) a^2(x).$$

Next, the operator is calculated as follows:

$$\begin{aligned}
\mathbb{L}_2 \Pi \varphi(u) &= \Pi Q_0 \mathbb{D}(x) R_0 Q_0 \mathbb{D}(x) \Pi \varphi(u) \\
&= \Pi q(x) P a(x) R_0 q(x) P a(x) \varphi''(u) \\
&= \Pi q(x) b(x) R_0 q(x) b(x) \varphi''(u) \\
&= \Pi B(x) R_0 B(x) \varphi''(u).
\end{aligned}$$

Hence
$$L_2\varphi(u) = \frac{1}{2}\sigma_1^2\varphi''(u),$$
where, by definition,
$$\sigma_1^2 := 2\int_E \pi(dx)C_0(x),$$
with:
$$C_0(x) := C(x)R_0C(x), \quad C(x) := b(x)/m(x),$$
$$b(x) := Pa(x) = \int_E P(x,dy)a(y),$$
that is exactly as in Theorem 3.5. Note that according to Appendix C, $\sigma_1^2 \geq 0$.

5.4.4 Random Evolutions with Markov Switching

The diffusion approximation principle for random evolution with Markov switching can be obtained from the results presented in Sections 5.3.2 and 5.3.3, for the semi-Markov random evolutions, by putting the "distance from exponential distribution" parameter $\mu(x)$ equal to 0 (see Remark 3.3, p. 83).

According to Propositions 5.8 and 5.9 we can formulate as a corollary the following result.

Proposition 5.10 *The diffusion approximation principle for the Markov random evolution given by a solution of the evolutionary equation (5.121), with the switching Markov process $x(t), t \geq 0$, defined by the generator (5.125), is realized by a solution of the singular perturbation problem for the generator (see Proposition 3.3)*
$$\mathbb{L}^\varepsilon = \varepsilon^{-2}Q + \varepsilon^{-1}\Gamma(x) + \Gamma_1(x).$$

The limit generator is given by (see Proposition 5.2)
$$\mathbb{L}\Pi = \Pi\Gamma_1(x)\Pi + \Pi\Gamma(x)R_0\Gamma(x)\Pi.$$

Thus we obtain the preliminary "blank-cheque" to construct the limit generator in diffusion approximation scheme for stochastic systems with ergodic Markov switching considered in Section 3.4 in the following form

$$\mathbb{L} = \widehat{\mathbb{\Gamma}}_1 + \widehat{\mathbb{\Gamma}}_0,$$

where:

$$\widehat{\mathbb{\Gamma}}_1 := \int_E \pi(dx) \mathbb{\Gamma}_1(x), \quad \widehat{\mathbb{\Gamma}}_0 := \int_E \pi(dx) \mathbb{\Gamma}(x) R_0 \mathbb{\Gamma}(x).$$

The diffusion approximation principle for jump random evolution with Markov switching coincides with the analogous one for the semi-Markov jump random evolution. We have to keep in mind that, in case of switching Markov process, $q(x)$ is the true intensity of the exponential distribution of renewal times. Hence, for example, the parameter is

$$q = \int_E \pi(dx) q(x),$$

without transformation from the equality

$$q(x) = 1/m(x), \quad m(x) = \mathbb{E}\theta_x,$$

which was used in the case of switching semi-Markov processes.

5.5 Diffusion Approximation with Equilibrium

The main problem in constructing the diffusion approximation principle for stochastic systems with equilibrium considered in Section 3.5 is the representation of the generator of the centered and normalized process in a suitable asymptotic form. Certainly, the situations considered in Sections 3.5.1 and 3.5.2 are completely different. The centered and normalized process (3.69) is Markovian, which essentially simplify the problem. While, the centered and normalized process (3.81) has to be extended to the Markov process by two components: the deterministic shift-process $\widehat{\xi}(t), t \geq 0$, defined by a solution of the evolutionary equation (3.80), and the switching Markov process $x(t), t \geq 0$, defined by the generator (3.79).

5.5.1 *Locally Independent Increment Processes*

The generator of the centered and normalized process (3.69) is constructed by using the generator (3.67) and the following relation (see (3.70))

$$\varepsilon\eta^\varepsilon(t/\varepsilon) = \rho + \varepsilon\zeta^\varepsilon(t). \tag{5.148}$$

Lemma 5.1 *The generator of the Markov process*

$$\zeta^\varepsilon(t) = \eta^\varepsilon(t/\varepsilon) - \varepsilon^{-1}\rho, \quad t \geq 0, \tag{5.149}$$

is represented as follows

$$\mathbb{L}^\varepsilon_\zeta \varphi(u) = \mathbb{L}^\varepsilon(\rho + \varepsilon u)\varphi(u), \tag{5.150}$$

with the generators

$$\mathbb{L}^\varepsilon(\rho + \varepsilon u)\varphi(u) = \varepsilon^{-2} \int_{\mathbb{R}^d} [\varphi(u + \varepsilon u) - \varphi(u)]\Gamma_\varepsilon(\rho + \varepsilon u, dv). \tag{5.151}$$

PROOF. The statement of Lemma 5.1 provides the equality (5.148) and the representation of the generator

$$\mathbb{L}^\varepsilon \varphi(u) = \varepsilon^{-1} \int_{\mathbb{R}^d} [\varphi(u + \varepsilon u) - \varphi(u)]\Gamma_\varepsilon(u, dv), \tag{5.152}$$

of the initial stochastic system described by the Markov process with locally independent increments $\eta^\varepsilon(t), t \geq 0, \varepsilon > 0$ (see Section 3.5.1).

Now conditions (3.71)-(3.73) of Theorem 3.6 provide the following asymptotic representation of generator (5.150)

$$\mathbb{L}^\varepsilon_\zeta \varphi(u) = \mathbb{L}^0 \varphi(u) + \theta^\varepsilon \varphi(u), \tag{5.153}$$

with the negligible term

$$\|\theta^\varepsilon \varphi\| \to 0, \quad \varepsilon \to 0, \quad \varphi \in C^2(\mathbb{R}^d). \tag{5.154}$$

Here, the generator $\mathbb{L}^0$, given by (3.74), defines the limit diffusion process $\zeta^0(t), t \geq 0$, in Theorem 3.6.

The verification of convergence

$$\zeta^\varepsilon(t) \Rightarrow \zeta^0(t), \quad \varepsilon \to 0, \tag{5.155}$$

can be obtained by using the Skorokhod limit theorem (see Chapter 6, and Appendix A).

It is worth noticing that in Theorem 3.7 the additional term $\widehat{C}(v)$ in diffusion coefficient, from the jump part of the stochastic additive functionals

5.5. DIFFUSION APPROXIMATION WITH EQUILIBRIUM

(see (3.87)), is ignored in order to simplify the construction of the diffusion approximation principle for Markov stochastic systems with equilibrium (without balance condition).

5.5.2 Stochastic Additive Functionals

As was mentioned in the introduction of Section 5.4.4, the centered and normalized stochastic additive functional (see relation (3.81))

$$\zeta^\varepsilon(t) = \varepsilon^{-1}[\xi^\varepsilon(t) - \widehat{\xi}(t)], \quad t \geq 0, \tag{5.156}$$

with the time re-scaled switching Markov process is as follows (see (3.82))

$$\xi^\varepsilon(t) = \xi_0^\varepsilon + \int_0^t \eta^\varepsilon(ds; x(s/\varepsilon^2)), \quad t \geq 0, \tag{5.157}$$

where the family of Markov processes $\eta^\varepsilon(t;x), t \geq 0, \varepsilon > 0, x \in E$, is given by the generators (see (3.83))

$$\mathbb{\Gamma}_\varepsilon(x)\varphi(u) = g_\varepsilon(u;x)\varphi'(u). \tag{5.158}$$

For simplification we dropped the remaining term in (3.83).

The decisive step in the asymptotic analysis of the considered problem is the construction of the generator of the three component Markov process

$$\zeta^\varepsilon(t), \quad \widehat{\xi}(t), \quad x_t^\varepsilon := x(t/\varepsilon^2), \quad t \geq 0. \tag{5.159}$$

Lemma 5.2 *The generator of the Markov process (5.159) is represented as follows*

$$\mathbb{L}_\zeta^\varepsilon \varphi(u,v,x) = [\varepsilon^{-2}Q + \mathbb{\Gamma}_\varepsilon(v+\varepsilon u;x)]\varphi(u,v,x)$$
$$+ \widehat{\mathbb{\Gamma}}(v)\varphi(u,v,x). \tag{5.160}$$

Here Q is the generator of the switching Markov process $x(t), t \geq 0$,

$$\widehat{\mathbb{\Gamma}}(v)\varphi(v) := \widehat{g}(v)\varphi'(v). \tag{5.161}$$

The generator $\mathbb{\Gamma}_\varepsilon$ is defined by

$$\mathbb{\Gamma}_\varepsilon(v+\varepsilon u;x)\varphi(u) := [g_\varepsilon(v+\varepsilon u;x) - \widehat{g}(v)]\varphi'(u). \tag{5.162}$$

PROOF. The representation (5.162) provides the following equality (compare with (5.156))

$$\xi^\varepsilon(t) = \widehat{\xi}(t) + \varepsilon \zeta^\varepsilon(t), \tag{5.163}$$

that is, under conditions:

$$\widehat{\xi}(t) = v, \quad \zeta^\varepsilon(t) = u, \quad \text{we have } \xi^\varepsilon(t) = v + \varepsilon u, \tag{5.164}$$

which are, exactly those used in (5.162).

The first and last terms on the right hand side of (5.160) are obtained by using independence of processes $\widehat{\xi}(t)$ and x_t^ε from $\zeta^\varepsilon(t)$.

The asymptotic representation (5.165) is obtained by using a solution of the singular perturbation problem in Proposition 5.2 for the truncated operator

$$\mathbb{L}_0^\varepsilon \varphi(u, v, x) = \varepsilon^{-2} Q + \varepsilon^{-1} Q_1(x) + Q_2(x),$$

where:

$$Q_1(x)\varphi(u,v) = \widetilde{g}(u;x)\varphi'_u(u,v), \quad \widetilde{g}(u;x) := g(u;x) - \widehat{g}(u),$$
$$Q_2(x)\varphi(u,v) = g(u,v)\varphi'_u(u,v).$$

Now the conditions of Theorem 3.7 (Section 3.5.2) and Proposition 5.2 provide the following asymptotic representation of the generator (5.160)

$$\mathbb{L}_0^\varepsilon \varphi^\varepsilon(u, v, x) = \widehat{\mathbb{L}}\varphi(u, v) + \theta^\varepsilon(x)\varphi(u, v), . \tag{5.165}$$

with the negligible term

$$\|\theta^\varepsilon(x)\varphi\| \to 0, \quad \varepsilon \to 0.$$

The generator $\widehat{\mathbb{L}}$ is given in Theorem 3.7 in the following form

$$\widehat{\mathbb{L}}\varphi(u,v) = g(u,v)\varphi'_u(u,v) + \frac{1}{2}B(v)\varphi''_{uu}(u,v) + \widehat{g}(v)\varphi'_v(u,v),$$

that is the limit generator of diffusion process $\widehat{\zeta}(t), t \geq 0$, and the equilibrium process $\widehat{\xi}(t), t \geq 0$.

Representations (5.165), and (5.165) are obtained by using Condition D2' of Theorem 3.7.

5.5.3 Stochastic Evolutionary Systems with Semi-Markov Switching

The diffusion approximation principle for stochastic evolutionary systems with semi-Markov switching can be constructed as in the previous sections by using the suitable asymptotic form of the extended compensating operator of the centered and normalized process considered in Section 3.5.3.

5.5. DIFFUSION APPROXIMATION WITH EQUILIBRIUM

We constructed the generators of the centered and normalized processes

$$\zeta_x^\varepsilon(t) := \varepsilon^{-1}[U_x^\varepsilon(t) - \widehat{U}(t)], \quad t \geq 0, \quad x \in E,$$

where the evolutionary systems $U_x^\varepsilon(t), t \geq 0, x \in E$, are defined by a solution of the equations

$$\frac{d}{dt} U_x^\varepsilon(t) = a_\varepsilon(U_x^\varepsilon(t); x), \quad x \in E.$$

Using the relation

$$U_x^\varepsilon(t) = \widehat{U}(t) + \varepsilon \zeta_x^\varepsilon(t),$$

it is easy to obtain the generator of the two component process $\zeta_x^\varepsilon(t), \widehat{U}(t), t \geq 0$, that is

$$\mathbb{A}^\varepsilon(v; x)\varphi(u, v) = \varepsilon^{-1}\widetilde{a}_\varepsilon(v + \varepsilon u; x)\varphi_u'(u, v), \tag{5.166}$$

where, by definition,

$$\widetilde{a}_\varepsilon(v; x) := a_\varepsilon(v; x) - \widehat{a}(v).$$

Under the condition of Theorem 3.8, the following asymptotic representation of generator (5.166) holds

$$\mathbb{A}^\varepsilon(v; x)\varphi(u, v) = \varepsilon^{-1}\widetilde{\mathbb{A}}(v; x)\varphi(u, v) + \mathbb{A}_1(v; x)\varphi(u, v)$$
$$+ \theta^\varepsilon(v; x)\varphi(u, v). \tag{5.167}$$

By definition we have:

$$\widetilde{\mathbb{A}}(v; x)\varphi(u) = \widetilde{a}(v; x)\varphi'(u), \tag{5.168}$$
$$\mathbb{A}_1(v; x)\varphi(u) = b(v, u; x)\varphi'(u), \widetilde{a}(v; x) := a(v; x) - \widehat{a}(v),$$

the generator of the average component $\widetilde{U}(t), t \geq 0$, is the following

$$\widehat{\mathbb{A}}(v)\varphi(v) = \widehat{a}(v)\varphi'(v), \tag{5.169}$$

and the negligible remaining term is

$$\|\theta^\varepsilon(v; x)\varphi\| \to 0, \quad \varepsilon \to 0,$$

with $\varphi \in C^{2,1}(\mathbb{R}^d \times \mathbb{R}^d)$.

Let us now define the extended compensating operator.

Lemma 5.3 *The extended compensating operator of the three-component process $\zeta^\varepsilon(t), \widehat{U}(t), x_t^\varepsilon := x(t/\varepsilon^2), t \geq 0$, is determined by the relation*

$$\mathbb{L}^\varepsilon \varphi(u,v,x) = \varepsilon^{-2} q(x) \left[\int_0^\infty F_x(dt) A_{\varepsilon^2 t}^\varepsilon(v;x) \widehat{A}_{\varepsilon^2 t}(v) P\varphi(u,v,x) - \varphi(u,v,x) \right]. \tag{5.170}$$

Here, the semigroups A_t^ε and $\widehat{A}_t$, $t \geq 0$, are defined by the generators (5.166)-(5.168) and (5.169).

The next step in the asymptotic analysis is to construct the asymptotic expansion of the compensating operator (5.170) with respect to ε.

Lemma 5.4 *The extentended compensating operator (5.170) has the following asymptotic representation on the test functions $\varphi \in C^{3,2}(\mathbb{R}^d \times \mathbb{R}^d)$*

$$\mathbb{L}^\varepsilon \varphi(u,v,x) = \varepsilon^{-2} Q \varphi(\cdot,\cdot,x) + \varepsilon^{-1} \widetilde{\mathbb{A}}(v;x) P\varphi(u,\cdot,\cdot) + Q_2(v;x) P\varphi(u,v,\cdot) \\ + \theta_l^\varepsilon(v;x)\varphi, \tag{5.171}$$

with:

$$Q_2(v;x)\varphi(u,v) := \mathbb{L}_0(v;x)\varphi(u,v) + \widehat{\mathbb{A}}(v)\varphi(u,v), \tag{5.172}$$

$$\mathbb{L}_0(v;x)\varphi(u) := b(v,u;x)\varphi'(u) + \frac{1}{2}B(v;x)\varphi''(u), \tag{5.173}$$

$$B(v;x) := \mu_2(x) a(v;x) a^*(v;x), \tag{5.174}$$

$$\mu_2(x) := m_2(x)/m(x), \quad m_2(x) := \int_0^\infty t^2 F_x(dt). \tag{5.175}$$

PROOF. The compensating operator is transformed as follows

$$\mathbb{L}^\varepsilon = \varepsilon^{-2} Q + \varepsilon^{-2} q(x) [\mathbb{F}^\varepsilon(x) - I] P. \tag{5.176}$$

Now, the following algebraic identity is used

$$ab - 1 = (a-1) + (b-1) + (a-1)(b-1). \tag{5.177}$$

Setting

$$a := A_{\varepsilon^2 t}^\varepsilon(v;x), \quad b := \widehat{A}_{\varepsilon^2 t}, \tag{5.178}$$

the terms in (5.176) with (5.177) and (5.178) are transformed by using the

5.5. DIFFUSION APPROXIMATION WITH EQUILIBRIUM

integral equation for semigroup:

$$F_a^\varepsilon(x) := \int_0^\infty F_x(dt)[A_{\varepsilon^2 t}^\varepsilon(v;x) - I]$$

$$= \varepsilon^2 \mathbb{\Gamma}^\varepsilon(x,v) \int_0^\infty \overline{F}_x(t) A_{\varepsilon^2 t}^\varepsilon(v;x) dt$$

$$= \varepsilon^2 m(x)\mathbb{\Gamma}^\varepsilon(x,v) + \varepsilon^4 [\mathbb{\Gamma}^\varepsilon(x,v)]^2 \int_0^\infty \overline{F}_x(t) A_{\varepsilon^2 t}^\varepsilon(v;x) dt$$

$$= \varepsilon^2 m(x)\mathbb{\Gamma}^\varepsilon(x,v) + \varepsilon^4 \frac{m_2(x)}{2}[\mathbb{\Gamma}^\varepsilon(x,v)]^2 + \varepsilon^6 [\mathbb{\Gamma}^\varepsilon(x,v)]^3 F_{a3}^\varepsilon(x),$$

where:

$$\overline{F}_x^{(k+1)}(t) := \int_t^\infty \overline{F}_x^{(k)}(s) ds, \quad \overline{F}_x^{(1)}(t) = \overline{F}_x(t),$$

$$F_{a3}^\varepsilon(x) := \int_0^\infty \overline{F}_x^3(t) A_{\varepsilon^2 t}^\varepsilon(v;x) dt.$$

Taking into account (5.166) the following expansion is obtained

$$F_a^\varepsilon(x) = \varepsilon m(x)\mathbb{A}_\varepsilon(v,x) + \varepsilon^2 \frac{m_2(x)}{2}[\mathbb{A}_\varepsilon(v,x)]^2 + \varepsilon^2 \theta_a^\varepsilon(x,v), \quad (5.179)$$

with the negligible term

$$\theta_a^\varepsilon(x,v) := \varepsilon[\mathbb{\Gamma}^\varepsilon(x,v)]^3 F_{a3}^\varepsilon(x),$$

on the test functions $\varphi \in C_0^3(\mathbb{R}^d)$.

Similarly, the asymptotic expansion can be obtained for the next two terms in (5.177)-(5.178)

$$F_b^\varepsilon(x) := \int_0^\infty F_x(dt)[\widehat{A}_{\varepsilon^2 t}(v) - I]$$

$$= \varepsilon^2 m(x)\widehat{\mathbb{A}}(v) + \varepsilon^2 \theta_b^\varepsilon(x), \quad (5.180)$$

with the negligible term

$$\theta_b^\varepsilon(x) := [\widehat{\mathbb{A}}(v)]^3 F_{b2}^\varepsilon(x), \quad F_{b2}^\varepsilon(x) := \int_0^\infty \overline{F}_x^{(2)}(t) \widehat{A}_{\varepsilon^2 t}(v) dt,$$

on the test functions $\varphi \in C_0^2(\mathbb{R}^d)$.

Finally, we analyze the third term:

$$F_{ab}^\varepsilon(x) := \int_0^\infty F_x(dt)[A_{\varepsilon^2 t}^\varepsilon(v;x) - I][\widehat{A}_{\varepsilon^2 t}^\varepsilon(v) - I]$$
$$= \mathbb{A}^\varepsilon(v,x)\widehat{\mathbb{A}}^\varepsilon(v)F_{ab1}^\varepsilon(x),$$

where:

$$F_{ab1}^\varepsilon(x) := \int_0^\infty F_x(dt)[\int_0^{\varepsilon^2 t} A_s^\varepsilon(v,x)ds \int_0^{\varepsilon^2 t} \widehat{A}_s^\varepsilon(v)ds]$$
$$= 2\varepsilon^4 \int_0^\infty F_x^{(2)}(dt) A_{\varepsilon^2 t}^\varepsilon(v;x)\widehat{A}_{\varepsilon^2 t}^\varepsilon(v) + \varepsilon^4 \theta_{ab}^\varepsilon(x).$$

Hence, by (5.167) and (5.169), we get

$$F_{ab}^\varepsilon(x) = -\varepsilon^3 m_2(x)\mathbb{A}_\varepsilon(v;x)\widehat{\mathbb{\Gamma}}(v) + \varepsilon^2 \theta_{ab}^\varepsilon(x) = \varepsilon^2 \theta_{ab1}^\varepsilon(x), \quad (5.181)$$

with the negligible term $\theta_{ab1}^\varepsilon(x)$ on the test functions $\varphi \in C^3(\mathbb{R}^d)$.

Gathering expressions (5.179)-(5.181), the asymptotic extension (5.171) for the compensating operator is obtained. □

The final step in the diffusion approximation principle for the stochastic systems with the semi-Markov switching is to determine the limit generator in Theorem 3.8 calculated by using a solution of the singular perturbation problem for the truncated operator

$$\mathbb{L}_0^\varepsilon \varphi(u,v,x) = [\varepsilon^{-2}Q + \varepsilon^{-1}\widetilde{\mathbb{A}}(v;x)P + Q_2(v;x)P]\varphi(u,v,x). \quad (5.182)$$

According to Proposition 5.2 we get the following result.

Lemma 5.5 *A solution of singular perturbation problem for the generator (3.111)*

$$\mathbb{L}^\varepsilon \varphi^\varepsilon(u,v,x) = \mathbb{L}\varphi(u,v) + \theta_L^\varepsilon(u,v,x), \quad (5.183)$$

on the test functions $\varphi^\varepsilon(u,v,x) = \varphi(u,v) + \varepsilon\varphi_1(u,v,x) + \varepsilon^2\varphi_2(u,v,x)$, *and negligible term* $\theta_L^\varepsilon(u,v,x)$, *is realized by the generator* $\mathbb{L}$ *given in Theorem 3.8, formulas (3.98)-(3.101).*

PROOF. According to Proposition 5.2, Section 5.2, the limit generator in (5.183) is represented as follows

$$\mathbb{L}\Pi = \Pi\widetilde{\mathbb{A}}(v;x)PR_0\widetilde{\mathbb{A}}(v;x)P\Pi + \Pi\mathbb{L}_0(v;x)P\Pi + \Pi\widehat{\mathbb{A}}(v)P\Pi,$$

5.5. DIFFUSION APPROXIMATION WITH EQUILIBRIUM

where the projector Π is defined as follows

$$\Pi\varphi(x) = \int_E \pi(dx)\varphi(x).$$

Let us calculate:

$$\begin{aligned}\mathbb{L}_1\Pi &= \Pi\widetilde{\mathbb{A}}(v;x)PR_0\widetilde{\mathbb{A}}(v;x)P\Pi\varphi(u)\\ &= \Pi\widetilde{\mathbb{A}}(v;x)PR_0\widetilde{\mathbb{A}}(v;x)\varphi(u)\\ &= \Pi\widetilde{\mathbb{A}}(v;x)PR_0\widetilde{a}(v;x)\varphi'(u).\end{aligned}$$

By the definition of the potential operator R_0 (Section 5.2), we have

$$QR_0 = R_0Q = \Pi - I,$$

or

$$q(x)[P-I]R_0 = \Pi - I,$$

hence, $PR_0 = R_0 + m(x)[\Pi - I]$.
So, we can write:

$$\begin{aligned}\mathbb{L}_1\Pi &= \Pi\widetilde{\mathbb{A}}(v;x)[R_0 - m(x)I]\widetilde{a}(v;x)\varphi'(u)\\ &= \Pi\widetilde{\mathbb{A}}(v;x)R_0\widetilde{a}(v;x)\varphi'(u) - \Pi\widetilde{a}(v;x)m(x)\widetilde{a}^*(v;x)\varphi''(u)\\ &= \Pi\widetilde{a}(v;x)R_0\widetilde{a}(v;x)\varphi''(u) - \Pi m(x)\widetilde{a}(v;x)\widetilde{a}^*(v;x)\varphi''(u).\end{aligned}$$

Hence, the first term, on the right hand side of the latter equality, is

$$\mathbb{L}_1\varphi(u) = \frac{1}{2}B_0(v)\varphi''(u) - \frac{1}{2}\widehat{A}_0(v)\varphi''(u), \tag{5.184}$$

where:

$$B_0(v) := 2\int_E \pi(dx)\widetilde{a}(v;x)R_0\widetilde{a}(v;x),$$

$$\widehat{A}_0(v) := 2\int_E \pi(dx)m(x)\widetilde{a}(v;x)\widetilde{a}^*(v;x).$$

The second term is:

$$\begin{aligned}\Pi\mathbb{L}_0(x,v)P\Pi\varphi(u) &= \Pi b(v,u;x)\varphi'(u) + \frac{1}{2}\Pi B_1(v;x)\varphi''(u)\\ &= \mathbb{L}_0(v)\varphi(u),\end{aligned}$$

where

$$\mathbb{L}_0(v)\varphi(u) = b(v,u)\varphi'(u) + \frac{1}{2}B_1(v)\varphi''(u), \quad (5.185)$$

and

$$B_1(v) := \int_E \pi(dx)\mu_2(x)B_1(v;x).$$

The functions $B_1(v;x)$, $b(v,u)$, $b(v,u;x)$ and $\mu_2(x)$ are defined respectively in (3.116), (3.99), (3.115) and (3.117).

Hence, putting together (5.184) and (5.185) we obtain the generator $\mathbb{L}$ of Theorem 3.8. □

5.6 Merging and Averaging in Split State Space

5.6.1 *Preliminaries*

Here, the phase merging principles are formulated in split state space for semi-Markov processes (Section 5.2.2) and for stochastic systems in average scheme (Section 5.6.3).

As in the previous sections of Chapter 5, the phase merging principles are constructed by using the corresponding propositions from Section 5.2 and the asymptotic representation of the extended compensating operators for the stochastic systems in split state space considered in Chapter 4.

The new additional particularity connected with the switching semi-Markov processes considered in the series scheme also with the small parameter series $\varepsilon \to 0$, $(\varepsilon > 0)$ disappears.

The split phase space $(E, \mathcal{E})$ is considered as follows (see Section 4.2.1):

$$E = \cup_{k=1}^{N} E_k, \quad E_k \cap E_{k'} = \emptyset, \quad k \neq k'. \quad (5.186)$$

The stochastic kernel $P(x, dy)$ is coordinated with split (5.186) in the following way:

$$P(x, E_k) = \mathbf{1}_k(x) := \begin{cases} 1, & x \in E_k \\ 0, & x \notin E_k. \end{cases}$$

The semi-Markov kernel

$$Q^\varepsilon(x, B, t) = P^\varepsilon(x, B)F_x(t), \quad (5.187)$$

5.6. MERGING AND AVERAGING IN SPLIT STATE SPACE

depends on the parameter series ε as follows

$$P^\varepsilon(x,B) = P(x,B) + \varepsilon P_1(x,B).$$

The perturbing kernel $P_1(x,B)$ is a signed kernel with additional conservative condition

$$P_1(x,E) = 0,$$

which is basic in the *ergodic merging principle* of Section 4.2.1, and

$$P(x,E) = -p(x), \quad p(x) \geq 0, \quad x \in E, \qquad (5.188)$$

is a bounded function, in the *absorbing merging principle*.

The embedded Markov chain $x_n^0, n \geq 0$, defined by the stochastic kernel $P(x,B)$, is supposed to be uniformly ergodic with stationary distributions $\rho_k(B), B \in \mathcal{E}_k, 1 \leq k \leq N$, satisfying the equations:

$$\rho_k(B) = \int_{E_k} \rho_k(dx) P(x,B), \quad \rho_k(E_k) = 1.$$

Moreover, the associated support Markov process $x^0(t), t \geq 0$, defined by the generator

$$Q\varphi(x) = q(x) \int_E P(x,dy)[\varphi(y) - \varphi(x)], \qquad (5.189)$$

with $q(x) := 1/m(x)$, $m(x) := \int_0^\infty \overline{F}_x(t)dt$, $x \in E$, also is supposed to be uniformly ergodic with stationary distributions $\pi_k(B), B \in \mathcal{E}_k$, satisfying the relations:

$$\pi_k(dx) = \rho_k(dx)m(x)/m_k, \quad m_k := \int_{E_k} \rho_k(dx)m(x),$$

or, in the equivalent form:

$$\pi_k(dx)q(x) = q_k \rho_k(dx), \quad q_k = 1/m_k.$$

The projector Π on the Banach space $\mathbf{B}$ is defined as follows:

$$\Pi\varphi(x) = \sum_{k=1}^N \widehat{\varphi}_k \mathbf{1}_k(x), \quad \widehat{\varphi}_k := \int_{E_k} \pi_k(dx)\varphi(x), \quad 1 \leq k \leq N.$$

Remark 5.3. In the absorbing split case (5.188), the additional assumption

$$\widehat{p}_k := \int_{E_k} \pi_k(dx)p(x) > 0, \quad \text{for some } 1 \leq k \leq N,$$

will be used. That means that absorption with positive probability in the split state space takes place.

5.6.2 Semi-Markov Processes in Split State Space

The phase merging principle for the semi-Markov processes $x^\varepsilon(t), t \geq 0$, in the series scheme with the small series parameter $\varepsilon \to 0$, $(\varepsilon > 0)$ in the split state space (5.186) and the semi-Markov kernel (5.187) satisfying the conditions of Theorem 4.1 (Section 4.2.1) is formulated by using the compensating operator of the coupled process $x^\varepsilon(t/\varepsilon), \widehat{x}^\varepsilon(t) := v(x^\varepsilon(t/\varepsilon)), t \geq 0$ (see Section 4.2.1) given in the following form

$$\mathbb{L}^\varepsilon \varphi(x,k) = [\varepsilon^{-1}Q + Q_1]\varphi(x,k), \qquad (5.190)$$

on the test functions $\varphi(x,k), k \in E, k \in \widehat{E} = \{1, ..., N\}$.

Here Q is the generator (5.189) of the associated Markov process, and Q_1 the operator

$$Q_1\varphi(x) := q(x) \int_E P_1(x, dy)\varphi(y). \qquad (5.191)$$

The representation (5.190)-(5.191) can be obtained from the definition of the compensating operator in the following form (see Section 1.3.4)

$$\mathbb{L}^\varepsilon \varphi(x,u,t) = \varepsilon^{-1} q(x) \left[\int_0^\infty F_x(ds) P^\varepsilon(x,dy) \varphi(y, v(y), t + \varepsilon s) - \varphi(x, v(x), t) \right]$$

and then considering on the test functions $\varphi(x,k)$ which do not depend on t.

It is worth noticing that the compensating operator (5.190) coincides with the generator (4.9) (Section 4.2.1) of the Markov process.

Hence the phase merging principle for the semi-Markov processes gives the same result as for the Markov processes.

Proposition 5.11 *The limit generator $\widehat{Q}$ of the limit Markov process $\widehat{x}(t), t \geq 0$, in Theorem 4.1, is defined by a solution of the singular perturbation problem given in Proposition 5.1 for the compensating operator*

$$\mathbb{L}^\varepsilon \varphi(u,v) = [\varepsilon^{-1}Q + Q_1]\varphi(u,v),$$

5.6. MERGING AND AVERAGING IN SPLIT STATE SPACE

of the coupled process $x^\varepsilon(t/\varepsilon), v(x^\varepsilon(t/\varepsilon)), t \geq 0$.

According to Proposition 5.1 the limit generator $\widehat{Q}$ is calculated as follows (compare with (4.17), Section 4.2.1)

$$\widehat{Q}\Pi = \Pi Q_1 \Pi,$$

that is exactly as in Theorem 4.1.

Corollary 5.2 *In the absorbing case (5.188) the generating matrix $\widehat{Q} = (\widehat{q}_{kr}; 1 \leq k, r \leq N)$ has the same representation (4.17), Section 4.2.1, but the conservative relation does not take place, precisely,*

$$\sum_{r \neq k} \widehat{p}_{kr} = 1 - \widehat{p}_{k0},$$

where $\widehat{p}_{k0}, k \in \widehat{E}$, are the absorbing probabilities of the limit merged Markov process $\widehat{x}(t), t \geq 0$, considered on the extended merged state space $E_0 = \{0; 1, ..., N\}$.

Note that it is exactly what was formulated in Theorem 4.2 (Section 4.2.2).

The *double merging scheme* considered in Section 4.2.3 leads to the following representation of the compensating operator of the coupled process:

$$x^\varepsilon(t/\varepsilon^2), \quad \widehat{\widehat{x}}^\varepsilon(t/\varepsilon^2) := \widehat{\widehat{v}}(x^\varepsilon(t/\varepsilon^2)), \quad t \geq 0,$$

$$\mathbb{L}^\varepsilon \varphi(x, v) = [\varepsilon^{-2} Q + \varepsilon^{-1} Q_1 + Q_2]\varphi(x, v), \tag{5.192}$$

where (see (4.23))

$$Q_i \varphi(x) = q(x) P_i \varphi(x), \quad P_i \varphi(x) := \int_E P_i(x, dy) \varphi(y), \quad i = 1, 2.$$

The stochastic kernel $P(x, dy)$ is coordinated with the split state space (4.22) as follows:

$$P(x, E_k^r) = \mathbf{1}_k^r(x) := \begin{cases} 1, & x \in E_k^r \\ 0, & x \notin E_k^r. \end{cases}$$

It is worth noticing that the split state space (4.22) and uniform ergodicity of the associated Markov process $x^0(t), t \geq 0$, with generator (4.26) and stationary distributions $\pi_k^r(dx), 1 \leq k \leq N, 1 \leq r \leq N_k$, in every class E_k^r provides the projector on the Banach space $\mathbf{B}$,

$$\Pi \varphi(x) := \sum_{k=1}^{N} \sum_{r=1}^{N_k} \widehat{\varphi}_k^r \mathbf{1}_k^r, \quad \widehat{\varphi}_k^r := \int_{E_k^r} \pi_k^r(dx) \varphi(x).$$

In the same way, the first convergence in (4.27) and uniform ergodicity of the limit merged Markov process $\widehat{x}(t), t \geq 0$, on the merged state space $\widehat{E} = \{\widehat{E}_k^r; 1 \leq k \leq N, 1 \leq r \leq N_k\}$ with the stationary distributions $\widehat{\pi}_k = (\pi_k^r; 1 \leq r \leq N_k), 1 \leq k \leq N$, provides the projector in Banach space **B**

$$\widehat{\Pi}\varphi(x) = \sum_{k=1}^{N}\sum_{r=1}^{N_k} \widehat{\pi}_k^r \widehat{\varphi}_k^r \mathbf{1}_k^r(x) =: \sum_{k=1}^{N} \widehat{\widehat{\varphi}}_k \mathbf{1}_k(x), \qquad (5.193)$$

$$\widehat{\widehat{\varphi}}_k := \sum_{r=1}^{N_k} \widehat{\pi}_k^r \widehat{\varphi}_k^r, \quad 1 \leq k \leq N.$$

According to Theorem 4.2 and using additional assumption ME4 (Section 4.2.1), the limit merged Markov process $\widehat{x}(t), t \geq 0$, defined by the generating matrix $\widehat{Q}$ is determined by the following relation

$$\widehat{Q}\Pi = \Pi Q_1 \Pi.$$

Hence, the generator $\widehat{Q}$ is reducible-invertible with the projector $\widehat{\Pi}$ on the null space $\mathcal{N}_{\widehat{Q}}$. This means that the asymptotic representation (5.192) of the compensating operator can be analyzed by using Proposition 5.3. The double merging principle is given in the following proposition, exactly as in Theorem 4.3.

Proposition 5.12 *The generator $\widehat{\widehat{Q}}_2$ of the double merged Markov process $\widehat{\widehat{x}}(t), t \geq 0$, is determined by the relations (see (5.28))*

$$\widehat{\widehat{Q}}_2 \widehat{\Pi} = \widehat{\Pi} \widehat{Q}_2 \widehat{\Pi}, \quad \widehat{Q}_2 \Pi = \Pi Q_2 \Pi. \qquad (5.194)$$

5.6.3 Average Stochastic Systems

The phase merging principles for the stochastic systems with split and merging of the switching semi-Markov process, considered in Section 4.3, are constructed by using the asymptotic representation of the extended compensating operator of stochastic system and a solution of the suitable singular perturbation problem from Section 5.2.1.

The new situation in the analysis of the compensating operator disappears due to the dependency of the generator of the associated Markov process on the series parameter ε.

According to Assumptions A1-A3, Section 4.3, and additional stated assumptions in Section 4.3.1, the compensating operator of the continuous

5.6. MERGING AND AVERAGING IN SPLIT STATE SPACE

random evolution in the split and merging scheme can be written in the following form (compare with (5.121), Section 5.3.4):

$$\mathbb{L}^\varepsilon \varphi(u,v,x) = \varepsilon^{-1} q(x)[F_\varepsilon(x) P^\varepsilon - I]\varphi(u,v,x),$$
$$F_\varepsilon(x) := \int_0^\infty F_x(dt) \Gamma_{\varepsilon s}(x), \qquad (5.195)$$
$$P^\varepsilon = P + \varepsilon P_1.$$

It is easy to obtain the following asymptotic representation for generator (5.195) (compare to (5.79))

$$\mathbb{L}^\varepsilon = \varepsilon^{-1} Q + [Q_1 + \boldsymbol{\Gamma}(x) P] + \theta^\varepsilon(x) \qquad (5.196)$$

where Q is the generator of the associated Markov process $x^0(t), t \geq 0$, and

$$Q_1 \varphi(x) := q(x) \int_E P_1(x, dy) \varphi(y).$$

Proposition 5.13 *The generator $\widehat{\mathbb{L}}$ of the limit coupled Markov process $\widehat{x}(t), \widehat{U}(t), t \geq 0$, can be represented in the following form*

$$\widehat{\mathbb{L}} = \widehat{Q}_1 + \widehat{\boldsymbol{\Gamma}}. \qquad (5.197)$$

PROOF. According to Proposition 5.1 the solution of the singular perturbation problem for generator (5.196) has the following representation

$$\widehat{\mathbb{L}} \Pi = \Pi[Q_1 + \boldsymbol{\Gamma}(x) P] \Pi = [\widehat{Q}_1 + \widehat{\boldsymbol{\Gamma}}] \Pi.$$

□

The generator (5.197) of the coupled Markov process $\widehat{x}(t), \widehat{U}(t), t \geq 0$, defines the limit stochastic system $\widehat{U}(t)$ given by a solution of the evolutionary equation

$$\left| \begin{array}{l} \frac{d}{dt} \widehat{U}(t) = \widehat{a}(\widehat{U}(t); \widehat{x}(t)), \\ \widehat{U}(0) = \xi(0), \end{array} \right.$$

with the average velocity

$$\widehat{a}(u; k) = \int_{E_k} \pi_k(dx) a(u; x), \quad 1 \leq k \leq N,$$

that is exactly as in Theorem 4.4, Section 4.3.1.

Using Theorem 4.5, Section 4.3.2, we can solve a similar problem for merging of the semi-Markov processes.

5.7 Diffusion Approximation with Split and Merging

The phase merging principles for the stochastic systems with split and merging of the switching semi-Markov process considered in Section 4.4 are constructed as in the previous Section 5.6.2, by using the asymptotic expansion of the extended compensating operator of the stochastic systems and a solution of the singular perturbation problem from Section 5.2.

In the diffusion approximation scheme with split and merging of state space the new special feature disappears owing to the dependency of the switching semi-Markov process on the series parameter $\varepsilon \to 0, (\varepsilon > 0)$.

The diffusion approximation scheme in Sections 4.4.2-4.4.5 are considered with Markov switching, which simplifies the asymptotic analysis. The generators of the Markov stochastic systems are considered instead of the compensating operator. Generalization for switching semi-Markov processes can be also obtained following the analysis considered in Section 5.7.1.

5.7.1 Ergodic Split and Merging

According to Conditions D1-D3, in Section 4.4, and additional assumptions of Theorem 4.7, Section 4.4.1, the compensating operator of the continuous random evolution in split and merging scheme is represented in the following form (compare with (5.123), Section 5.4.2)

$$\mathbb{L}^\varepsilon \varphi(u,x) = \varepsilon^{-2} q(x)[F_\varepsilon(x) P^\varepsilon \varphi(u,x) - \varphi(u,x)]. \quad (5.198)$$

Here, by definition:

$$F_\varepsilon(x) := \int_0^\varepsilon F_x(ds) \Gamma^\varepsilon_{\varepsilon^2 s}(x), \quad (5.199)$$

and:

$$P^\varepsilon \varphi(x) := \int_E P^\varepsilon(x, dy) \varphi(y), \quad (5.200)$$

$$P^\varepsilon(x, B) = P(x, B) + \varepsilon^2 P_1(x, B).$$

Proposition 5.14 *The compensating operator (5.198)-(5.200) acting on the test functions $\varphi \in C^3(\mathbb{R}^d \times E)$ has the following asymptotic representation (compare with Proposition 3.2, Section 3.2.1)*

$$\mathbb{L}^\varepsilon \varphi(u,x) = [\varepsilon^{-2} Q + \varepsilon^{-1} \mathbb{\Gamma}(x) P + \widetilde{Q}_2(x) + \theta^\varepsilon(x)] \varphi(u,x),$$

where
$$\widetilde{Q}_2(x) = [\mathbb{\Gamma}_1(x) + \mu_2(x)\mathbb{\Gamma}^2(x)]P + Q_1.$$

PROOF. In representation (3.10) put $Q^\varepsilon = Q + \varepsilon^2 Q_1$, with $Q_1 = qP_1$. □

Now, the solution of the singular perturbation problem, given in Proposition 5.2, for the truncated operator
$$\mathbb{L}_0^\varepsilon = \varepsilon^{-2}Q + \varepsilon^{-1}\mathbb{\Gamma}(x)P + \widetilde{Q}_2(x),$$
gives the limit generator $\widehat{\mathbb{L}}$ of the coupled Markov process $\widehat{\xi}(t), \widehat{x}(t), t \geq 0$, in the form given in Theorem 4.7, Section 4.4.1.

Calculation of the limit generator $\widehat{\mathbb{L}}$ almost coincides with calculation of the limit generator in Section 5.4.2.

As a result we obtain the following construction of the limit generator (compare with (5.130))
$$\widehat{\mathbb{L}} = \widehat{Q}_1 + \widehat{\mathbb{\Gamma}}_0 + \widehat{\mathbb{\Gamma}}_{00}(k) + \widehat{\mathbb{B}}(k), \quad k \in \widehat{E}, \qquad (5.201)$$

with the operators depending on the state of the limit merged Markov process $\widehat{x}(t), t \geq 0$, that is:
$$\widehat{\mathbb{\Gamma}}_0(k) := \int_{E_k} \pi_k(dx)\mathbb{\Gamma}_0(x), \quad \mathbb{\Gamma}_0(x) = \mathbb{\Gamma}(x)R_0\mathbb{\Gamma}(x),$$
$$\widehat{\mathbb{\Gamma}}_{00}(k) := \int_{E_k} \mu(x)\mathbb{\Gamma}^2(x), \quad \widehat{\mathbb{B}}(k) := \int_{E_k} \pi_k(dx)\mathbb{B}(x), \qquad (5.202)$$

and $\widehat{Q}_1$ defined by $\widehat{Q}_1\Pi = \Pi Q_1 \Pi$.

Formulas (5.201)-(5.202) allow us to construct the limit generator for the stochastic systems considered in diffusion approximation scheme with split and merging of the state space of the switching semi-Markov process.

The limit generator $\mathbb{L}$ of the limit diffusion process $\widehat{\xi}(t), t \geq 0$, switched by the merged Markov process $\widehat{x}(t), t \geq 0$, in Theorem 4.7, is calculated by formulas (5.201)-(5.202), setting:
$$\mathbb{\Gamma}(x)\varphi(u) = a(u;x)\varphi'(u),$$
$$\mathbb{B}(x)\varphi(u) = a_1(u;x)\varphi'(u) + \frac{1}{2}C_0(u;x)\varphi''(u).$$

5.7.2 Split and Double Merging

According to conditions D1-D2 in Section 4.4 and additional Condition BC2, Section 4.4.3, the generator of the random evolution

$\xi^\varepsilon(t), x^\varepsilon(t/\varepsilon^2), t \geq 0$, described in Theorem 4.9 is represented in the following form

$$\mathbb{L}^\varepsilon = \varepsilon^{-3}Q + \varepsilon^{-2}Q_1 + \varepsilon^{-1}\mathbb{A}(x) + \mathbb{B}(x) + \theta_l^\varepsilon(x), \qquad (5.203)$$

where by definition:

$$\mathbb{A}(x)\varphi(u) := a(u;x)\varphi'(u), \qquad (5.204)$$

$$\mathbb{B}(x)\varphi(u) := a_1(u;x)\varphi'(u) + C_0(u;x)\varphi''(u), \qquad (5.205)$$

with negligible term $\|\theta_l^\varepsilon(x)\varphi\| \to 0$, as $\varepsilon \to 0$, $\varphi \in C^3(\mathbb{R}^d)$.

Proposition 5.15 *The generator $\mathbb{L}$ of the limit diffusion process in Theorem 4.9 is calculated by using a solution of the singular perturbation problem for the generator (5.203) given in Proposition 5.4 in the following form*

$$\mathbb{L} = \widehat{\widehat{\mathbb{B}}} + \widehat{\widehat{\mathbb{A}\widehat{R}_0\mathbb{A}}}. \qquad (5.206)$$

The calculation of $\mathbb{L}$ in the formula (5.206) using (5.204) and (5.205) leads to the representation of the limit generator in Theorem 4.9.

5.7.3 Double Split and Merging

Conditions MD1-MD4, Section 4.2.3 and Conditions D1-D3, BC3, Section 4.4.4, in Theorem 4.10, provide the generator $\mathbb{L}^\varepsilon$ of the Markov process

$$\xi^\varepsilon(t), \quad x_t^\varepsilon := x(t/\varepsilon^3), \quad \widehat{\widehat{x}}^\varepsilon_t := \widehat{\widehat{v}}(x_t^\varepsilon), \quad t \geq 0,$$

represented in the following form

$$\mathbb{L}^\varepsilon \varphi(u, k, x) = [\varepsilon^{-3}Q + \varepsilon^{-2}Q_1(x) + \varepsilon^{-1}\mathbb{A}(x)$$
$$+ [Q_2 + \mathbb{B}(x) + \theta_l^\varepsilon(x)]]\varphi(u, k, x), \qquad (5.207)$$

where, by definition:

$$\mathbb{A}(x)\varphi(u) := a(u;x)\varphi'(u), \qquad (5.208)$$

$$\mathbb{B}(x)\varphi(u) := a_1(u;x)\varphi'(u) + \frac{1}{2}C_0(u;x)\varphi''(u),$$

with the negligible term

$$\|\theta_l^\varepsilon(x)\varphi\| \to 0, \quad \varepsilon \to 0, \quad \varphi \in C^3(\mathbb{R}^d).$$

5.7. DIFFUSION APPROXIMATION WITH SPLIT AND MERGING

Proposition 5.16 *The generator* $\mathbb{L}$ *of the limit diffusion process* $\widehat{\widehat{\xi}}(t)$, $t \geq 0$, *switched by the twice merged Markov process* $\widehat{\widehat{x}}(t), t \geq 0$, *is defined by a solution of the regular perturbed problem for the truncated operator*

$$\mathbb{L}_0^\varepsilon = \varepsilon^{-3}Q + \varepsilon^{-2}Q_1(x) + \varepsilon^{-1}\mathbb{A}(x) + [Q_2 + \mathbb{B}(x)], \quad (5.209)$$

in the following form

$$\mathbb{L} = \widehat{\widehat{Q}}_2 + \widehat{\widehat{\mathbb{B}}} + \widehat{\widehat{\mathbb{A}R_0\mathbb{A}}}, \quad (5.210)$$

where the generator $\widehat{\widehat{Q}}_2$ of the twice merged Markov process $\widehat{\widehat{x}}(t), t \geq 0$, is given in Condition MD4, Section 4.2.3. The potential $\widehat{R}_0$ is defined for the generator $\widehat{Q}_1$ as follows $\widehat{Q}_1\widehat{R}_0 = \widehat{R}_0\widehat{Q}_1 = \widehat{\Pi} - I$. The twice average operators in (5.210) are calculated by

$$\widehat{\widehat{\mathbb{B}\Pi}} = \widehat{\Pi}\widehat{\mathbb{B}\Pi}, \quad \widehat{\mathbb{B}}\Pi = \Pi\mathbb{B}(x)\Pi, \quad (5.211)$$

and analogously,

$$\widehat{\widehat{\mathbb{A}\Pi}} = \widehat{\Pi}\widehat{\mathbb{A}}_0\widehat{\Pi}, \quad \widehat{\mathbb{A}}_0 = \widehat{\mathbb{A}R_0\mathbb{A}}, \quad \widehat{\mathbb{A}\Pi} = \Pi\mathbb{A}(x)\Pi. \quad (5.212)$$

PROOF. The formulas (5.210)-(5.212) are obtained straightforwardly from Proposition 5.4, Section 5.2.

Calculation by formulas (5.210)-(5.212), with (5.208), give the representation of the limit generator $\mathbb{L}$ in (4.47) of Theorem 4.10. □

5.7.4 Double Split and Double Merging

Under the conditions of Theorem 4.11, Section 4.4.5, and taking into account the conditions of Theorem 4.3 we can calculate the generator $\mathbb{L}^\varepsilon$ of the coupled Markov processes $\xi^\varepsilon(t), x_t^\varepsilon := x^\varepsilon(t/\varepsilon^4), t \geq 0$, with the first component $\xi^\varepsilon(t), t \geq 0$, given in Theorem 4.11 in the following form

$$\mathbb{L}^\varepsilon \varphi(u,x) = [\varepsilon^{-4}Q + \varepsilon^{-3}Q_1(x) + \varepsilon^{-2}Q_2(x) + \varepsilon^{-1}\mathbb{A}(x) + \mathbb{B}(x)]\varphi(u,x)$$
$$+ \theta_l^\varepsilon(x)\varphi, \quad (5.213)$$

where, by definition, Q is the generator of the support Markov process $x^0(t), t \geq 0$, given by the generator (4.26) and $Q_i(x) := q(x)P_i(x,B), i = 1, 2$, (see Section 4.2). The operators are (compare with Section 5.7.2):

$$\mathbb{A}(x)\varphi(u) := a(u;x)\varphi'(u),$$

$$\mathbb{B}(x)\varphi(u) := a_1(u;x)\varphi'(u) + C_0(u;x)\varphi''(u),$$

with the negligible term: $\|\theta_l^\varepsilon(x)\varphi\| \to 0$, as $\varepsilon \to 0$, for $\varphi \in C^3(\mathbb{R}^d)$.

Proposition 5.17 *The generator* $\mathbb{L}$ *of the limit diffusion process* $\widehat{\widehat{\xi}}(t)$, $t \geq 0$, *is calculated by using a solution of the singular perturbation problem for the generator (5.213) given in Proposition 5.5 in the following form*

$$\mathbb{L} = \widehat{\widehat{\mathbb{B}}} + \widehat{\widehat{\mathbb{A}}}\widehat{\widehat{R_0}}\widehat{\widehat{\mathbb{A}}}. \tag{5.214}$$

The calculation in formula (5.214) leads to the representation of the limit generator in the form given in Theorem 4.11.

PROOF. According to the conditions of Theorem 4.3, we can verify that Conditions (i)-(iii) of Proposition 5.5 really hold by using conditions of Theorem 4.3. The contracted operator $\widehat{Q}_1$, defined by the relation $\widehat{Q}_1 \Pi = \Pi Q_1(x)\Pi$, defines the ergodic merged Markov process $\widehat{x}(t), t \geq 0$. That is, the generator $\widehat{Q}_1$ is reducible-invertible with the projector $\widehat{\Pi}$ as in Assumption (ii) of Proposition 5.5. The second convergence in Theorem 4.3 induces that the limit twice merged Markov process $\widehat{\widehat{x}}(t), t \geq 0$, is ergodic, whose stationary distribution defines the projector $\widehat{\widehat{\Pi}}$ on the null space $\mathcal{N}_{\widehat{\widehat{Q}}_2}$ of its generator $\widehat{\widehat{Q}}_2$ as in Assumption (iii) of Proposition 5.5. Assumption (i) in Proposition 5.5 is obviously valid.

Hence the solution of the singular perturbation problem given in Proposition 5.5 gives the preliminary representation of the limit generator in Theorem 4.10. □

Chapter 6

Weak Convergence

6.1 Introduction

The present chapter presents the entire proofs of weak convergence. Actually, the algorithmic part of the proofs, concerning convergence of the generators, providing finite-dimensional distribution convergence, was given in the previous Chapter 5. Here we prove the tightness of the probability measures of stochastic processes. Of course, in a Polish space, as is our case here, relative compactness and tightness are equivalent.

The proof of relative compactness, is given via the scheme of proof of Stroock and Varadhan[173] for space $\mathbf{C}[0, \infty)$. More precisely, we use the extension of Sviridenko to space $\mathbf{D}[0, \infty)$, that is the compact containment condition and nonnegative submartingale (see Theorem 6.2).

The proofs concerning specific systems are based on the pattern limit theorems given in Section 6.2; for stochastic systems with Markov switching see Theorem 6.3; with asymptotic merging for averaging see Theorem 6.4; for diffusion approximation see Theorem 6.5; for semi-Markov switching see Theorem 6.6. Theorem 6.7 gives an alternative proof to the previous theorems, based on the continuous-time extended Markov renewal process.

Appendix A gives general definitions and basics on weak convergence concerning this chapter.

6.2 Preliminaries

Phase merging, averaging, diffusion and Poisson approximation algorithms for stochastic systems with Markov and semi-Markov switching are based on limit theorems for stochastic processes in series scheme.

The processes arising in applications and being of interest to us now are

Markovian (or have an embedded Markov chain) with values in a *standard state space* $(E, \mathcal{E})$.

Recall that, if two probability measures are the limits of the same sequence in a common standard state space, then they are equal. Every probability measure on a standard state space is tight [52].

The functional space $\mathbf{D}_E[0, \infty)$ of right continuous functions $x : \mathbb{R}_+ := [0, \infty) \to E$, with left limit (cadlag functions) is considered as the space of sample paths of stochastic processes. It is well known that for a standard state space E, the space $\mathbf{D}_E[0, \infty)$, with the Skorokhod topology, is also a *standard state space*.

Of course, the space $\mathbf{C}_E[0, \infty)$ of continuous functions in uniform topology is also considered as a space of simple paths of stochastic processes, especially for limit processes [56].

The main type of convergence of stochastic processes is the *weak convergence of finite-dimensional distributions*, that is for the family of stochastic processes $x^\varepsilon(t), t \geq 0, \varepsilon > 0$, in the series scheme with the small series parameter $\varepsilon > 0, \varepsilon \to 0$, there is a process $x(t), t \geq 0$, such that

$$\lim_{\varepsilon \to 0} \mathbb{E}\varphi(x^\varepsilon(t_1), \cdots, x^\varepsilon(t_N)) = \mathbb{E}\varphi(x(t_1), \cdots, x(t_N)), \qquad (6.1)$$

for any test function $\varphi(x_1, \cdots, x_N) \in C(E^N)$, the space of real-valued bounded continuous functions on E^N, and for any finite set $\{t_1, \cdots, t_N\} \in S$, where S is a dense set in $\mathbb{R}_+$. We will denote this type of convergence by

$$x^\varepsilon(t) \stackrel{D}{\Rightarrow} x(t), \quad \varepsilon \to 0.$$

A more general type of convergence of stochastic processes is the *weak convergence of associated measures*, that is, the probability distributions (see Appendix A)

$$P_\varepsilon(B) := \mathbb{P}(x^\varepsilon(\cdot) \in B), \quad B \in \mathcal{D}_E,$$

where $\mathcal{D}_E$ is the Borel σ-algebra on $\mathbf{D}_E[0, \infty)$. So, the following limit takes place

$$\lim_{\varepsilon \to 0} \mathbb{E}\varphi(x^\varepsilon(\cdot)) = \mathbb{E}\varphi(x(\cdot)), \qquad (6.2)$$

for all $\varphi \in C(\mathbf{D}_E)$, the space of continuous real-valued functions on $\mathbf{D}_E[0, \infty)$ in the Skorokhod topology. We will denote the weak convergence

6.2. PRELIMINARIES

of processes and of associated measures as follows

$$x^\varepsilon \Rightarrow x, \quad P_\varepsilon \Rightarrow P, \quad \varepsilon \to 0.$$

The connection between the above two types of convergence of stochastic processes is realized by means of *relative compactness* of the family of associated measures $(P_\varepsilon, \varepsilon > 0)$, that is, for any sequence, say (P_{ε_n}) of $(P_\varepsilon, \varepsilon > 0)$ there exists a weak convergent subsequence $(P_{\varepsilon'_n}) \subset (P_{\varepsilon_n})$.

Theorem 6.1 ([45]) *Let $x^\varepsilon(t), t \geq 0, \varepsilon > 0$, and $x(t), t \geq 0$, be processes with simple paths in $\mathbf{D}_E[0, \infty)$.*

(a) If $x^\varepsilon \Rightarrow x$ as $\varepsilon \to 0$, then $x^\varepsilon \stackrel{D}{\Rightarrow} x$, for the set $S = \{t > 0 : \mathbb{P}(x(t) = x(t-)) = 1\}$.

(b) If $x^\varepsilon, \varepsilon > 0$, is relatively compact and there exists a dense set $S \subset \mathbb{R}_+$ such that $x^\varepsilon \stackrel{D}{\Rightarrow} x$, as $\varepsilon \to 0$, on the set S, then $x^\varepsilon \Rightarrow x$, as $\varepsilon \to 0$.

By Prohorov's theorem [16], relative compactness of a family of probability measures on $\mathbf{D}_E[0, \infty)$ is equivalent to tightness of this family.

Many different kinds of procedures are available for verifying the relative compactness for a family of probability measures associated with given stochastic processes. In the case of Markov processes the martingale characterization approach seems to be the most effective one [100], in particular the one developed by Strook and Varadhan [173] is effective in the verification of relative compactness. It is worth noticing that the relative compactness conditions formulated in [173] for stochastic processes with sample paths in the space $\mathbf{C}_E[0, \infty)$ are valid for the space $\mathbf{D}_E[0, \infty)$.

Theorem 6.2 ([173]) *Let the family of $\mathbb{R}^d$-valued stochastic processes, $x^\varepsilon(t), t \geq 0, \varepsilon > 0$, such that:*

H1: *For all nonnegative functions $\varphi \in C_0^\infty(\mathbb{R}^d)$ there exists a constant $A_\varphi \geq 0$ such that $(\varphi(x^\varepsilon(t)) + A_\varphi t, \mathcal{F}_t^\varepsilon)$ is a nonnegative submartingale.*

H2: *Given a nonnegative $\varphi \in C_0^\infty(\mathbb{R}^d)$, the constant A_φ can be chosen so that it does not depend on the translates of φ.*

Then, under the initial condition

$$\lim_{c \to \infty} \sup_{0 < \varepsilon \leq \varepsilon_0} \mathbb{P}(|x^\varepsilon(0)| \geq c) = 0,$$

the family of associated probability measures $P_\varepsilon, \varepsilon > 0$, is relatively compact.

In order to verify the weak convergence of a family of stochastic processes on $\mathbf{D}_E[0,\infty)$, we have to establish the relative compactness and the weak convergence of finite-dimensional distributions.

Both these problems for a family of Markov processes in $\mathbf{D}_E[0,\infty)$ can be solved by using the martingale characterization of Markov processes and convergence of generators.

The particularity of the fast time-scaling switching processes is that the convergence of generators or compensating operators cannot be obtained in a direct way because the generators are considered in a singular perturbed form. But, as was shown in Chapter 5, the phase merging and averaging algorithms as well as diffusion and Poisson approximation schemes can be obtained by using a solution of the singular perturbation problem for reducible-invertible operators. Such an approach will be used in the following.

Another approach to verifying the relative compactness of a family of Markov processes consists in using the martingale characterization and the compactness conditions for square integrable martingales [132].

The uniqueness of the limit measure follows from the uniqueness of solution of the martingale problem [45].

6.3 Pattern Limit Theorems

6.3.1 Stochastic Systems with Markov Switching

The stochastic systems with Markov switching in series scheme with the small series parameter $\varepsilon > 0$, defined by the coupled Markov process (see Sections 3.2-3.3)

$$\xi^\varepsilon(t), \quad x^\varepsilon(t), \quad t \geq 0, \varepsilon > 0,$$

can be characterized by the martingale

$$\mu_t^\varepsilon = \varphi(\xi^\varepsilon(t), x^\varepsilon(t)) - \int_0^t \mathbb{L}^\varepsilon \varphi(\xi^\varepsilon(s), x^\varepsilon(s))ds. \tag{6.3}$$

The generators $\mathbb{L}^\varepsilon, \varepsilon > 0$, have the common domain of definition $\mathcal{D}(\mathbb{L})$, which is supposed to be dense in $C_0^2(\mathbb{R}^d \times E)$.

The limit Markov process $\xi(t), t \geq 0$, is considered on $\mathbb{R}^d$, characterized

6.3. PATTERN LIMIT THEOREMS

by the martingale

$$\mu_t = \varphi(\xi(t)) - \int_0^t \mathbb{L}\varphi(\xi(s))ds, \tag{6.4}$$

where the closure $\overline{\mathcal{D}(\mathbb{L})}$ of the domain $\mathcal{D}(\mathbb{L})$ of the generator $\mathbb{L}$ is a convergence-determining class (see Appendix A).

Theorem 6.3 ([100]) *Let the following conditions hold for a family of Markov processes $\xi^\varepsilon(t), t \geq 0, \varepsilon > 0$:*

C1: *There exists a family of test functions $\varphi^\varepsilon(u,x)$ in $C_0^2(\mathbb{R}^d \times E)$, such that*

$$\lim_{\varepsilon \to 0} \varphi^\varepsilon(u,x) = \varphi(u), \quad \text{uniformly on } u, x.$$

C2: *The following convergence holds*

$$\lim_{\varepsilon \to 0} \mathbb{L}^\varepsilon \varphi^\varepsilon(u,x) = \mathbb{L}\varphi(u), \quad \text{uniformly on } u, x.$$

The family of functions $\mathbb{L}^\varepsilon \varphi^\varepsilon, \varepsilon > 0$, is uniformly bounded, and $\mathbb{L}\varphi$ and $\mathbb{L}^\varepsilon \varphi^\varepsilon$ belong to $C(\mathbb{R}^d \times E)$.

C3: *The quadratic characteristics of the martingales (6.3) have the representation*

$$\langle \mu^\varepsilon \rangle_t = \int_0^t \zeta^\varepsilon(s)ds,$$

where the random functions $\zeta^\varepsilon, \varepsilon > 0$, satisfy the condition

$$\sup_{0 \leq s \leq T} \mathbb{E}|\zeta^\varepsilon(s)| \leq c < +\infty. \tag{6.5}$$

C4: *The convergence of the initial values holds, that is,*

$$\xi^\varepsilon(0) \xrightarrow{P} \widehat{\xi}(0), \quad \varepsilon \to 0,$$

and

$$\sup_{\varepsilon > 0} \mathbb{E}|\xi^\varepsilon(0)| \leq C < +\infty.$$

Then the weak convergence

$$\xi^\varepsilon(t) \Longrightarrow \xi(t), \quad \varepsilon \to 0,$$

takes place. The limit Markov process $\xi(t), t \geq 0$, is characterized by the martingale (6.4).

In the particular case where Condition (6.5) is replaced by

$$\sup_{0\leq t\leq T} \mathbb{E}\,|\zeta^\varepsilon(t)| \to 0, \quad \varepsilon \to 0,$$

the limit process $\xi(t), t \geq 0$, is given by the solution of the deterministic equation

$$\varphi(\xi(t)) - \int_0^t \mathbb{L}\varphi(\xi(s))ds = \varphi(\xi(0)),$$

or, in the equivalent form

$$\frac{d}{dt}\varphi(\xi(t)) = \mathbb{L}\varphi(\xi(t)). \tag{6.6}$$

PROOF. Condition C3 of the theorem means that the quadratic characteristics of the martingale (6.3) is relatively compact (see [70,100]), that is, there exists a sequence $\varepsilon_n \to 0$, such that the weak convergence

$$\mu_t^{\varepsilon_n} \Longrightarrow \mu_t^0, \quad n \to \infty,$$

takes place. Now, in order to prove that the martingale characterization of the limit process is (6.4), let us calculate:

$$\begin{aligned}
\mathbb{E}\mu_t^\varepsilon &= \mathbb{E}[\varphi^\varepsilon(\xi^\varepsilon(t), x^\varepsilon(t)) - \int_0^t \mathbb{L}^\varepsilon\varphi^\varepsilon(\xi^\varepsilon(s), x^\varepsilon(s))ds] \\
&= \mathbb{E}[\varphi^\varepsilon(\xi^\varepsilon(t), x^\varepsilon(t)) - \varphi(\xi^\varepsilon(t))] \\
&\quad + \mathbb{E}[\varphi(\xi^\varepsilon(t)) - \int_0^t \mathbb{L}\varphi(\xi^\varepsilon(s))ds] \\
&\quad + \mathbb{E}\int_0^t [\mathbb{L}\varphi(\xi^\varepsilon(t)) - \mathbb{L}^\varepsilon\varphi^\varepsilon(\xi^\varepsilon(s), x^\varepsilon(s))ds].
\end{aligned}$$

The first and third terms on the right hand side tend to zero, as $\varepsilon \to 0$, by Conditions C1 and C2 of the theorem. Due to the relative compactness of the family of stochastic processes $\xi^\varepsilon(t), t \geq 0, \varepsilon > 0$, the following convergence

$$\varphi(\xi^{\varepsilon_n}(t)) - \int_0^t \mathbb{L}\varphi(\xi^{\varepsilon_n}(s))ds \Rightarrow \mu_t, \quad n \to \infty, \tag{6.7}$$

takes place. Now, due to Condition C3, we calculate:

$$\begin{aligned}
\mathbb{E}\mu_t^{\varepsilon_n} &= \mathbb{E}\varphi^{\varepsilon_n}(\xi^{\varepsilon_n}(0), x) \\
&= \mathbb{E}\varphi(\xi^{\varepsilon_n}(0)) + \mathbb{E}[\varphi^{\varepsilon_n}(\xi^{\varepsilon_n}(0), x) - \varphi(\xi^{\varepsilon_n}(0))].
\end{aligned}$$

6.3. PATTERN LIMIT THEOREMS

Hence, the following convergence holds

$$\mathbb{E}\mu_t^{\varepsilon_n} \longrightarrow \mathbb{E}\varphi(\xi(0)), \quad n \to \infty.$$

Consequently, we get that the limit process $\xi(t), t \geq 0$, is characterized by the martingale (6.4), with

$$\mathbb{E}\mu_t - \mathbb{E}\varphi(\zeta(0)).$$

□

The following convergence theorem is an adaptation of Theorem 8.2, p. 226 and Theorem 8.10, p. 234, in [45], to our conditions with the solution of singular perturbation problem.

Theorem 6.4 ([45]) *Suppose that for the generator* $\mathbb{L}$ *of the coupled Markov process* $\xi(t), \widehat{x}(t), t \geq 0$, *on the state space* $\mathbb{R}^d \times V$, *with* V *a finite set, there is at most one solution of the martingale problem in* $\mathbf{D}_{\mathbb{R}^d \times V}[0, \infty)$, *and that the closure of the domain* $\mathcal{D}(\mathbb{L})$ *is a convergence-determining class.*

Suppose that the family of Markov processes $\xi^\varepsilon(t/\varepsilon), x_t^\varepsilon, t \geq 0, \varepsilon > 0$ *on* $\mathbb{R}^d \times E$ *defined by the generators* $\mathbb{L}^\varepsilon, \varepsilon > 0$, *with domains* $\mathcal{D}(\mathbb{L}^\varepsilon)$ *dense in* $C_0^2(\mathbb{R}^d \times E)$, *satisfies the following conditions:*

C1: *The family of probability measures* $(P^\varepsilon, \varepsilon > 0)$ *associated to the processes* $(\xi^\varepsilon(t), v(x^\varepsilon(t/\varepsilon)), t \geq 0, \varepsilon > 0)$ *is relatively compact.*

C2: *There exists a collection of functions* $\varphi^\varepsilon(u, x) \in C(\mathbb{R}^d \times E)$, *such that the following uniform convergence takes place*

$$\lim_{\varepsilon \to 0} \varphi^\varepsilon(u, x) = \varphi(u, v(x)) \in C(\mathbb{R}^d \times V) \tag{6.8}$$

and such that for every $T > 0$

$$\lim_{\varepsilon \to 0} \sup_{0 \leq t \leq T} \mathbb{E}|\varphi^\varepsilon(u, x^\varepsilon(t/\varepsilon))| < +\infty. \tag{6.9}$$

C3: *The uniform convergence of generators*

$$\lim_{\varepsilon \to 0} \mathbb{L}^\varepsilon \varphi^\varepsilon(u, x) = \mathbb{L}\varphi(u, v(x)), \tag{6.10}$$

takes place, and the functions $\mathbb{L}^\varepsilon \varphi^\varepsilon$, $\varepsilon > 0$, *are uniformly bounded on* $\varepsilon > 0$, *and* $\mathbb{L}\varphi \in C(\mathbb{R}^d \times V)$.

C4: *The convergence in probability of the initial values holds, that is,*

$$(\xi^\varepsilon(0), v(x^\varepsilon(0))) \xrightarrow{P} (\xi(0), \widehat{x}(0)), \quad \varepsilon \to 0,$$

with uniformly bounded expectation

$$\sup_{\varepsilon>0} \mathbb{E}\,|\xi^\varepsilon(0)| \leq c < +\infty.$$

Then the weak convergence in $\mathbf{D}_{\mathbb{R}^d \times V}[0,\infty)$

$$(\xi^\varepsilon(t), v(x^\varepsilon(t))) \Longrightarrow (\xi(t), \widehat{x}(t)), \quad \varepsilon \to 0,$$

takes place.

The limit Markov process $\xi(t), \widehat{x}(t), t \geq 0$, is defined by the generator $\mathbb{L}$.

Remark 6.1. The main algorithmic conditions (8.53) and (8.54) of Theorem 8.10 in [45] are represented in conditions of Theorem 6.4, respectively (6.8) and (6.10). Additional conditions of boundedness (8.51) and (8.52) correspond to additional conditions C2 and C3. All other conditions of Theorem 8.10 are represented in the convergence Theorem 6.4 in the same form.

We will use also the following theorem which is a compilation of Theorem 9.4, p. 145, and Corollary 8.6, p. 231 in [45] under our conditions in diffusion approximation schemes.

Theorem 6.5 ([45]) Let us consider the family of coupled Markov processes

$$\xi^\varepsilon(t), \ x^\varepsilon(t/\varepsilon^2), \quad t \geq 0, \ \varepsilon > 0, \tag{6.11}$$

with state space $\mathbb{R}^d \times E$, and generators $\mathbb{L}^\varepsilon, \varepsilon > 0$, with domains $\mathcal{D}(\mathbb{L}^\varepsilon)$ dense in $C(\mathbb{R}^d \times E)$. Let $\xi(t), \widehat{x}(t), t \geq 0$, be a Markov process with state space $\mathbb{R}^d \times V$, and generator $\mathbb{L}$ with domain $\mathcal{D}(\mathbb{L})$, and let $\overline{\mathcal{D}(\mathbb{L})}$ be a convergence class. Consider also the test functions

$$\varphi^\varepsilon(u,x) = \varphi(u, v(x)) + \varepsilon\varphi_1(u,x).$$

Suppose that the following conditions are fulfilled:

C1: The family of processes $\xi^\varepsilon(t), t \geq 0, \varepsilon > 0$, satisfies the compact containment condition

$$\lim_{c \to \infty} \sup_{\varepsilon > 0} \mathbb{P}\left(\sup_{0 \leq t \leq T} |\xi^\varepsilon(t)| > c\right) = 0.$$

C2: There exists a collection of functions $\varphi^\varepsilon \in C(\mathbb{R}^d \times E)$, such that, for any $T > 0$, we have

$$\lim_{\varepsilon \to 0} \mathbb{E}\left[\sup_{0 \le t \le T} \left|\varphi^\varepsilon(\xi^\varepsilon(t), v(x^\varepsilon(t/\varepsilon^2))) - \varphi(\xi^\varepsilon(t), v(x^\varepsilon(t/\varepsilon^2)))\right|\right] = 0 \tag{6.12}$$

and

$$\sup_{\varepsilon > 0} \mathbb{E}\left[\|\mathbb{L}^\varepsilon \varphi^\varepsilon\|_{\infty,T}\right] < +\infty, \tag{6.13}$$

where $\|\varphi\|_{\infty,T} = \sup_{0 \le t \le T} |\varphi(\xi(t), v(x(t/\varepsilon^2)))|$.

C3: The convergence in probability of the initial values holds, that is,

$$(\xi^\varepsilon(0), v(x^\varepsilon(0))) \xrightarrow{P} (\xi(0), \widehat{x}(0)), \quad \varepsilon \to 0,$$

with uniformly bounded expectation

$$\sup_{\varepsilon > 0} \mathbb{E}\,|\xi^\varepsilon(0)| \le c < +\infty.$$

Then, the weak convergence

$$(\xi^\varepsilon(t), v(x^\varepsilon(t/\varepsilon^2))) \Longrightarrow (\xi(t), \widehat{x}(t)), \quad \varepsilon \to 0,$$

takes place.

6.3.2 Stochastic Systems with Semi-Markov Switching

The stochastic systems with semi-Markov switching in the series scheme are determined by the embedded Markov renewal process (see Section 3.2)

$$\xi_n^\varepsilon := \xi^\varepsilon(\tau_n^\varepsilon), \quad x_n^\varepsilon := x^\varepsilon(\tau_n^\varepsilon), \quad n \ge 0, \tag{6.14}$$

which can be characterized by the martingale

$$\mu_{n+1}^\varepsilon = \varphi(\xi_{n+1}^\varepsilon, x_{n+1}^\varepsilon) - \sum_{k=1}^n \theta_{k+1}^\varepsilon \mathbb{L}^\varepsilon \varphi(\xi_k^\varepsilon, x_k^\varepsilon), \quad n \ge 0, \tag{6.15}$$

with respect to the filtration $\mathcal{F}_n^\varepsilon := \sigma(\xi_k^\varepsilon, x_k^\varepsilon, \tau_k^\varepsilon; 0 \le k \le n), n \ge 0$.

The compensating operator of the extended Markov renewal process (6.14) in the average approximation scheme is defined by (see Section 3.3)

$$\mathbb{L}^\varepsilon = \varepsilon^{-1} Q + \mathbb{\Gamma}(x) P + \varepsilon \theta_2^\varepsilon(x), \tag{6.16}$$

on the test functions $\varphi(u, x)$ in $C_0^2(\mathbb{R}^d \times E)$.

In the diffusion approximation scheme (see Section 5.4), it is defined by

$$\mathbb{L}^\varepsilon = \varepsilon^{-2}Q + \varepsilon^{-1}\mathbb{\Gamma}(x)P + Q_2(x)P + \varepsilon\theta_3^\varepsilon(x), \tag{6.17}$$

on the test functions $\varphi(u,x)$ in $C_0^3(\mathbb{R}^d \times E)$.

The limit Markov process $\xi(t), t \geq 0$, is supposed to be characterized by the martingale

$$\mu_t = \varphi(\xi(t)) - \int_0^t \mathbb{L}\varphi(\xi(s))ds, \tag{6.18}$$

where the closure $\overline{\mathcal{D}(\mathbb{L})}$ of the domain $\mathcal{D}(\mathbb{L})$ of the generator $\mathbb{L}$ is a convergence-determining class.

Theorem 6.6 *Let the following conditions hold:*

C1: *The family of stochastic processes $\xi^\varepsilon(t), t \geq 0, \varepsilon > 0$, is relatively compact.*

C2: *There exists a family of test functions $\varphi^\varepsilon(u,x)$ in $C_0^3(\mathbb{R}^d \times E)$, such that*

$$\lim_{\varepsilon \to 0} \varphi^\varepsilon(u,x) = \varphi(u), \quad \text{uniformly on } u, x.$$

C3: *The following uniform convergence holds*

$$\lim_{\varepsilon \to 0} \mathbb{L}^\varepsilon \varphi^\varepsilon(u,x) = \mathbb{L}\varphi(u), \quad \text{uniformly on } u, x.$$

The family of functions $\mathbb{L}^\varepsilon \varphi^\varepsilon, \varepsilon > 0$, is uniformly bounded, and $\mathbb{L}^\varepsilon \varphi^\varepsilon$ and $\mathbb{L}\varphi$ belong to $C(\mathbb{R}^d \times E)$.

C4: *The convergence of the initial values holds, that is,*

$$\xi^\varepsilon(0) \xrightarrow{P} \widehat{\xi}(0), \quad \varepsilon \to 0,$$

and

$$\sup_{\varepsilon > 0} \mathbb{E}|\xi^\varepsilon(0)| \leq C < +\infty.$$

Then the weak convergence

$$\xi^\varepsilon(t) \Longrightarrow \xi(t), \quad \varepsilon \to 0, \tag{6.19}$$

takes place. The limit process $\xi(t), t \geq 0$, is characterized by the martingale (6.18).

6.3. PATTERN LIMIT THEOREMS

In the particular case where the martingale is constant $\mu_t \equiv \mu_0 = \text{const.}$, then the limit process $\xi(t), t \geq 0$, is given by the solution of the deterministic equation

$$\varphi(\xi(t)) - \int_0^t \mathbb{L}\varphi(\xi(s))ds = \varphi(\xi(0)), \qquad (6.20)$$

or, in an equivalent form

$$\frac{d}{dt}\varphi(\xi(t)) = \mathbb{L}\varphi(\xi).$$

PROOF. Let us introduce the following random variables

$$\nu^\varepsilon(t) := \max\{n \geq 0 : \tau_n^\varepsilon \leq t\},$$
$$\nu_+^\varepsilon(t) := \nu^\varepsilon(t) + 1, \quad \tau_+^\varepsilon(t) := \tau_{\nu_+^\varepsilon(t)}^\varepsilon, \quad \tau^\varepsilon(t) := \tau_{\nu^\varepsilon(t)}^\varepsilon.$$

Recall that the time-scaled semi-Markov process in the average scheme is considered as $x^\varepsilon(s) := x(s/\varepsilon)$, and in the diffusion approximation scheme as $x^\varepsilon(s) := x(s/\varepsilon^2)$.

Note also that the random variables $\nu_+^\varepsilon(t)$ are stopping times with respect to the filtration

$$\mathcal{F}_t^\varepsilon := \sigma(\xi^\varepsilon(s), x^\varepsilon(s), \tau^\varepsilon(s); 0 \leq s \leq t), \quad t \geq 0. \qquad (6.21)$$

For the proof of theorem we need the following lemma.

In what follows, we consider the *embedded stochastic system* with piecewise trajectories as follows

$$\xi_+^\varepsilon(t) := \xi^\varepsilon(\tau_+^\varepsilon(t)), \quad \xi_\tau^\varepsilon(t) := \xi^\varepsilon(\tau^\varepsilon(t)), \quad t \geq 0.$$

Lemma 6.1 *The process*

$$\zeta_t^\varepsilon(t) = \varphi(\xi_+^\varepsilon(t), x_+^\varepsilon(t)) - \int_0^{\tau_+^\varepsilon(t)} \mathbb{L}^\varepsilon\varphi(\xi_\tau^\varepsilon(s), x^\varepsilon(s))ds, \qquad (6.22)$$

has the martingale property

$$\mathbb{E}[\zeta^\varepsilon(t) - \zeta^\varepsilon(s) \mid \mathcal{F}_s^\varepsilon] = 0, \quad 0 \leq s \leq t \leq T. \qquad (6.23)$$

PROOF. It is worth noticing that

$$\zeta^\varepsilon(t) = \zeta^\varepsilon(\tau^\varepsilon(t)) = \mu_{\nu_+^\varepsilon(t)}^\varepsilon, \quad \text{for} \quad \tau^\varepsilon(t) \leq t < \tau_+^\varepsilon(t),$$

and $\quad \zeta^\varepsilon(\tau_n^\varepsilon) = \mu_{n+1}^\varepsilon, \quad n \geq 0.$

The truncated random variables $\nu^\varepsilon_+(t) \wedge N$, for all positive integer value of $N > 0$, are finite. Hence the following martingale property takes place[174],

$$\mathbb{E}[\mu^\varepsilon_{\nu^\varepsilon_+(t) \wedge N} - \mu^\varepsilon_{\nu^\varepsilon_+(s) \wedge N} \mid \mathcal{F}^\varepsilon_s] = 0. \tag{6.24}$$

Taking the limit in (6.24) for $N \to \infty$, we get that for N large enough, $\nu^\varepsilon_+(t) \wedge N = \nu^\varepsilon_+(t)$. Hence, we obtain

$$\mathbb{E}[\mu^\varepsilon_{\nu^\varepsilon_+(t)} - \mu^\varepsilon_{\nu^\varepsilon_+(s)} \mid \mathcal{F}^\varepsilon_s] = 0. \tag{6.25}$$

Since, by construction, $\zeta^\varepsilon(t) = \zeta^\varepsilon(\tau^\varepsilon_+(t)) = \mu^\varepsilon_{\nu^\varepsilon_+(t)}$, the martingale property (6.23) of the process (6.22) is proved. $\square$

Now, for verifying that the weak convergence (6.19), in Theorem 6.6, holds, we have to estimate the expectation of the following process

$$\mu^\varepsilon_t = \varphi(\xi^\varepsilon(t)) - \int_0^t \mathbb{L}\varphi(\xi^\varepsilon(s))ds,$$

by using the conditions of Theorem 6.6, and the martingale property (6.23) of the piecewise process (6.22).

Let us calculate:

$$\begin{aligned}
\mathbb{E}[\mu^\varepsilon_t] &= \mathbb{E}[\varphi(\xi^\varepsilon(t)) - \int_0^t \mathbb{L}\varphi(\xi^\varepsilon(s))ds] \\
&= \mathbb{E}[\varphi(\xi^\varepsilon(t)) - \varphi^\varepsilon(\xi^\varepsilon(t), x^\varepsilon(t))] \\
&\quad + \mathbb{E}[\varphi^\varepsilon(\xi^\varepsilon(t), x^\varepsilon(t)) - \varphi^\varepsilon(\xi^\varepsilon_+(t), x^\varepsilon_+(t))] \\
&\quad + \mathbb{E}[\varphi^\varepsilon(\xi^\varepsilon_+(t), x^\varepsilon_+(t)) - \int_0^{\tau^\varepsilon_+(t)} \mathbb{L}^\varepsilon\varphi^\varepsilon(\xi^\varepsilon_\tau(s), x^\varepsilon(s))ds] \\
&\quad + \mathbb{E}[\int_0^{\tau^\varepsilon_+(t)} \mathbb{L}^\varepsilon\varphi^\varepsilon(\xi^\varepsilon_\tau(s), x^\varepsilon(s))ds - \int_0^t \mathbb{L}^\varepsilon\varphi^\varepsilon(\xi^\varepsilon_\tau(s), x^\varepsilon(s))ds] \\
&\quad + \mathbb{E}\int_0^t \mathbb{L}^\varepsilon[\varphi^\varepsilon(\xi^\varepsilon_\tau(s), x^\varepsilon(s)) - \varphi^\varepsilon(\xi^\varepsilon(s), x^\varepsilon(s))]ds \\
&\quad + \mathbb{E}\int_0^t [\mathbb{L}^\varepsilon\varphi^\varepsilon(\xi^\varepsilon(s), x^\varepsilon(s)) - \mathbb{L}\varphi(\xi^\varepsilon(s))]ds.
\end{aligned}$$

First note that the third term of the sum is exactly equal to $\zeta^\varepsilon(t)$ (see (6.22)), and the fourth term is equal to

$$\mathbb{E}\int_t^{\tau^\varepsilon_+(t)} \mathbb{L}^\varepsilon\varphi^\varepsilon(\xi^\varepsilon_\tau(s), x^\varepsilon(s))ds \longrightarrow 0, \quad \varepsilon \to 0,$$

thanks to the property of the renewal moments

$$\mathbb{E}[\tau_+^\varepsilon(t) - t] \to 0, \quad \varepsilon \to 0,$$

uniformly on every finite time interval $[0, T]$ (see Appendix C).

The first two terms tend to zero by Condition C2 of Theorem 6.6. The fifth and sixth terms tend to zero by Condition C3 of Theorem 6.6 and finally the dominated convergence theorem allows us to get the limit with expectation.

Now we have to estimate the third term by using the martingale property (6.23) of the process (6.22) as follows:

$$\mathbb{E}\zeta^\varepsilon(t) = \mathbb{E}\zeta^\varepsilon(0) = \mathbb{E}\varphi^\varepsilon(\xi^\varepsilon(0), x)$$
$$= \mathbb{E}\varphi(\xi^\varepsilon(0)) + \mathbb{E}[\varphi^\varepsilon(\xi^\varepsilon(0), x) - \varphi(\xi^\varepsilon(0))].$$

The last term tends to zero by Condition C2 of Theorem 6.2. In conclusion we obtain

$$\mathbb{E}\mu_t^\varepsilon = \mathbb{E}\varphi(\xi^\varepsilon(0)) + \theta^\varepsilon,$$

where the negligible term $\theta^\varepsilon \to 0$, as $\varepsilon \to 0$.

Hence, by Condition C4, we get

$$\mathbb{E}\mu_t = \mathbb{E}\varphi(\xi(0)).$$

Now, by Condition C1 of Theorem 6.6, there exists a sequence $\varepsilon_n \to 0$, such that the weak convergence (6.19) takes place and simultaneously the limit process $\xi(t), t \geq 0$, is characterized by the martingale (6.18). $\square$

6.3.3 Embedded Markov Renewal Processes

We present here an alternative approach to the previous pattern limit theorems based on the continuous time embedded Markov renewal process.

In the *average scheme*, the scaled in continuous time embedded Markov renewal process is considered in the following form

$$\xi_t^\varepsilon := \xi_{[t/\varepsilon]}^\varepsilon, \quad x_t^\varepsilon := x_{[t/\varepsilon]}^\varepsilon, \quad t \geq 0, \qquad (6.26)$$

where $[t]$ means the integer part of the real number $t \geq 0$.

The corresponding coupled random evolution is characterized by the martingale

$$\mu_t^\varepsilon = \varphi(\xi_t^\varepsilon, x_t^\varepsilon) - \varepsilon \sum_{k=0}^{[t/\varepsilon]-1} \mathbb{L}^\varepsilon \varphi(\xi_k^\varepsilon, x_k^\varepsilon), \quad t \geq 0, \qquad (6.27)$$

where the generator of the coupled Markov renewal process (6.26) is defined by (see Section 3.3)

$$\mathbb{L}^\varepsilon = \varepsilon^{-1}Q + \mathbb{\Gamma}(x)P + \varepsilon\theta_1^\varepsilon(x), \qquad (6.28)$$

where

$$Q := P - I, \quad P\varphi(x) := \int_E P(x, dy)\varphi(y). \qquad (6.29)$$

The negligible term is

$$\theta_1^\varepsilon(x) = \mathbb{\Gamma}^{(2)}(x)F_\varepsilon^{(2)}(x)P. \qquad (6.30)$$

In the *diffusion approximation scheme*, the scaled in continuous time embedded Markov renewal process is considered in the following form

$$\xi_t^\varepsilon := \xi_{[t/\varepsilon^2]}^\varepsilon, \quad x_t^\varepsilon := x_{[t/\varepsilon^2]}^\varepsilon, \quad t \geq 0. \qquad (6.31)$$

The corresponding coupled random evolution is characterized by the martingale

$$\mu_t^\varepsilon = \varphi(\xi_t^\varepsilon, x_t^\varepsilon) - \varepsilon^2 \sum_{k=0}^{[t/\varepsilon^2]-1} \mathbb{L}^\varepsilon \varphi(\xi_k^\varepsilon, x_k^\varepsilon), \quad t \geq 0, \qquad (6.32)$$

where the generator of the coupled Markov renewal process (6.31) is represented in the following asymptotic form (see Section 5.4)

$$\mathbb{L}^\varepsilon = \varepsilon^{-2}Q + \varepsilon^{-1}\mathbb{\Gamma}(x)P + Q_2(x)P + \varepsilon\theta_3^\varepsilon(x). \qquad (6.33)$$

The limit Markov process $\xi_t, t \geq 0$, is supposed to be characterized by the martingale

$$\mu_t = \varphi(\xi_t) - \int_0^t \mathbb{L}\varphi(\xi_s)ds, \quad t \geq 0, \qquad (6.34)$$

with the generator $\mathbb{L}$ whose closure $\overline{\mathcal{D}(\mathbb{L})}$ of domain $\mathcal{D}(\mathbb{L})$ is a convergence-determining class.

Theorem 6.7 *Let the following conditions hold:*

6.3. PATTERN LIMIT THEOREMS

C1: The family of embedded Markov renewal processes $\xi_t^\varepsilon, x_t^\varepsilon, t \geq 0, \varepsilon > 0$, is relatively compact.

C2: There exists a family of test functions $\varphi^\varepsilon(u,x)$ in $C_0^\infty(\mathbb{R}^d \times E)$, such that

$$\lim_{\varepsilon \to 0} \varphi^\varepsilon(u,x) = \varphi(u), \quad \text{uniformly on } u, x.$$

C3: The following convergence holds

$$\lim_{\varepsilon \to 0} \mathbb{L}^\varepsilon \varphi^\varepsilon(u,x) = \mathbb{L}\varphi(u), \quad \text{uniformly on } u, x.$$

The family of functions $\mathbb{L}^\varepsilon \varphi^\varepsilon, \varepsilon > 0$, is uniformly bounded, and $\mathbb{L}^\varepsilon \varphi^\varepsilon$ and $\mathbb{L}\varphi$ belong to $C(\mathbb{R}^d \times E)$.

C4: The convergence of the initial values holds, that is,

$$\xi_0^\varepsilon \xrightarrow{P} \xi_0, \quad \varepsilon \to 0,$$

and

$$\sup_{\varepsilon > 0} \mathbb{E}|\xi_0^\varepsilon| \leq C < +\infty.$$

Then the weak convergence

$$\xi_t^\varepsilon \Longrightarrow \xi_t, \quad \varepsilon \to 0, \tag{6.35}$$

takes place. The limit process process $\xi_t, t \geq 0$ is characterized by the martingale (6.34).

In the particular case, where the martingale is constant $\mu_t \equiv \mu_0 = \text{const.}$, the limit process is determined by a solution of the deterministic equation

$$\varphi(\xi_t) - \int_0^t \mathbb{L}\varphi(\xi_s)ds = \mathbb{E}\xi_0,$$

or, in equivalent form

$$\frac{d}{dt}\varphi(\xi_t) = \mathbb{L}\varphi(\xi_t).$$

PROOF. In order to verify the weak convergence (6.35), we have to estimate the expectation of the following process

$$\zeta_t^\varepsilon = \varphi(\xi_t^\varepsilon) - \int_0^t \mathbb{L}\varphi(\xi_s^\varepsilon)ds, \quad t \geq 0, \tag{6.36}$$

by using the conditions of Theorem 6.7, and the martingale property of the piecewise process (6.27).

Let us calculate:

$$\mathbb{E}\zeta_t^\varepsilon = \mathbb{E}[\varphi(\xi_t^\varepsilon) - \int_0^t \mathbb{L}\varphi(\xi_s^\varepsilon)ds]$$
$$= \mathbb{E}[\varphi(\xi_t^\varepsilon) - \varphi^\varepsilon(\xi_t^\varepsilon, x_t^\varepsilon)]$$
$$+ \mathbb{E}[\varphi^\varepsilon(\xi_t^\varepsilon, x_t^\varepsilon) - \varepsilon \sum_{k=0}^{[t/\varepsilon]-1} \mathbb{L}^\varepsilon \varphi^\varepsilon(\xi_k^\varepsilon, x_k^\varepsilon)]$$
$$+ \mathbb{E}\varepsilon \sum_{k=0}^{[t/\varepsilon]-1} [\mathbb{L}^\varepsilon \varphi^\varepsilon(\xi_k^\varepsilon, x_k^\varepsilon) - \mathbb{L}\varphi(\xi_k^\varepsilon)]$$
$$+ \mathbb{E}\left[\varepsilon \sum_{k=0}^{[t/\varepsilon]-1} \mathbb{L}\varphi(\xi_k^\varepsilon) - \int_0^t \mathbb{L}\varphi(\xi_s^\varepsilon)ds\right].$$

Due to Condition C2 of theorem, the first and third terms of the sum tend to zero as $\varepsilon \to 0$. The forth term also tends to zero, since in the square brackets we have the difference between the integral and the corresponding integral sum. The second term is exactly the martingale (6.27).

In conclusion we obtain:

$$\mathbb{E}\zeta_t^\varepsilon = \mathbb{E}\mu_t^\varepsilon + \theta^\varepsilon$$
$$= \mathbb{E}\mu_0^\varepsilon + \theta^\varepsilon$$
$$= \mathbb{E}\varphi^\varepsilon(\xi_0^\varepsilon, x) + \theta^\varepsilon$$
$$= \mathbb{E}\varphi(\xi_0) + \mathbb{E}[\varphi^\varepsilon(\xi_0^\varepsilon, x) - \varphi(\xi_0)] + \theta^\varepsilon.$$

Finally,

$$\mathbb{E}\zeta_t^\varepsilon = \mathbb{E}\varphi(\xi_0) + \theta^\varepsilon,$$

with the negligible term

$$\theta^\varepsilon \to 0, \quad \varepsilon \to 0.$$

Hence, the limit process in (6.36),

$$\zeta_t^\varepsilon \Longrightarrow \mu_t, \quad \varepsilon \to 0,$$

is the martingale (6.34). $\square$

6.4 Relative Compactness

In this sections the relative compactness of the family of stochastic systems $\xi^\varepsilon(t), t \geq 0, \varepsilon > 0$, is realized by using the Stroock and Varadhan criteria formulated in Theorem 6.2.

6.4.1 Stochastic Systems with Markov Switching

The stochastic systems with Markov switching in the series scheme with the small parameter $\varepsilon > 0, \varepsilon \to 0$, is characterized by the martingale

$$\mu_t^\varepsilon = \varphi(\xi^\varepsilon(t), x^\varepsilon(t)) - \int_0^t \mathbb{L}^\varepsilon \varphi(\xi^\varepsilon(s), x^\varepsilon(s))ds, \tag{6.37}$$

where the generators $\mathbb{L}^\varepsilon, \varepsilon > 0$, have the common domain of definition $\mathcal{D}(\mathbb{L})$, is supposed to be dense in $C_0^2(\mathbb{R}^d \times E)$.

Lemma 6.2 *Let the generators $\mathbb{L}^\varepsilon, \varepsilon > 0$, have the following estimation*

$$|\mathbb{L}^\varepsilon \varphi(u)| \leq C_\varphi, \tag{6.38}$$

for any real-valued nonnegative function $\varphi \in C_0^2(\mathbb{R}^d)$, where the constant C_φ depends on the norm of φ, but not on $\varepsilon > 0$, nor on shifts of φ [45]. *Suppose that the compact containment condition holds*

$$\lim_{l \to \infty} \sup_{\varepsilon > 0} \mathbb{P}\left(\sup_{0 \leq t \leq T} |\xi^\varepsilon(t)| > l \right) = 0. \tag{6.39}$$

Then the family of stochastic processes $\xi^\varepsilon(t), t \geq 0, \varepsilon > 0$, is relatively compact.

PROOF. Let us consider the process

$$\eta^\varepsilon(t) := \varphi(\xi^\varepsilon(t)) + C_\varphi t, \quad t \geq 0,$$

and prove that it is an $\mathcal{F}_t^\varepsilon$-nonnegative submartingale. To see this, let us calculate, by using the martingale characterization (6.37), for $0 \leq s \leq t$:

$$\mathbb{E}[\eta^\varepsilon(t) \mid \mathcal{F}_s^\varepsilon] = \mathbb{E}[\varphi(\xi^\varepsilon(t)) \mid \mathcal{F}_s^\varepsilon] + C_\varphi s$$

$$= \mathbb{E}[\int_0^s \mathbb{L}^\varepsilon \varphi(\xi^\varepsilon(u))du \mid \mathcal{F}_s^\varepsilon] + C_\varphi + \mu_s^\varepsilon$$

$$+ \mathbb{E}[\int_s^t (\mathbb{L}^\varepsilon \varphi(\xi^\varepsilon(u)) + C_\varphi)du \mid \mathcal{F}_s^\varepsilon].$$

So, the following equality takes place

$$\mathbb{E}[\eta^\varepsilon(t) \mid \mathcal{F}_s^\varepsilon] = \eta^\varepsilon(s) + \mathbb{E}[\int_s^t (\mathbb{L}^\varepsilon \varphi(\xi^\varepsilon(u)) + C_\varphi) du \mid \mathcal{F}_s^\varepsilon],$$

where the last term is nonnegative due to the estimation (6.54). Hence

$$\mathbb{E}[\eta^\varepsilon(t) \mid \mathcal{F}_s^\varepsilon] \geq \eta^\varepsilon(s), \quad \text{for } s < t.$$

Now, we can see that both hypotheses of Theorem 6.2 are valid. □
We need the following lemma for the proof of Lemma 6.4 below.

Lemma 6.3 *(Lemma 3.2, p. 174, in [45]) Let $x(t)$, $t \geq 0$, be a Markov process defined by the generator $\mathbb{L}$, and $\mathcal{G}_t \supset \mathcal{F}_t^x$. Then for any fixed $\lambda \in \mathbb{R}$ and $\varphi \in \mathcal{D}(\mathbb{L})$*

$$e^{-\lambda t} \varphi(x(t)) + \int_0^t e^{-\lambda s}[\lambda \varphi(x(s)) - \mathbb{L}\varphi(x(s))]ds$$

is a $\mathcal{G}_t$-martingale.

Lemma 6.4 *Let the generators $\mathbb{L}^\varepsilon, \varepsilon > 0$, have the following estimation for $\varphi_0(u) = \sqrt{1+u^2}$,*

$$\mathbb{L}^\varepsilon \varphi_0(u) \leq C_l \varphi_0(u), \quad |u| \leq l,$$

where the constant C_l depends on the function φ_0, but not on $\varepsilon > 0$, and

$$\mathbb{E}\left[|\xi^\varepsilon(0)|\right] \leq b < +\infty. \tag{6.40}$$

Then the compact containment condition holds

$$\lim_{\ell \to \infty} \sup_{\varepsilon > 0} \mathbb{P}^\varepsilon \left(\sup_{0 \leq t \leq T} |\xi^\varepsilon(t)| \geq \ell \right) = 0. \tag{6.41}$$

PROOF. Since $\varphi_0'(u) = u/\sqrt{1+u^2}$, and $\varphi''(u) = (1+u^2)^{-3/2}$, and $\varphi_0(u) \leq 1 + |u|$, we have

$$|\varphi_0'(u)| \leq 1 \leq \varphi_0(u), \quad |\varphi_0''(u)| \leq 1 \leq \varphi_0(u), \quad u \in \mathbb{R}.$$

Let us define the stopping time τ_ℓ^ε, by

$$\tau_\ell^\varepsilon = \begin{cases} \inf\{t \in [0,T] : |\xi^\varepsilon(t)| \geq \ell\} \\ T \end{cases} \quad \text{if } |\xi^\varepsilon(t)| \leq \ell, \text{for all } t \in [0,T]. \tag{6.42}$$

6.4. RELATIVE COMPACTNESS

By Lemma 6.3, applied to a stopping time instead of a fixed time t, we have that

$$e^{-C_l t \wedge \tau_\ell^\varepsilon} \varphi_0(\xi^\varepsilon(t \wedge \tau_\ell^\varepsilon)) + \int_0^{t \wedge \tau_\ell^\varepsilon} e^{-C_l s} \left[C_l \varphi_0(\xi^\varepsilon(s)) - \mathbb{L}^\varepsilon \varphi_0(\xi^\varepsilon(s)) \right] ds, \tag{6.43}$$

is a martingale. We get, for $s \leq t \wedge \tau_\ell^\varepsilon$,

$$C_l \varphi_0(\xi^\varepsilon(s)) - \mathbb{L}^\varepsilon \varphi_0(\xi^\varepsilon(s)) \geq 0, \tag{6.44}$$

and from (6.43) we obtain

$$\mathbb{E}\left[e^{-C_l t \wedge \tau_\ell^\varepsilon} \varphi_0(\xi^\varepsilon(t \wedge \tau_\ell^\varepsilon)) \right] \leq \mathbb{E}\mu_t^\varepsilon = \mathbb{E}\mu_0^\varepsilon = \mathbb{E}\varphi_0(\xi^\varepsilon(0)). \tag{6.45}$$

The convexity of φ_0 and the inequality $\varphi_0(u) \geq 1$, together provide the estimation:

$$p_\ell^\varepsilon := \mathbb{P}^\varepsilon \left(\sup_{0 \leq t \leq T} |\xi^\varepsilon(t)| \geq \ell \right)$$
$$\leq \mathbb{P}^\varepsilon \left(\varphi_0(\xi^\varepsilon(\tau_\ell^\varepsilon)) \geq \varphi_0(\ell) \right),$$

and Chebichev's inequality yields

$$p_\ell^\varepsilon \leq \mathbb{E}\left[\varphi_0(\xi^\varepsilon(\tau_\ell^\varepsilon)) \right] / \varphi_0(\ell). \tag{6.46}$$

From inequality (6.46), together with inequality $e^{-\tau_\ell^\varepsilon} \geq e^{-T}$ (since $\tau_\ell^\varepsilon \leq T$), we obtain

$$p_\ell^\varepsilon \leq e^{C_l T} \mathbb{E}\left[e^{-C_l \tau_\ell^\varepsilon} \varphi_0(\xi^\varepsilon(\tau_\ell^\varepsilon)) \right] / \varphi_0(\ell), \tag{6.47}$$

and from (6.45) we get

$$p_\ell^\varepsilon \leq e^{C_l T} \mathbb{E}\left[\varphi_0(\xi^\varepsilon(0)) \right] / \varphi_0(\ell). \tag{6.48}$$

Now, by the inequalities $\sqrt{1+u^2} \leq 1 + |u|$ and (6.40), we get

$$p_\ell^\varepsilon \leq e^{C_l T}(b+1)/\varphi_0(\ell) \longrightarrow 0, \quad \ell \to \infty. \tag{6.49}$$

$\square$

Corollary 6.1 *Let the generators* $\mathbb{L}^\varepsilon, \varepsilon > 0$, *have the following estimation*

$$|\mathbb{L}^\varepsilon \varphi(u)| \leq C_\varphi, \tag{6.50}$$

for any real-valued nonnegative function $\varphi \in C_0^2(\mathbb{R}^d)$, where the constant C_φ depends on the norm of φ, and for $\varphi_0(u) = \sqrt{1+u^2}$,

$$\mathbb{L}^\varepsilon \varphi_0(u) \leq C_l \varphi_0(u), \quad |u| \leq l,$$

where the constant C_l depends on the function φ_0, but not on $\varepsilon > 0$.
Then the family of processes $\xi^\varepsilon(t), t \geq 0, \varepsilon > 0$, is relatively compact.

6.4.2 Stochastic Systems with Semi-Markov Switching

The stochastic systems with semi-Markov switching in the series scheme with the small series parameter $\varepsilon > 0, \varepsilon \to 0$, is characterized by the process (see Lemma 6.1)

$$\zeta^\varepsilon(t) = \varphi(\xi_+^\varepsilon(t), x^\varepsilon(t)) - \int_0^{\tau_+^\varepsilon(t)} \mathbb{L}^\varepsilon \varphi(\xi_\tau^\varepsilon(s), x^\varepsilon(s))ds, \qquad (6.51)$$

where the compensating operators $\mathbb{L}^\varepsilon, \varepsilon > 0$, have the following asymptotic representation (see Proposition 3.1)

$$\mathbb{L}^\varepsilon \varphi(u, x) = \varepsilon^{-1} Q\varphi + \mathbb{\Gamma}(x) P\varphi + \varepsilon \theta_2^\varepsilon(x)\varphi, \qquad (6.52)$$

on the test functions $\varphi(u, x)$ in $C_0^2(\mathbb{R}^d \times E)$.

The remaining term is

$$\theta_2^\varepsilon(x) = \mathbb{\Gamma}^2(x) F_\varepsilon^{(2)}(x) Q_0.$$

The process $\zeta^\varepsilon(t)$ in (6.51) has the martingale property (see Lemma 6.1).

The relative compactness of the family of the stochastic systems $\xi^\varepsilon(t), t \geq 0, \varepsilon > 0$, is realized by using the Stroock and Varadhan criteria, formulated in Theorem 6.2.

Lemma 6.5 *Let the compensating operators $\mathbb{L}^\varepsilon, \varepsilon > 0$, have the following estimation*

$$|\mathbb{L}^\varepsilon \varphi(u)| \leq C_\varphi, \qquad (6.53)$$

for any real-valued nonnegative functions $\varphi(u)$ in $C_0^\infty(\mathbb{R}^d)$, where the constant C_φ depends only on the norm of φ, but not on ε nor on shifts of φ[45].

Let the compact containment condition (6.39) holds, and

$$\mathbb{E}|\xi^\varepsilon(0)| \leq c < +\infty.$$

Then the family of the stochastic systems $\xi_+^\varepsilon(t), t \geq 0, \varepsilon > 0$, is relatively compact.

PROOF. The proof of this lemma is similar as the proof of Lemma 6.2. □

Corollary 6.2 *Let the compensating operator (6.52) has the following estimation*

$$|\mathbb{L}^\varepsilon \varphi(u)| \leq C_\varphi, \qquad (6.54)$$

for any real-valued nonnegative function $\varphi \in C_0^3(\mathbb{R}^d)$, where the constant C_φ depends only on the norm of φ, and for $\varphi_0(u) = \sqrt{1+u^2}$,

$$\mathbb{L}^\varepsilon \varphi_0(u) \leq C_l \varphi_0(u), \quad |u| \leq l,$$

where the constant C_l depends on the function φ_0, but not on $\varepsilon > 0$. Then the family of processes $\xi^\varepsilon(t), t \geq 0, \varepsilon > 0$, is relatively compact.

6.4.3 Compact Containment Condition

The relative compactness of the family of stochastic processes $\xi^\varepsilon(t)$, $t \geq 0, \varepsilon > 0$, with simple paths in the space $\mathbf{D}_{\mathbb{R}^d}[0, \infty)$, is obtained provided that the *compact containment condition* holds [45]

$$\lim_{l \to \infty} \sup_{\varepsilon > 0} \mathbb{P}\left(\sup_{0 \leq t \leq T} |\xi^\varepsilon(t)| > l\right) = 0, \qquad (6.55)$$

together with the additional condition that $\varphi(\xi^\varepsilon(t))$ is relatively compact for each test function $\varphi(u)$ in a dense set, say H, in $C(E)$, in the topology of uniform convergence on compact set.

The unique solution of the martingale problem for the limit generator of Markov process together with Condition (6.55) provides the weak convergence of the processes (see Theorem 9.1, Ch. 3 in [45]).

Lemma 6.6 *The family of processes $\xi^\varepsilon(t)$, $t \geq 0$, $0 < \varepsilon \leq \varepsilon_0$, characterized by the compensating operator (6.52), with bounded initial value $\mathbb{E}|\xi^\varepsilon(0)| \leq b < \infty$, satisfies the compact containment condition (see [45])*

$$\lim_{\ell \to \infty} \sup_{0 < \varepsilon \leq \varepsilon_0} \mathbb{P}\left(\sup_{0 \leq t \leq T} |\xi^\varepsilon(t)| \geq \ell\right) = 0. \qquad (6.56)$$

PROOF. We will use the function $\varphi_0(u) = \sqrt{1+u^2}$. The asymptotic representation (6.52) for the compensating operator and the following properties of $\varphi_0(u)$:

$$|\varphi_0'(u)| \leq 1 \leq \varphi_0(u), \quad |\varphi_0''(u)| \leq \varphi_0(u), \qquad (6.57)$$

yield the inequality

$$\mathbb{L}^\varepsilon \varphi_0(u) \leq C_\ell \varphi_0(u), \quad \text{for} \quad |u| \leq \ell.$$

Let us use now the process $\zeta_c^\varepsilon(t)$, $t \geq 0$, defined by (see Lemma 7.8)

$$\zeta_c^\varepsilon(t) = e^{-c\tau_+^\varepsilon(t)} \varphi(\xi^\varepsilon(\tau_+^\varepsilon(t)), x^\varepsilon(\tau_+^\varepsilon(t)), x^\varepsilon(\tau_+^\varepsilon(t)), \tau_+^\varepsilon(t) + J^\varepsilon(t)).$$

First, the following inequality is satisfied

$$J^\varepsilon(t) := \int_0^{\tau_+^\varepsilon(t)} \left[e^{-cs} c\varphi_0(\xi^\varepsilon(\tau_+^\varepsilon(t))) - e^{-c\tau^\varepsilon(s)} \mathbb{L}\varphi_0(\xi^\varepsilon(\tau^\varepsilon(s))) \right] ds \geq 0, \qquad (6.58)$$

for any large enough $c > 0$.

Using the martingale property of the process $\zeta_c^\varepsilon(t)$ and inequality (6.58), we obtain the following inequality

$$\mathbb{E} e^{-c\tau_+^\varepsilon(T)} \varphi_0(\xi^\varepsilon(\tau_+^\varepsilon(T))) \leq \mathbb{E}\varphi_0(\xi^\varepsilon(0)). \qquad (6.59)$$

The left hand side in (6.59) is estimated as follows:

$$\Psi^\varepsilon(T) := \mathbb{E} e^{-c\tau_+^\varepsilon(T)} \varphi_0(\xi^\varepsilon(\tau_+^\varepsilon(T))) \qquad (6.60)$$
$$= e^{-cT} \mathbb{E} e^{-c\gamma^\varepsilon(T)} \varphi_0(\xi^\varepsilon(\tau_+^\varepsilon(T)))$$
$$= e^{-cT} \mathbb{E} e^{-c\gamma^\varepsilon(T)} \varphi_0(\xi^\varepsilon(T))$$
$$+ e^{-cT} \mathbb{E} e^{-c\gamma^\varepsilon(T)} [\varphi_0(\xi^\varepsilon(\tau_+^\varepsilon(T))) - \varphi_0(\xi^\varepsilon(T)))].$$

The second term is estimated using property (6.57):

$$\Psi_2^\varepsilon(T) := e^{-cT} \mathbb{E} e^{-c\gamma^\varepsilon(T)} (\varphi_0(\xi^\varepsilon(\tau_+^\varepsilon(T))) - \varphi_0(\xi^\varepsilon(T)))]$$
$$\leq e^{-cT} \mathbb{E} e^{-c\gamma^\varepsilon(T)} b\gamma^\varepsilon(T) \varphi_0(\xi^\varepsilon(T))]. \qquad (6.61)$$

Hence,

$$\Psi_2^\varepsilon(T) \leq e^{-cT} \mathbb{E} \left[\varphi_0(\xi^\varepsilon(T)) \mathbb{E}[e^{-c\gamma^\varepsilon(T)} b\gamma^\varepsilon(T) \mid \mathcal{F}_T^\varepsilon] \right].$$

We will use below the following property of random sojourn times $\gamma^\varepsilon(T)$ (see Appendix C), for all $\delta > 0$,

$$\mathbb{P}_x \left(\max_{0 \leq t \leq T} \gamma^\varepsilon(T) \geq \delta \right) \longrightarrow 0, \quad \varepsilon \to 0. \qquad (6.62)$$

6.4. RELATIVE COMPACTNESS

Note that the function $d(s) = bse^{-cs}$ is bounded in $s \in \mathbb{R}_+$

$$0 \le d(s) \le \ell < +\infty. \tag{6.63}$$

Let us estimate:

$$\begin{aligned}\mathbb{E}[d(\gamma^\varepsilon(T)) \mid \mathcal{F}_T^\varepsilon] &= \mathbb{E}[d(\gamma^\varepsilon(T))[I(\gamma^\varepsilon(T) \ge \delta) + I(\gamma^\varepsilon(T) < \delta)] \mid \mathcal{F}_T^\varepsilon] \\ &\le \ell \mathbb{P}(\gamma^\varepsilon(T) \ge \delta_{\varepsilon_0}) + d(\delta_{\varepsilon_0}) \\ &= \gamma_{\varepsilon_0} \to 0, \quad \varepsilon_0 \to 0.\end{aligned} \tag{6.64}$$

Similarly, using property (6.62), we estimate:

$$\begin{aligned}\mathbb{E}[e^{-c\gamma^\varepsilon(T)} \mid \mathcal{F}_T^\varepsilon] &= \mathbb{E}\left[e^{-c\gamma^\varepsilon(T)}[I(\gamma^\varepsilon(T) \ge \delta) + I(\gamma^\varepsilon(T) < \delta)] \mid \mathcal{F}_T^\varepsilon\right] \\ &\ge e^{-c\delta}\mathbb{P}(\gamma^\varepsilon(T) < \delta) \\ &= e^{-c\delta}[1 - \mathbb{P}(\gamma^\varepsilon(T) \ge \delta)].\end{aligned} \tag{6.65}$$

By (6.62), we have

$$\mathbb{E}[e^{-c\gamma^\varepsilon(T)} \mid \mathcal{F}_T^\varepsilon] \ge h > 0, \quad \text{for} \quad 0 < \varepsilon \le \varepsilon_0.$$

Inequality (6.59) can be now transformed into the following

$$he^{-cT}\mathbb{E}\varphi_0(\xi^\varepsilon(T)) \le \mathbb{E}\varphi_0(\xi^\varepsilon(0)). \tag{6.66}$$

The convexity of $\varphi_0(u) = \sqrt{1+u^2}$, and the inequality $\varphi_0(u) \ge 1$, provide the estimation:

$$\begin{aligned}p_l^\varepsilon &= \mathbb{P}(\sup_{0 \le t \le T} |\xi^\varepsilon(t)| \ge l) \\ &\le \mathbb{P}(\varphi_0(\xi^\varepsilon(\tau_l^\varepsilon)) \ge \varphi_0(l)),\end{aligned}$$

and Chebishev's inequality yields

$$p_l^\varepsilon \le \mathbb{E}\varphi_0(\xi^\varepsilon(\tau_l^\varepsilon))/\varphi_0(l).$$

Now, Inequality (6.66) is used:

$$\begin{aligned}p_l^\varepsilon &\le he^{cT}\mathbb{E}\varphi_0(\xi^\varepsilon(0))/\varphi_0(l) \\ &\le he^{cT}\mathbb{E}|\xi^\varepsilon(0)|/\varphi_0(l) \\ &\le he^{cT}(b+1)/\varphi_0(l) \to 0, \quad l \to \infty.\end{aligned}$$

□

6.5 Verification of Convergence

The verification of convergence of stochastic systems with semi-Markov switching in the average merging scheme (see Theorem 3.1) is based on the determination of the pattern limit Theorem 6.6 conditions, by using the explicit representation of the solution of the singular perturbation problem given in Proposition 5.7.

First, the explicit representation of the remaining term in (5.65)-(5.67) and Condition A2 in Section 3.3.1 (Theorem 3.1) provide that Condition (6.53) in Lemma 6.5 is valid.

Next, the compact containment condition (6.56) is realized by Lemma 6.6. So, Condition C1 of Theorem 6.6 is true. Conditions C2 and C3 also are true from the same explicit representation (5.65)-(5.67) of the limit generator and the remaining term and, of course, Conditions A2 and A3 in Section 3.3.1 (Theorem 3.1).

The characterization of the limit process in (6.20) with the limit generator $\mathbb{L} = \widehat{\Gamma}$, $\widehat{\Gamma}\varphi(u) = \widehat{g}(u)\varphi'(u)$, (see (5.59)), due to Condition C4, completes the proof of Theorem 3.1.

Verification of weak convergence of stochastic additive functionals (3.56) in Theorem 3.4 is achieved following an analogous scheme to Theorem 3.1. First, Conditions C2 and C3 of Theorem 6.6 are obtained by using asymptotic representation (3.10) in Proposition 3.2. Next, compact containment condition (6.39) is realized by Lemma 6.6. Condition (6.53) in Lemma 6.5 is proved for the perturbed test functions $\varphi^\varepsilon(u, x) = \varphi(u) + \varepsilon\varphi_1(u, x)$, such that

$$|\mathbb{L}^\varepsilon \varphi^\varepsilon(u, x)| \leq C_\varphi,$$

where the constant C_φ depends only on $\varphi(u)$, but not on ε nor on shifts of φ.

The characterization of the limit generator $\mathbb{L}$, in Proposition 5.8, completes the proof of Theorem 3.4.

The proof of Theorem 3.3 can be obtained as a particular case of that of Theorem 3.4.

The verification of weak convergence of Theorems 3.2 and 3.5 is obtained similarly by using Propositions 3.4 and 5.9 respectively.

6.5. VERIFICATION OF CONVERGENCE

The convergence in distribution of the coupled Markov process $\zeta^\varepsilon(t), \widehat{\xi}(t), t \geq 0, \varepsilon > 0$, is made by the Pattern Limit Theorem 6.3. For the weak convergence, we propose to the interested reader to calculate the square characteristic of the martingale characterization of the coupled Markov process $\zeta^\varepsilon(t), \widehat{\xi}(t), t \geq 0$, and verify the relative compactness of the family $\zeta^\varepsilon(t), t \geq 0, \varepsilon > 0$, as $\varepsilon \to 0$.

Theorem 3.6 can be considered as a particular case of Theorem 3.7.

The weak convergence in Theorems 4.7–4.11 is based on the solutions of the singular perturbation problems given in Propositions 5.14–5.17 and on the Pattern Limit Theorem 6.4 in average merging scheme and Theorem 6.5 in diffusion approximation scheme. The switching semi-Markov processes is considered in Theorem 6.6. The verification of relative compactness is made by using the estimations of generator (or compensating operator) on the test functions given in Lemmas 6.2 and 6.4. The relative compactness of the processes on the series scheme is shown by using Stroock-Varadhan approach given in Theorem 6.2.

Chapter 7

Poisson Approximation

7.1 Introduction

The Poisson approximation merging scheme is represented here for two kinds of stochastic systems: *impulsive processes* with Markov switching (Sections 7.2.1 and 7.2.2) and *stochastic additive functionals* with semi-Markov switching (Section 7.2.3). The average and diffusion approximation merging principles are constructed for stochastic systems in the series scheme with the small series parameter $\varepsilon \to 0$, $(\varepsilon > 0)$ normalizing the values of jumps. In the Poisson approximation scheme, the jump values of the stochastic system are split into two parts: a small jump taking values with probabilities close to one and a big jump taken values with probabilities tending to zero together with the series parameter $\varepsilon \to 0$.

So, in the Poisson merging principle the probabilities (or intensities) of jumps are normalized by the series parameter ε. The main assumption in the Poisson merging principle is the asymptotic representation of the probability measure on the *measure-determining class* of functions $\varphi \in C_3(\mathbb{R})$, which are real-valued, bounded, and such that [70] $\varphi(u)/u^2 \to 0$, $|u| \to 0$, (see Appendix B).

The techniques of proofs developed here are quite different from those used in the previous chapters for diffusion and average approximations. The proofs of theorems in the present chapter make use of semimartingale theory. Theorems 7.1 and 7.2 concern impulsive process, with and without state space merging of switching Markov process. Theorem 7.3 concerns additive functionals with semi-Markov switching. The main framework of proofs is that of Theorems VIII.2.18, and IX.3.27 in [70] (see Appendix B, Theorems B.1 and B.2). But the main point here is to prove convergence of predictable characteristics of semimartingales which are integral functionals of switching Markov processes. This is done by techniques given in Chapters 5 and 6.

The Poisson merging principle is constructed similarly to the average approximation principle (see Section 5.4) with some special devices. As usual, there are four different schemes: the continuous and jump random evolutions considered with Markov and semi-Markov switching.

The associated continuous random evolution in the Poisson approximation scheme is given by the family of generators $\Gamma_\varepsilon(x), x \in E$, which defines the switched Markov processes with locally independent increments $\eta^\varepsilon(t;x), t \geq 0, x \in E$, with values in $\mathbb{R}^d, d \geq 1$, and the switching Markov renewal process $x_n^\varepsilon, \tau_n^\varepsilon, n \geq 0$, which determines the states $x_n^\varepsilon \in E$, and the renewal times by the transition probabilities given by the Markov kernel.

The starting point of construction of the Poisson approximation principle is the compensating operator of the continuous random evolution in series scheme (see Section 5.3). For reasons of easier understanding by the reader, we consider, in the first part, Markov switching, but the same approach can be used for semi-Markov switching when we replace the generator by the compensating operator.

7.2 Stochastic Systems in Poisson Approximation Scheme

7.2.1 *Impulsive Processes with Markov Switching*

Let $x(t)$, $t \geq 0$, be a Markov jump process on a standard state space $(E, \mathcal{E})$ defined by the generator

$$Q\varphi(x) = q(x) \int_E P(x, dy)[\varphi(y) - \varphi(x)]. \qquad (7.1)$$

The semi-Markov kernel

$$Q(x, B, t) = P(x, B)(1 - e^{-q(x)t}), \quad x \in E, \ B \in \mathcal{E}, \ t \geq 0, \qquad (7.2)$$

defines the associated Markov renewal process $x_k, \tau_k, k \geq 0$, where $x_k, k \geq 0$, is the embedded Markov chain defined by the stochastic kernel

$$P(x, B) = \mathbb{P}(x_{k+1} \in B \mid x_k = x),$$

and $\tau_k, k \geq 0$, is the point process of jump times defined by the distribution function of sojourn times $\theta_{k+1} = \tau_{k+1} - \tau_k, \ k \geq 0$,

$$\mathbb{P}(\theta_{k+1} \leq x \mid x_k = x) = 1 - e^{-q(x)t}.$$

7.2. POISSON APPROXIMATION SCHEME

We suppose that the Markov process $x(t), t \geq 0$, is uniformly ergodic with stationary distribution $\pi(B)$, $B \in \mathcal{E}$. Thus the embedded Markov chain x_k, $k \geq 0$, is uniformly ergodic too, with stationary distribution $\rho(B)$, $B \in \mathcal{E}$, connected by the following relations

$$\pi(dx)q(x) = q\rho(dx), \quad q := \int_E \pi(dx)q(x). \tag{7.3}$$

In the sequel we will suppose that

$$0 < q_0 \leq q(x) \leq q_1 < +\infty, \quad x \in E. \tag{7.4}$$

The *impulsive process* with Markov switching is defined by

$$\xi^\varepsilon(t) := \xi^\varepsilon(0) + \sum_{k=1}^{\nu(t/\varepsilon)} \alpha_k^\varepsilon(x_k), \quad t \geq 0, \tag{7.5}$$

where $\nu(t) = \max\{k : \tau_k \leq t\}$ is the counting process of jumps. The family of random variables $\alpha_k^\varepsilon(x)$, $k \geq 1$, $x \in E$, is considered in the series scheme with a small series parameter $\varepsilon > 0$, and is defined by the following distribution functions on the real line $\mathbb{R}$

$$\Phi_x^\varepsilon(u) = \mathbb{P}(\alpha_k^\varepsilon(x) \leq u), \quad u \in \mathbb{R}, \ x \in E. \tag{7.6}$$

Analogous results can be obtained for the impulsive processes in $\mathbb{R}^d$, $d \geq 1$.

In the sequel, we will suppose that for any fixed sequence (z_k) in E, the sequence $\alpha_k^\varepsilon(z_k), k \geq 1$, is constituted of independent random variables.

Let the following conditions hold.

A1: The switching jump Markov process $x(t)$, $t \geq 0$, is uniformly ergodic with the stationary distributions (7.3).

A2: The family of random variables $\alpha_k^\varepsilon(x), k \geq 1, x \in E$, is uniformly square integrable, that is,

$$\sup_{\varepsilon > 0} \sup_{x \in E} \int_{|u|>c} u^2 \Phi_x^\varepsilon(du) \longrightarrow 0, \quad c \to \infty.$$

A3: Approximation of mean values

$$\int_\mathbb{R} u \Phi_x^\varepsilon(du) = \varepsilon[a(x) + \theta_a^\varepsilon(x)], \quad \int_\mathbb{R} u^2 \Phi_x^\varepsilon(du) = \varepsilon[c(x) + \theta_c^\varepsilon(x)],$$

with $\sup_{x \in E} |a(x)| \leq a < \infty$, and $\sup_{x \in E} |c(x)| \leq c < +\infty$.

A4: Poisson approximation condition

$$\int_{\mathbb{R}} g(u)\Phi_x^\varepsilon(du) = \varepsilon[\Phi_x(g) + \theta_g^\varepsilon(x)], \quad g \in C_3(\mathbb{R}),$$

and $\sup_{x \in E} |\Phi_x(g)| \leq \Phi(g) < \infty$.

A5: Square-integrability condition

$$\sup_{x \in E} \int_{\mathbb{R}} u^2 \Phi_x(du) < +\infty.$$

where the measure $\Phi_x(du)$ is defined on the measure-determining class $C_3(\mathbb{R})$, by the relation

$$\Phi_x(g) = \int_{\mathbb{R}} g(u)\Phi_x(du), \quad g \in C_3(\mathbb{R}).$$

The negligible terms $\theta_a^\varepsilon(x)$, $\theta_c^\varepsilon(x)$ and $\theta_g^\varepsilon(x)$ in the above conditions satisfy

$$\sup_{x \in E} |\theta_\cdot^\varepsilon(x)| \to 0, \quad \varepsilon \to 0.$$

Theorem 7.1 *Under Assumptions A1-A5, the impulsive process (7.5) converges weakly to the compound Poisson process*

$$\xi^0(t) := \sum_{k=1}^{\nu_0(t)} \alpha_k^0 + tqa_0, \quad t \geq 0. \tag{7.7}$$

The distribution function $\Phi^0(u)$ of the i.i.d. random variables α_k^0, $k \geq 1$, is defined on the measure-determining class $C_3(\mathbb{R})$ of functions g by the relation

$$\mathbb{E}g(\alpha_k^0) = \int_{\mathbb{R}} g(u)\Phi^0(du) = \widehat{\Phi}(g)/\widehat{\Phi}(1), \quad g \in C_3(\mathbb{R}), \tag{7.8}$$

where:

$$\widehat{\Phi}(g) := \int_E \rho(dx)\Phi_x(g), \quad \widehat{\Phi}(1) := \int_E \rho(dx)\Phi_x(\mathbb{R}). \tag{7.9}$$

The counting Poisson process $\nu_0(t)$ is defined by the intensity

$$q_0 := q\widehat{\Phi}(1). \tag{7.10}$$

The drift parameter a_0 is defined by

$$a_0 = \widehat{a} - \widehat{\Phi}(1)\mathbb{E}\alpha_1^0, \quad \widehat{a} := \int_E \rho(dx)a(x). \tag{7.11}$$

7.2. POISSON APPROXIMATION SCHEME

The following corollary gives an adaptation of the above theorem in the case of finite valued random variables $\alpha_k^\varepsilon(x)$.

Corollary 7.1 *The impulsive process (7.5) with a finite number of jump values:*

$$\mathbb{P}(\alpha_k^\varepsilon(x) = a_m) = \varepsilon p_m(x), \quad 1 \leq m \leq M,$$

$$\mathbb{P}(\alpha_k^\varepsilon(x) = \varepsilon a_0) = 1 - \varepsilon p_0(x), \quad (7.12)$$

$$p_0(x) = \sum_{m=1}^{M} p_m(x),$$

converges weakly to the compound Poisson process (7.7) determined by the distribution function of jumps:

$$\mathbb{P}(\alpha_k^0 = a_m) = p_m^0, \quad 1 \leq m \leq M,$$

where:

$$p_m^0 = \widehat{p}_m/\widehat{p}_0, \quad \widehat{p}_m = \int_E \rho(dx) p_m(x), \quad 1 \leq m \leq M, \quad (7.13)$$

and $\widehat{p}_0 := \sum_{m=1}^{M} \widehat{p}_m$.
The intensity of the counting Poisson process $\nu_0(t)$, $t \geq 0$, *is defined by*

$$q_0 := q\widehat{p}_0, \quad (7.14)$$

the drift parameter a_0 is given in (7.12).

Remark 7.1. Assumptions A3 and A4 together split jumps into two parts. The first part gives the deterministic drift, and the second part gives the jumps of the limit Poisson process. The small jumps of the initial process characterized by the function $a(x)$ in A3, are transformed into deterministic drift $\widehat{a}$ for the limit process.

Remark 7.2. The stochastic exponential process for the impulsive process

(7.5) is defined as follows [106]

$$\mathcal{E}(\lambda\xi^\varepsilon)_t := \prod_{k=1}^{\nu(t/\varepsilon)} [1 + \lambda\alpha_k^\varepsilon(x_k)], \quad t \geq 0. \tag{7.15}$$

The weak limit of the process (7.15), as $\varepsilon \to 0$, is

$$\mathcal{E}(\lambda\xi^0)_t := \prod_{k=1}^{\nu^0(t)} [1 + \lambda\alpha_k^0]\varepsilon^{-tqa_0}, \quad t \geq 0. \tag{7.16}$$

▷ **Example 7.1.** Let us consider a two state ergodic Markov process $x(t), t \geq 0$, with generating matrix Q, and the transition matrix P of the embedded Markov chain

$$Q = \begin{pmatrix} -\lambda & \lambda \\ \mu & -\mu \end{pmatrix}, \quad P = \begin{pmatrix} 0 & 1 \\ 1 & 0 \end{pmatrix}.$$

Thus, the stationary distributions of $x(t), t \geq 0$, and $x_n, n \geq 0$, are respectively:

$$\pi = \left(\frac{\mu}{\lambda+\mu}, \frac{\lambda}{\lambda+\mu}\right), \quad \rho = \left(\frac{1}{2}, \frac{1}{2}\right).$$

Now, suppose that, for each $\varepsilon > 0$, the random variables $\alpha_k^\varepsilon(x)$, $x = 1, 2, k \geq 1$, take values in $\{\varepsilon a_0, a_1\}$ with probabilities depending on the state x, $\Phi_x^\varepsilon(\varepsilon a_0) = \mathbb{P}(\alpha_k^\varepsilon = \varepsilon a_0) = 1 - \varepsilon p_x$ and $\Phi_x^\varepsilon(a_1) = \mathbb{P}(\alpha_k^\varepsilon = a_1) = \varepsilon p_x$, for $x \in E$.

We have

$$\int g(u)\Phi_x^\varepsilon(du) = \varepsilon[g(a_1)p_x + \theta_g^\varepsilon(x)],$$

where $\theta_g^\varepsilon(x) := \varepsilon a_0^2 g(\varepsilon a_0)(1 - \varepsilon p_x)/\varepsilon^2 a_0^2 = \varepsilon a_0^2 \cdot o(1) = o(\varepsilon)$, for $\varepsilon \to 0$, and

$$\int u\Phi_x^\varepsilon(du) = \varepsilon[(a_0 + a_1 p_x) + \theta^\varepsilon(x)],$$

where $\theta^\varepsilon(x) = -\varepsilon a_0 p_x$.

For the limit process, we have $\mathbb{P}(\widehat{\alpha}^0 = a_1) = 1$, thus

$$\xi^0(t) = qa_0 t + a_1 \nu^0(t),$$

with $\mathbb{E}\nu^0(t) = q_0 t$, $q = \lambda + \mu$, $q_0 = q\widehat{p}_0 = q(p_1 + p_2)/2$.

Let us now take: $\lambda = \mu = 0.01; p_1 = 0.5; p_2 = 0.6; a_1 = 100; a_0 = -2; \varepsilon = 0.1$. Then we get $q_0 = 0.0165$, and Fig. 7.1 gives two trajectories in the time interval $[0, 4500]$, one for the initial process and the other for the limit process.

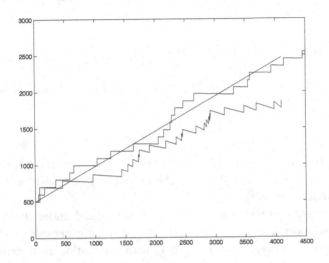

Fig. 7.1 Trajectories of the initial and limit processes, and of the drift

7.2.2 Impulsive Processes in an Asymptotic Split Phase Space

Now the switching Markov process $x^\varepsilon(t)$, $t \geq 0$, is considered in the series scheme with a small series parameter $\varepsilon > 0$, on an asymptotic split state space:

$$E = \bigcup_{v \in V}^{N} E_v, \quad E_v \bigcap E_{v'} = \emptyset, \quad v \neq v', \qquad (7.17)$$

where $(V, \mathcal{V})$ is the factor compact measurable space.

The generator is given by the relation

$$Q^\varepsilon \varphi(x) = \int_E Q_\varepsilon(x, dy)[\varphi(y) - \varphi(x)]. \qquad (7.18)$$

The transition kernel Q_ε has the following representation

$$Q_\varepsilon(x, B) = q(x) P^\varepsilon(x, B) = Q(x, B) + \varepsilon Q_1(x, B), \qquad (7.19)$$

with the stochastic kernel P^ε representation

$$P^\varepsilon(x, B) = P(x, B) + \varepsilon P_1(x, B). \tag{7.20}$$

The stochastic kernel $P(x, B)$ is coordinated to the split state space (7.17) as follows:

$$P(x, E_v) = \mathbf{1}_v(x) := \begin{cases} 1, & x \in E_v \\ 0, & x \notin E_v. \end{cases} \tag{7.21}$$

In the sequel we suppose that the signed kernel P_1 is of bounded variation, that is,

$$|P_1|(x, E) < +\infty. \tag{7.22}$$

According to (7.20) and (7.21), the Markov process $x^\varepsilon(t)$, $t \geq 0$, spends a long time in every class E_v, and the probability of transition from one class to another is $O(\varepsilon)$.

The state space merging scheme (7.17) is realized under the condition that the support Markov process $x(t)$, $t \geq 0$, with generator (7.1) is uniformly ergodic in every class E_v, $v \in V$, with the stationary distributions

$$\pi_v(dx)q(x) = q_v \rho_v(dx), \quad q_v := \int_{E_v} \pi_v(dx)q(x). \tag{7.23}$$

Let us define the merging function

$$v(x) = v, \quad x \in E_v. \tag{7.24}$$

By the state merging scheme, the merged Markov process converges weakly (see Section 4.2),

$$v(x^\varepsilon(t/\varepsilon)) \Longrightarrow \widehat{x}(t), \quad \varepsilon \to 0, \tag{7.25}$$

to the merged Markov process $\widehat{x}(t)$, $t \geq 0$, defined on the merged state space V by the generating kernel

$$\widehat{Q}(v, B_\Gamma) = \int_{E_v} \pi_v(dx) Q(x, B_\Gamma), \quad B_\Gamma = \bigcup_{v \in \Gamma} E_v, \; \Gamma \in \mathcal{V}. \tag{7.26}$$

The counting process of jumps, denoted by $\widehat{\nu}(t), t \geq 0$, can be obtained as the following limit

$$\varepsilon \nu^\varepsilon(t/\varepsilon) \Longrightarrow \widehat{\nu}(t), \quad \varepsilon \to 0.$$

7.2. POISSON APPROXIMATION SCHEME

Theorem 7.2 *Under conditions A1-A5, in the state space merging scheme the impulsive process with Markov switching in series scheme*

$$\xi^\varepsilon(t) := \sum_{k=1}^{\nu^\varepsilon(t/\varepsilon)} \alpha_k^\varepsilon(x_k^\varepsilon), \quad t \geq 0, \tag{7.27}$$

converges weakly to the additive semimartingale $\xi^0(t)$, $t \geq 0$,

$$\xi^0(t) = \int_0^t \xi^0_{\widehat{x}(s)}(ds), \tag{7.28}$$

or, in the equivalent increment form

$$\xi^0(t) = \sum_{k=1}^{\widehat{\nu}(t)} \xi^0_{\widehat{x}_{k-1}}(\widehat{\theta}_k) + \xi^0_{\widehat{x}(t)}(\widehat{\gamma}(t)). \tag{7.29}$$

The compound Poisson processes $\xi_v^0(t)$ *are defined by the generators*

$$\mathbb{A}(v)\varphi(u) = q_v^0 \int_{\mathbb{R}} [\varphi(u+z) - \varphi(u)]\Phi_v(dz),$$

and $\nu_v^0(t)$ *are the counting Poisson processes characterized by the intensity* $q_v^0 = q_v\widehat{\Phi}_v(1)$, *or, in an explicit form*

$$\xi_v^0(t) = \sum_{k=1}^{\nu_v^0(t)} \alpha_{vk}^0 + tq_va_v^0, \quad v \in V,$$

for fixed $v \in V$, *where* $\alpha_{vk}^0, k \geq 1$, *are i.i.d. random variables with common distribution function defined by*

$$\Phi_v^0(g) = \widehat{\Phi}_v(g)/\widehat{\Phi}_v(1).$$

The drift parameter is given by

$$a_v^0 = c(v) = \widehat{a}(v) - \widehat{\Phi}_v(1)\mathbb{E}\alpha_{v1}^0.$$

Remark 7.3. In applications, the limit semimartingale (7.28) can be considered in the following form

$$\xi^0(t) = \sum_{k=1}^{\widehat{\nu}(t)} \widehat{\theta}_k c(\widehat{x}_{k-1}) + \widehat{\gamma}(t)c(\widehat{x}(t)) + \mu_0(t), \tag{7.30}$$

where $\mu_0(t)$ is a martingale fluctuation. The predictable term in (7.30) is a linear deterministic drift between jumps of the merged switching Markov process $\widehat{x}(t)$, $t \geq 0$.

7.2.3 Stochastic Additive Functionals with Semi-Markov Switching

Let us consider an additive functional with semi-Markov switching depending on the small series parameter $\varepsilon > 0$, namely

$$\xi^\varepsilon(t) = \xi^\varepsilon(0) + \int_0^t \eta^\varepsilon(ds; x(s/\varepsilon)), \tag{7.31}$$

where $\eta^\varepsilon(t;x)$, $t \geq 0, x \in E, \varepsilon > 0$, is a family of Markov jump processes in the series scheme defined by the generators

$$\mathbb{\Gamma}_\varepsilon(x)\varphi(u) = \varepsilon^{-1} \int_{\mathbb{R}^d} [\varphi(u+v) - \varphi(u)] \Gamma_\varepsilon(dv; x), \quad x \in E, \tag{7.32}$$

switched by the semi-Markov process $x(t), t \geq 0$, defined on a standard state space $(E, \mathcal{E})$ by the semi-Markov kernel

$$Q(x, B, t) = P(x, B) F_x(t), \quad x \in E, B \in \mathcal{E}, t \geq 0, \tag{7.33}$$

which defines the associated Markov renewal process $x_n, \tau_n, n \geq 0$:

$$\begin{aligned} Q(x, B, t) &= \mathbb{P}(x_{n+1} \in B, \theta_{n+1} \leq t \mid x_n = x) \\ &= \mathbb{P}(x_{n+1} \in B \mid x_n = x) \mathbb{P}(\theta_{n+1} \leq t \mid x_n = x). \end{aligned} \tag{7.34}$$

Remark 7.4. Here we do not consider drift for processes $\eta^\varepsilon(t;x), t \geq 0$, as was the case in diffusion approximation (see Section 4.1), since only random jumps can be transformed into jumps of limit Poisson processes.

Let the following conditions hold.

C1: The switching semi-Markov process $x(t)$, $t \geq 0$, is uniformly ergodic with the stationary distribution:

$$\pi(dx) = \rho(dx) m(x)/m,$$

$$m(x) := \mathbb{E}\theta_x = \int_0^\infty \bar{F}_x(t) dt, \quad m := \int_E \rho(dx) m(x),$$

7.2. POISSON APPROXIMATION SCHEME

$$\rho(B) = \int_E \rho(dx) P(x, B), \quad \rho(E) = 1.$$

C2: Approximation of the mean jumps:

$$a_\varepsilon(x) = \int_{\mathbb{R}^d} v \Gamma_\varepsilon(dv; x) = \varepsilon[a(x) + \theta_a^\varepsilon(x)], \qquad (7.35)$$

and

$$c_\varepsilon(x) = \int_{\mathbb{R}^d} vv^* \Gamma_\varepsilon(dv; x) = \varepsilon[c(x) + \theta_c^\varepsilon(x)], \qquad (7.36)$$

and $a(x), c(x)$ are bounded, that is, $|a(x)| \leq a < +\infty, |c(x)| \leq c < +\infty$.

C3: Poisson approximation condition

$$\Gamma_g^\varepsilon(x) = \int_{\mathbb{R}^d} g(v) \Gamma_\varepsilon(dv; x) = \varepsilon[\Gamma_g(x) + \theta_g^\varepsilon(x)] \qquad (7.37)$$

for all $g \in C_3(\mathbb{R}^d)$, and the kernel $\Gamma_g(x)$ is bounded for all $g \in C_3(\mathbb{R}^d)$, that is,

$$|\Gamma_g(x)| \leq \Gamma_g.$$

The negligible terms in (7.35)-(7.37) satisfy the condition

$$\sup_{x \in E} |\theta^\varepsilon(x)| \to 0, \quad \varepsilon \to 0. \qquad (7.38)$$

C4: Uniform square-integrability

$$\lim_{c \to \infty} \sup_{x \in E} \int_{|v| \geq c} vv^* \Gamma(dv; x) = 0,$$

where the kernel $\Gamma(dv; x)$ is defined on the measure-determining class $C_3(\mathbb{R}^d)$ by the relation

$$\Gamma_g(x) = \int_{\mathbb{R}^d} g(v) \Gamma(dv; x), \quad g \in C_3(\mathbb{R}).$$

C5: Cramér's condition

$$\sup_{x \in E} \int_0^\infty e^{ht} F_x(dt) \leq H < +\infty, \quad h > 0.$$

Now, we get the following result.

Theorem 7.3 *Under Assumptions C1-C5, the additive functional (7.31) converges weakly to the Markov process $\xi^0(t)$, $t \geq 0$, defined by the generator*

$$\widehat{\Gamma}\varphi(u) = \widehat{a}\varphi'(u) + \int_{\mathbb{R}^d} [\varphi(u+v) - \varphi(u) - v\varphi'(u)]\widehat{\Gamma}(dv), \qquad (7.39)$$

where

$$\widehat{a} = \int_E \pi(dx)a(x), \qquad (7.40)$$

and

$$\widehat{\Gamma}(dv) = \int_E \pi(dx)\Gamma(dv;x). \qquad (7.41)$$

The generator of the limit process can also be represented as follows

$$\Gamma\varphi(u) = \widehat{a}_0\varphi'(u) + \int_{\mathbb{R}^d} [\varphi(u+v) - \varphi(u)]\widehat{\Gamma}(dv). \qquad (7.42)$$

The first term of the sum in (7.42) determines the deterministic drift and the second one gives the representation of the generator of the stochastic part of the limit process.

It is worth noticing that under the Poisson approximation condition $C3$, the small jumps of the initial process characterized by the function $a(x)$ in Condition $C2$, are transformed into a deterministic drift with velocity $\widehat{a}$ which is the average value of $a(x)$ with respect to the stationary distribution of the switching ergodic semi-Markov process.

Due to both the representations (7.39) and (7.42) of the limit generator, and the approximation conditions C2 and C3, the small jumps of the initial functional are transformed into the deterministic drift $U_0(t) = \widehat{a}_0 t$, where

$$\widehat{a}_0 = \widehat{a} - \widehat{b}, \quad \widehat{b} := \int_{\mathbb{R}^d} v\widehat{\Gamma}(dv). \qquad (7.43)$$

The big jumps of the initial functional (7.31) are distributed following the averaged distribution function

$$\widehat{F}(dv) := \widehat{\Gamma}(dv)/\widehat{\Gamma}(\mathbb{R}^d), \qquad (7.44)$$

with the intensity of jump moments $\widehat{\gamma} := \widehat{\Gamma}(\mathbb{R}^d)$. The limit Markov process has the representation $\xi^0(t) = U^0(t) + \zeta^0(t)$, where the jump Markov process

$\zeta^0(t)$ has the following generator

$$\widehat{\Gamma}_0\varphi(u) = \widehat{\gamma}\int_{\mathbb{R}^d}[\varphi(u+v)-\varphi(u)]\widehat{F}(dv).$$

7.3 Semimartingale Characterization

The stochastic systems in Poisson approximation scheme are here considered as additive semimartingales characterized by their predictable characteristics [70] (see Section 1.4).

The limit predictable characteristics determine the limit semimartingale for the corresponding stochastic system. So, the limit theorems for semimartingales are used [70] (see Appendix B).

Since Theorem 7.1 is a particular case of Theorem 7.2, we will prove only the latter one.

The setting for proving convergence in Theorem 7.2 is the following. Consider a family of switching Markov processes

$$\zeta^\varepsilon(t),\ x^\varepsilon(t/\varepsilon),\ t\geq 0, \qquad (7.45)$$

in the series scheme with a small series parameter $\varepsilon > 0$, on the product space $\mathbb{R}^d \times E$, defined by the generators $\mathbb{L}^\varepsilon, \varepsilon > 0$. The domains of definition $\mathcal{D}(\mathbb{L}^\varepsilon)$ are supposed to be dense in the space $C(\mathbb{R}^d \times E)$ of real-valued, bounded, continuous functions $\varphi(u,x)$, $u \in \mathbb{R}^d$, $x \in E$, with sup-norm $\|\varphi\| = \sup_{u\in\mathbb{R}^d,\ x\in E}|\varphi(u,x)|$.

The first switched component $\zeta^\varepsilon(t)$ takes values in the Euclidean space $\mathbb{R}^d$, $d \geq 1$. The second switching Markov component $x^\varepsilon(t/\varepsilon)$ is defined on the standard state space $(E,\mathcal{E})$ by the generator

$$Q^\varepsilon\varphi(x) = \int_E Q_\varepsilon(x,dy)[\varphi(y)-\varphi(x)],$$

in perturbed form, with the kernel:

$$Q_\varepsilon(x,dy) = q(x)\int_E[P(x,dy)+\varepsilon P_1(x,dy)][\varphi(y)-\varphi(x)]$$
$$= Q_0(x,dy) + \varepsilon Q_1(x,dy).$$

The switched Markov process $x^\varepsilon(t)$, $t \geq 0$, is considered on the asymptotic split state space (7.17). The merged state space V is defined by the merging function (7.24).

The limit Markov process

$$\zeta(t),\ \widehat{x}(t),\ t \geq 0, \qquad (7.46)$$

is considered on the product space $\mathbb{R}^d \times V$ and is defined by the generator $\mathbb{L}$, with domain $\mathcal{D}(\mathbb{L})$ dense in $C(\mathbb{R}^d \times V)$.

7.3.1 Impulsive Processes as Semimartingales

Let $\mathcal{F}_t^x := \sigma(x(s), 0 \leq s \leq t), t \geq 0$, be the natural filtration of the Markov process $x(t), t \geq 0$. Let us define also the filtration $\mathbb{F}^\varepsilon = (\mathcal{F}_t^\varepsilon, t \geq 0)$, $\mathcal{F}_t^\varepsilon := \sigma(x^\varepsilon(s), \alpha_k^\varepsilon(x_k), 0 \leq s \leq t, k \leq \nu^\varepsilon(t))$, and the discrete time filtration $\mathcal{F}_n^\varepsilon := \sigma(x_k^\varepsilon, \alpha_k^\varepsilon(x_k), 0 \leq s \leq t, k \leq n), n \geq 0$.

The semimartingale characterization of the impulsive process with Markov switching (7.27) is given by the predictable characteristics as follows.

Lemma 7.1 *Under Assumptions A1-A5, the predictable characteristics $(B^\varepsilon(t), C^\varepsilon(t), \Phi_g^\varepsilon(t))$ of the semimartingale*

$$\xi^\varepsilon(t) = \sum_{k=1}^{\nu^\varepsilon(t/\varepsilon)} \alpha_k^\varepsilon(x_k^\varepsilon), \quad t \geq 0, \qquad (7.47)$$

are defined as follows. The first predictable characteristic is

$$B^\varepsilon(t) = \varepsilon \sum_{k=1}^{\nu^\varepsilon(t/\varepsilon)} b(x_{k-1}^\varepsilon) + \theta_b^\varepsilon(t), \quad t \geq 0,$$

where $b(x) = Pa(x) = \int_E P(x, dy) a(y)$, $x \in E$, and the predictable measure

$$\Phi_g^\varepsilon(t) = \varepsilon \sum_{k=1}^{\nu^\varepsilon(t/\varepsilon)} P\Phi_{x_{k-1}^\varepsilon}(g) + \theta_c^\varepsilon(t; g), \quad t \geq 0, \qquad (7.48)$$

where $P\Phi_x(g) = \int_E P(x, dy) \Phi_y(g)$.
The modified second characteristic is

$$C^\varepsilon(t) = \varepsilon \sum_{k=1}^{\nu^\varepsilon(t/\varepsilon)} C(x_{k-1}^\varepsilon) + \theta^\varepsilon(t), \ t \geq 0, \qquad (7.49)$$

where

$$C(x) = \int_E P(x, dy) c(y), \quad c(x) := \int_\mathbb{R} u^2 \Phi_x(du).$$

7.3. SEMIMARTINGALE CHARACTERIZATION

The continuous part of the second predictable characteristic is $C_c^\varepsilon(t) \equiv 0$.

The negligible terms satisfy the following asymptotic conditions for every finite $T > 0$:

$$\sup_{0 \leq t \leq T} |\theta_c^\varepsilon(t)| \xrightarrow{P} 0, \quad \varepsilon \to 0,$$

$$\sup_{0 \leq t \leq T} |\theta_c^\varepsilon(t;g)| \xrightarrow{P} 0, \quad \varepsilon \to 0, \quad \text{for every } g \in C_3(\mathbb{R}). \quad (7.50)$$

PROOF. Concerning the semimartingale (7.47) we have (see Theorem II.3.11, in [70]),

$$B^\varepsilon(t) = \sum_{k=1}^{\nu^\varepsilon(t/\varepsilon)} \mathbb{E}[\alpha_k^\varepsilon(x_k^\varepsilon) \mid \mathcal{F}_{k-1}^\varepsilon]. \quad (7.51)$$

In particular

$$\mathbb{E}[\alpha_k^\varepsilon(x_k^\varepsilon) \mid \mathcal{F}_{k-1}^\varepsilon] = \int_E P^\varepsilon(x_{k-1}^\varepsilon, dy) \mathbb{E}\alpha_k^\varepsilon(y).$$

By Condition A3, we get:

$$\mathbb{E}[\alpha_k^\varepsilon(x_k^\varepsilon) \mid \mathcal{F}_{k-1}^\varepsilon] = \varepsilon \int_E [P(x_{k-1}^\varepsilon, dy) + \varepsilon P_1(x_{k-1}^\varepsilon, dy)][a(y) + \theta^\varepsilon(y)]$$

$$= \varepsilon [Pa(x_{k-1}^\varepsilon) + \theta^\varepsilon(x_{k-1}^\varepsilon)].$$

Now, relation (7.51) becomes

$$B^\varepsilon(t) = \varepsilon \sum_{k=1}^{\nu^\varepsilon(t/\varepsilon)} Pa(x_{k-1}^\varepsilon) + \theta^\varepsilon(t),$$

where

$$\theta^\varepsilon(t) = \varepsilon \sum_{k=1}^{\nu^\varepsilon(t/\varepsilon)} \theta^\varepsilon(x_{k-1}^\varepsilon),$$

which is a negligible term satisfying (7.50).

In the same way, as in (7.51), we get:

$$\Phi_g^\varepsilon(t) = \sum_{k=1}^{\nu^\varepsilon(t/\varepsilon)} \mathbb{E}[g(\alpha_k^\varepsilon(x_k^\varepsilon)) \mid \mathcal{F}_{k-1}^\varepsilon] = \sum_{k=1}^{\nu^\varepsilon(t/\varepsilon)} P\Phi_{x_{k-1}^\varepsilon}^\varepsilon(g),$$

where, by Condition A4, we have:

$$\Phi_x^\varepsilon(g) = \mathbb{E}[g(\alpha_k^\varepsilon(x))] = \int_\mathbb{R} g(u)\Phi_x^\varepsilon(du) = \varepsilon[\Phi_x(g) + \theta_x^\varepsilon(g)]. \quad (7.52)$$

Thus the predictable measure $\Phi_g^\varepsilon(t)$ is represented by (7.48).

By Condition A5, the modified second characteristic is represented as follows[70]:

$$C^\varepsilon(t) = \sum_{k=1}^{\nu^\varepsilon(t/\varepsilon)} \mathbb{E}\left[\alpha_k^\varepsilon(x_k^\varepsilon))^2 \mid \mathcal{F}_{k-1}^\varepsilon\right]$$

$$= \varepsilon \sum_{k=1}^{\nu^\varepsilon(t/\varepsilon)} c(x_{k-1}^\varepsilon) + \theta^\varepsilon(t).$$

□

Lemma 7.2 *The coupled Markov process*

$$B_0^\varepsilon(t) = \varepsilon \sum_{k=1}^{\nu^\varepsilon(t/\varepsilon)} b(x_{k-1}^\varepsilon), \quad x^\varepsilon(t/\varepsilon), \quad t \geq 0, \quad (7.53)$$

characterized by the martingale

$$\mu_t^\varepsilon = \varphi(B_0^\varepsilon(t), x^\varepsilon(t/\varepsilon)) - \int_0^t \mathbb{L}^\varepsilon \varphi(B_0^\varepsilon(s), x^\varepsilon(s/\varepsilon))ds, \quad t \geq 0,$$

with respect to the filtration $\mathcal{F}_{t/\varepsilon}^\varepsilon, t \geq 0$, *and* $\varphi \in \mathcal{D}(\mathbb{L}^\varepsilon)$, *is defined by the generator* $\mathbb{L}^\varepsilon$, *which is represented in explicit form as follows*

$$\mathbb{L}^\varepsilon \varphi(u, x) = \varepsilon^{-1} Q + [Q_1 + Q_0 \mathbb{B}^\varepsilon(x)]\,\varphi(u, x), \quad (7.54)$$

and in asymptotic form

$$\mathbb{L}^\varepsilon \varphi(u, x) = \varepsilon^{-1} Q\varphi(u, x) + [Q_1 + Q_0 \mathbb{B}(x)]\,\varphi(u, x) + Q_0 \theta^\varepsilon(x)\varphi(u, x), \quad (7.55)$$

where:

$$Q\varphi(\cdot, x) = q(x) \int_E P(x, dy)[\varphi(\cdot, y) - \varphi(\cdot, x)],$$

$$Q_1 \varphi(\cdot, x) = q(x) \int_E P_1(x, dy)\varphi(\cdot, y),$$

7.3. SEMIMARTINGALE CHARACTERIZATION

$$Q_0\varphi(\cdot,x) = q(x)\int_E P(x,dy)\varphi(\cdot,y),$$

$$\mathbb{B}^\varepsilon(x)\varphi(u,\cdot) = \varepsilon^{-1}[\varphi(u+\varepsilon b(x),\cdot) - \varphi(u,\cdot)],$$

$$\mathbb{B}(x)\varphi(u,\cdot) = b(x)\varphi'_u(u,\cdot),$$

and

$$\theta^\varepsilon(x)\varphi(u,\cdot) = \varepsilon^{-1}[\varphi(u+\varepsilon b(x),\cdot) - \varphi(u,\cdot) - \varepsilon b(x)\varphi'_u(u,\cdot)].$$

PROOF. For the proof of this lemma, see Section 5.3.3. □

A standard calculus, via a stochastic perturbation problem, gives us the following result (see Section 4.5).

Lemma 7.3 *The generator* $\mathbb{L}$ *of the limit coupled Markov process*

$$B(t) = \int_0^t b(\widehat{x}(s))ds, \quad \widehat{x}(t), \quad t \geq 0,$$

is given by

$$\mathbb{L}\varphi(u,v) = \widehat{b}(v)\varphi'_u(u,v) + \widehat{Q}_1\varphi(u,v) \qquad (7.56)$$

where $\widehat{b}(v) = q_v \int_{E_v} \rho_v(dx) b(x) = q_v \int_{E_v} \rho_v(dx) a(x) = q_v \widehat{a}(v).$

The generator of the coupled Markov process (7.45) is represented in the following singular perturbed form

$$\mathbb{L}^\varepsilon\varphi(u,x) = \varepsilon^{-1}[Q^\varepsilon + Q_1^\varepsilon]\varphi(u,x), \qquad (7.57)$$

where the operator Q_1^ε corresponds to the first switched component in (7.45).

PROOF OF THEOREM 7.2. The coupled Markov processes

$$B_0^\varepsilon(t), \quad x^\varepsilon(t/\varepsilon), \quad t \geq 0, \ \varepsilon > 0, \qquad (7.58)$$

in the series scheme are defined by the generators (7.54) in Lemma 7.2. For $\varphi(u,v(x)) \in C_0^2(\mathbb{R} \times V)$ the generator (7.54) is represented under the asymptotic form as

$$\mathbb{L}^\varepsilon = \varepsilon^{-1}Q + [Q_1 + Q_0\mathbb{B}(x)] + Q_0\theta^\varepsilon(x) \qquad (7.59)$$

with, for $\varphi(u,\cdot) = \varphi(u)$,

$$\mathbb{B}(x)\varphi(u) = b(x)\varphi'(u),$$

and the residual operator

$$\theta^\varepsilon(x)\varphi(u) = \varepsilon^{-1}[\varphi(u+\varepsilon b(x)) - \varphi(u) - \varepsilon b(x)\varphi'(u)]. \qquad (7.60)$$

For $\varphi(u) \in C_0^2(\mathbb{R})$ the following negligible condition holds

$$|\theta^\varepsilon(x)\varphi(u)| \leq \varepsilon b \,\|\varphi''\| \to 0, \quad \varepsilon \to 0. \qquad (7.61)$$

We will use a solution of the following singular perturbation problem given in Proposition 5.1 for the test functions

$$\varphi^\varepsilon(u,x) = \varphi(u,v(x)) + \varepsilon\varphi_1(u,x), \qquad (7.62)$$

that is,

$$\mathbb{L}^\varepsilon \varphi^\varepsilon = \mathbb{L}\widehat{\varphi} + \varepsilon\theta^\varepsilon, \qquad (7.63)$$

where $\widehat{\varphi} = \widehat{\varphi}(u,v) \in C_0^2(\mathbb{R} \times V)$.

The limit operator $\mathbb{L}$ is given by the contracted operator

$$\mathbb{L} = \widehat{Q}_1 + \widehat{Q_0 \mathbb{B}},$$

where the contracted operators $\widehat{Q}_1$ and $\widehat{Q_0 \mathbb{B}}$ are defined by:

$$\widehat{Q}_1 \Pi = \Pi Q_1 \Pi, \quad \text{and} \quad \widehat{Q_0 \mathbb{B}} \Pi = \Pi Q_0 \mathbb{B} \Pi.$$

Condition A3 is fulfilled for the coupled Markov process (7.58) (see Appendix C).

The limit generator

$$\mathbb{L}\varphi(u,v) = \widehat{Q}_1 \varphi(\cdot,v) + \widehat{b}(v)\varphi'_u(u,\cdot),$$

satisfies the preliminary conditions of Theorem 6.4 (see Theorem 2 in [174]).

The limit generator $\mathbb{L}$ given by relation (7.56) is obtained by a singular perturbation approach.

Then, from Lemma 7.3, the following weak convergence, in $\mathbf{D}_{\mathbb{R}\times V}[0,\infty)$,

$$(B^\varepsilon(t), v(x^\varepsilon(t/\varepsilon))) \Longrightarrow (B(t), \widehat{x}(t)), \quad \varepsilon \to 0,$$

takes place, where $B(t) = \int_0^t b(\widehat{x}(s))ds$ is a continuous trajectory stochastic process.

Due to the fact that the process $B(t)$ has continuous trajectories (a.s.), the weak convergence in $\mathbf{D}_\mathbb{R}[0,\infty)$ is equivalent to uniform convergence in probability in $\mathbf{C}_\mathbb{R}[0,\infty)$ on every finite time interval $[0,T]$ [16], that is

$$\sup_{0 \leq t \leq T} |B^\varepsilon(t) - B(t)| \xrightarrow{P} 0, \quad \varepsilon \to 0.$$

In the same way, we obtain the weak convergence in $\mathbf{D}_\mathbb{R}[0,\infty)$ of the predictable measures, that is, for any fixed function $g \in C_3(\mathbb{R})$,

$$\Phi_g^\varepsilon(t) \xrightarrow{P} \nu_t(g), \quad \varepsilon \to 0. \qquad (7.64)$$

The modified second characteristic $C^\varepsilon(t)$, given by (7.49) converges to

$$\widehat{C}(t) = \int_0^t \widehat{C}(\widehat{x}(s))ds,$$

as stated in Theorem 7.2.

All the conditions of the limit theorem, Theorem B.2 (see Appendix B) for semimartingales, are fulfilled. Due to Condition A2, in particular, the square integrability Condition B.1 holds true.

The strong domination hypothesis is valid with the dominating function $F_t = tF$, for some constant F, due to the boundedness of all functions $b(x)$, $C(x)$ and $\Phi_x(g)$ for $g \in C_3(\mathbb{R})$ and the known inequality [60]: $\varepsilon\nu^\varepsilon(t/\varepsilon) \leq ct$, $c > 0$ (see Appendix C). Condition A5 implies the condition on big jumps for the last predictable measure of Theorem B.2 (Appendix B). Conditions (iv) and (v) of Theorem B.2 are obviously fulfilled. The weak convergence of predictable characteristics (B.3), (B.4) and (B.5) (Theorem B.2, Appendix B) are proved in Lemmas 7.1-7.2. The last Condition (B.6) of Theorem B.2 holds due to the uniformly square integrable Condition A2.

Hence, we get the weak convergence in $\mathbf{D}_\mathbb{R}[0,\infty)$ of the impulsive process $\xi^\varepsilon(t)$ to the process $\xi^0(t)$ which is defined by the predictable characteristics, as stated in the above Theorem 7.2. $\square$

7.3.2 Stochastic Additive Functionals

For semi-Markov switching we need the compensating operator of the Markov renewal process. The additive functional (7.31) is first considered as an additive semimartingale defined by its predictable characteristics (see Section 1.4).

PROOF OF THEOREM 7.3 In order to prove this theorem, the following lemmas are needed.

Lemma 7.4 *Under the assumptions of Theorem 7.3, the predictable characteristics* $(B^\varepsilon(t), C^\varepsilon(t), \gamma^\varepsilon(t))$ *of the semimartingale*

$$\xi^\varepsilon(t) = \xi_0^\varepsilon + \int_0^t \eta^\varepsilon(ds; x(s/\varepsilon)),$$

are defined by the following relations:
- *the predictable process is*

$$B^\varepsilon(t) = \varepsilon \Big[\int_0^{t/\varepsilon} a(x(s))ds + \theta_b^\varepsilon(t)\Big], \quad t \geq 0;$$

- *the modified second characteristic is*

$$C^\varepsilon(t) = \varepsilon \Big[\int_0^{t/\varepsilon} C(x(s))ds + \theta_c^\varepsilon(t)\Big], \quad t \geq 0;$$

- *the predictable measure is*

$$\gamma_g^\varepsilon(t) = \varepsilon \Big[\int_0^{t/\varepsilon} \Gamma_g(x(s))ds + \theta_\gamma^\varepsilon(t)\Big], \quad t \geq 0,$$

where $C(x) = \int_{\mathbb{R}^d} vv^*\Gamma(dv;x)$, *and* θ_b^ε, θ_c^ε, *and* $\theta_\gamma^\varepsilon$ *satisfy the negligible condition*

$$\|\theta_\cdot^\varepsilon\| \to 0, \quad \varepsilon \to 0.$$

PROOF. The statements of this lemma are almost direct consequences of the factorization theorem (see Section 2.9). □

In what follows, we will study only the convergence of $(B_0^\varepsilon(t), C_0^\varepsilon(t), \gamma_0^\varepsilon(t;g))$, where:

$$B_0^\varepsilon(t) = \int_0^t a(x(s/\varepsilon))ds, \quad t \geq 0,$$

$$C_0^\varepsilon(t) = \int_0^t C(x(s/\varepsilon))ds, \quad t \geq 0,$$

$$\gamma_0^\varepsilon(t;g) = \int_0^t \Gamma_g(x(s/\varepsilon))ds, \quad t \geq 0.$$

In the sequel the process $A^\varepsilon(t)$ will denote one of the above predictable characteristics $B_0^\varepsilon(t), C_0^\varepsilon(t), \gamma_0^\varepsilon(t)$.

7.3. SEMIMARTINGALE CHARACTERIZATION

The extended Markov renewal process is considered as a three-component Markov chain

$$A_n^\varepsilon = A_0^\varepsilon(\tau_n^\varepsilon), \quad x_n^\varepsilon, \quad \tau_n^\varepsilon, \quad n \geq 0, \tag{7.65}$$

where $x_n^\varepsilon = x^\varepsilon(\tau_n^\varepsilon)$, $x^\varepsilon(t) := x(t/\varepsilon)$ and $\tau_{n+1}^\varepsilon = \tau_n^\varepsilon + \varepsilon \theta_n^\varepsilon$, $n \geq 0$, and

$$\mathbb{P}(\theta_{n+1}^\varepsilon \leq t \mid x_n^\varepsilon = x) = F_x(t) = \mathbb{P}(\theta_x \leq t).$$

We are using here the compensating operator of the extended Markov renewal process (7.65) (see Section 1.3.4).

Let $A_t(x)$, $t \geq 0$, $x \in E$, be a family of semigroups determined by the generators

$$\mathbb{A}(x)\varphi(u) = a(x)\varphi'(u). \tag{7.66}$$

Lemma 7.5 *The compensating operator of the extended Markov renewal process (7.65) can be defined by the relation (see Section 2.8)*

$$\mathbb{L}^\varepsilon \varphi(u,x,t) = \left[\int_0^\infty F_x(ds) A_{\varepsilon s}(x) \int_E P(x,dy)\varphi(u,y,t+\varepsilon s) - \varphi(u,x,t)\right]/\varepsilon m(x). \tag{7.67}$$

PROOF. The proof of this lemma follows directly from Definition 2.12. □

Lemma 7.6 *The extended Markov renewal process (7.65) is characterized by the martingale*

$$\mu_{n+1}^\varepsilon = \varphi(A_{n+1}^\varepsilon, x_{n+1}^\varepsilon, \tau_{n+1}^\varepsilon) - \sum_{k=0}^n \varepsilon \theta_{k+1}^\varepsilon \mathbb{L}^\varepsilon \varphi(A_k^\varepsilon, x_k^\varepsilon, \tau_k^\varepsilon), \quad n \geq 0. \tag{7.68}$$

PROOF. The martingale property of (7.68) follows from the homogeneous Markov property

$$\mathbb{E}[\varphi(A_{n+1}^\varepsilon, x_{n+1}^\varepsilon, \tau_{n+1}^\varepsilon) \mid A_n^\varepsilon = u, x_n^\varepsilon = x, \tau_n^\varepsilon = t]$$
$$= \mathbb{E}[\varphi(A_1^\varepsilon, x_1^\varepsilon, \tau_1^\varepsilon) \mid A_0^\varepsilon = u, x_0^\varepsilon = x, \tau_0^\varepsilon = t]. \tag{7.69}$$

□

In what follows, the martingale property will be used for the process

$$\zeta^\varepsilon(t) = \varphi\left(A^\varepsilon(\tau_+^\varepsilon(t)), x^\varepsilon(\tau_+^\varepsilon(t)), \tau_+^\varepsilon(t)\right)$$
$$- \int_0^{\tau_+^\varepsilon(t)} \mathbb{L}^\varepsilon \varphi\left(A^\varepsilon(\tau^\varepsilon(s)), x^\varepsilon(s), \tau^\varepsilon(s)\right) ds, \tag{7.70}$$

where $\tau_+^\varepsilon(t) := \tau_{\nu_+^\varepsilon(t)}$, $\nu_+^\varepsilon(t) := \nu^\varepsilon(t) + 1$.

Note that the following relations hold:
$$\zeta^\varepsilon(\tau_n) = \mu^\varepsilon_{n+1}, \quad n \geq 0, \tag{7.71}$$
and
$$\zeta^\varepsilon(t) = \zeta^\varepsilon(\tau^\varepsilon(t)), \quad \text{for} \quad \tau^\varepsilon(t) \leq t < \tau^\varepsilon_+(t). \tag{7.72}$$

The random numbers $\nu^\varepsilon_+(t)$ are stopping times for
$$\mathcal{F}^\varepsilon_n = \sigma(A^\varepsilon_k, x^\varepsilon_k, \tau^\varepsilon_k; \, 0 \leq k \leq n), \quad n \geq 0.$$

Lemma 7.7 *The process (7.72) has the martingale property*
$$\mathbb{E}[\zeta^\varepsilon(t) - \zeta^\varepsilon(s) \mid \mathcal{F}^\varepsilon_s] = 0, \quad \text{for} \quad 0 \leq s < t \leq T,$$
where
$$\mathcal{F}^\varepsilon_t := \sigma(A^\varepsilon(s), x^\varepsilon(s), \tau^\varepsilon(s); \, 0 \leq s \leq t).$$

PROOF. The proof of this lemma is based on relation (7.71) and on the martingale property of the sequence μ^ε_n, $n \geq 0$, defined in (7.68). □

Remark 7.5. Note that the process $\zeta^\varepsilon(t)$, $t \geq 0$, is not a martingale since it is not $\mathcal{F}^\varepsilon_t$-adapted.

The next lemma is basic in the proof of the *compact containment condition* for the additive functionals $A^\varepsilon_0(t)$, $t \geq 0$. (Compare with Lemma 3.2, Ch. 4, in [45]).

Lemma 7.8 *The process*
$$\zeta^\varepsilon_c(t) = e^{-c\tau^\varepsilon_+(t)} \varphi\left(A^\varepsilon(\tau^\varepsilon_+(t)), x^\varepsilon(\tau^\varepsilon_+(t)), \tau^\varepsilon_+(t)\right)$$
$$+ \int_0^{\tau^\varepsilon_+(t)} \left[e^{-cs} c\varphi\left(A^\varepsilon(\tau^\varepsilon_+(s)), x^\varepsilon(\tau^\varepsilon_+(s)), \tau^\varepsilon_+(s)\right) \right.$$
$$\left. - e^{-c\tau^\varepsilon(s)} \mathbf{L}^\varepsilon \varphi\left(A^\varepsilon(\tau^\varepsilon(s)), x^\varepsilon(s), \tau^\varepsilon(s)\right) \right] ds \tag{7.73}$$

has the martingale property for every $c \in \mathbb{R}$, that is,
$$\mathbb{E}[\zeta^\varepsilon_c(t) - \zeta^\varepsilon_c(s) \mid \mathcal{F}^\varepsilon_s] = 0, \quad \text{for} \quad 0 \leq s < t \leq T. \tag{7.74}$$

PROOF. The proof of this lemma is based on the representation
$$\zeta^\varepsilon_c(t) = e^{-c\tau^\varepsilon_+(t)} \zeta^\varepsilon(t) + c \int_0^{\tau^\varepsilon_+(t)} e^{-cs} \zeta^\varepsilon(s) ds,$$

7.3. SEMIMARTINGALE CHARACTERIZATION

where the process $\zeta^\varepsilon(s)$, $s \geq 0$, is defined by (7.70). □

The algorithm of Poisson approximation given in Theorem 7.3 provides the asymptotic representation of the compensating operator.

Lemma 7.9 *The compensating operator (7.67) applied to functions $\varphi \in C^2(\mathbb{R}) \times \mathbf{B}(E)$ has the asymptotic representation*

$$\mathbb{L}^\varepsilon \varphi(u, x) = \varepsilon^{-1} Q \varphi(u, x) + \mathbb{A}(x) P \varphi(u, x) + \varepsilon \theta^\varepsilon(x) P \varphi(u, x), \quad (7.75)$$

where:

$$Q\varphi(\cdot, x) = q(x) \int_E P(x, dy)[\varphi(\cdot, y) - \varphi(\cdot, x)], \quad q(x) := 1/m(x),$$

$$\mathbb{A}(x)\varphi(u, \cdot) = a(x)\varphi'_u(u, \cdot).$$

The negligible operator is defined as follows

$$\theta^\varepsilon(x)\varphi(u, \cdot) = \mathbb{A}^2(x)\mathbb{A}^\varepsilon(x)\varphi(u, \cdot),$$

where

$$\mathbb{A}^\varepsilon(x) = \int_0^\infty A_{\varepsilon s}(x)\overline{F}_x^{(2)}(s)ds, \quad \overline{F}_x^{(2)}(s) := \int_s^\infty \overline{F}_x(t)dt.$$

The asymptotic expansion (7.75) can be obtained as in Section 5.3.2.

PROOF OF THEOREM 7.3. Let us first consider a solution of singular perturbation problem for the compensating operator $\mathbb{L}^\varepsilon$ using the asymptotic representation (7.75).

Corollary 7.2 *According to Proposition 5.1, a solution of the singular perturbation problem*

$$\mathbb{L}^\varepsilon[\varphi(u) + \varepsilon\varphi_1(u, x)] = \mathbb{L}\varphi(u) + \theta_0^\varepsilon(x)\varphi(u), \quad (7.76)$$

whose negligible term in (7.76) satisfies

$$\|\theta_0^\varepsilon(x)\varphi\| \to 0, \quad (7.77)$$

is the limit operator which is defined as a contracted operator

$$\mathbb{L}\varphi(u) = \Pi\mathbb{A}(x)\Pi\varphi(u) = \Pi\widehat{\mathbb{A}}\varphi(u),$$

where

$$\widehat{\mathbb{A}} = \int_E \pi(dx)\mathbb{A}(x).$$

and given by
$$\mathbb{L}\varphi(u) = \widehat{a}\varphi'(u).$$

Hence, we get from (7.66)
$$\widehat{\mathbb{A}}\varphi(u) = \widehat{a}\varphi'(u), \quad \widehat{a} := \int_E \pi(dx)a(x).$$

Now, the proof of the theorem is achieved in the same way as in Theorem 7.2, by using Lemmas 7.6, 6.4, and Theorem B.1 in Appendix B. □

In this chapter we have considered processes with independent increments, impulsive processes and stochastic additive functionals in order to simplify proofs. Processes with locally independent increments, considered in Chapters 3, 4 and 5, can be studied in the same way.

A more general approximation by a Lévy process is given in Section 9.2.

Chapter 8

Applications I

In this chapter we present some applications of the previous results encountered in real application problems as the absorption time distributions, stationary phase merging, superposition of two independent renewal processes and semi-Markov random walks.

8.1 Absorption Times

Absorption times constitute a concept very useful in many applications: reliability, survival analysis, queuing theory, risk analysis, etc.

The phase merging effect can be observed for an "almost ergodic" Markov process with an absorbing state.

Let $x^\varepsilon(t)$, $t \geq 0, \varepsilon > 0$, be a family of Markov jump processes on a measurable phase space $E^0 = E \cup \{0\}$ with absorbing state 0 (see Fig. 8.1), considered in a series scheme with the small series parameter $\varepsilon > 0$, defined by the generator

$$Q^\varepsilon \varphi(x) = q(x) \int_E P^\varepsilon(x, dy)[\varphi(y) - \varphi(x)] \qquad (8.1)$$

on the Banach space **B** of real-valued bounded functions $\varphi(x)$ with sup-norm $\|\varphi\| := \sup_{x \in E} |\varphi(x)|$.

Fig. 8.1 Merging with absorption

The basic assumption is that the stochastic kernel P^ε has the following representation

$$P^\varepsilon(x, B) = P(x, B) - \varepsilon P_1(x, B),$$

where the stochastic kernel $P(x, B)$ on E defines the support Markov chain x_n, $n \geq 0$, uniformly ergodic with stationary distribution $\rho(B)$, $B \in \mathcal{E}$.

The perturbing kernel $P_1(x, B)$ provides the probability of absorption of the embedded Markov chain x_n^ε, $n \geq 0$:

$$\begin{aligned}\mathbb{P}(x_{n+1}^\varepsilon = 0 \mid x_n^\varepsilon = x) &= P^\varepsilon(x, \{0\}) \\ &= 1 - P^\varepsilon(x, E) \\ &= \varepsilon P_1(x, E) =: \varepsilon p(x).\end{aligned} \qquad (8.2)$$

The hitting time to absorption for the Markov process $x^\varepsilon(t)$, $t \geq 0$, is defined by

$$\zeta^\varepsilon := \inf\{t : x^\varepsilon(t) = 0\}.$$

Lemma 8.1 *The transition probabilities*

$$\Phi_\varepsilon(t, x; B) = \mathbb{P}(x^\varepsilon(t/\varepsilon) \in B, \varepsilon\zeta^\varepsilon > t \mid x^\varepsilon(0) = x), \qquad (8.3)$$

satisfy the evolutionary equation

$$\frac{d}{dt}\Phi_\varepsilon(t, x; B) = \varepsilon^{-1} Q^\varepsilon \Phi_\varepsilon(t, x; B) \qquad (8.4)$$

with initial value:

$$\Phi_\varepsilon(0, x; B) = \mathbf{1}_B(x) := \begin{cases} 1, & x \in B \\ 0, & x \notin B. \end{cases}$$

PROOF. By using the representation, on $\{x^\varepsilon(0) = x\}$, $x \in E$,

$$\begin{aligned}\mathbf{1}(x^\varepsilon(t/\varepsilon) \in B, \varepsilon\zeta^\varepsilon > t) &= \mathbf{1}(x^\varepsilon(t/\varepsilon) \in B, \varepsilon\zeta^\varepsilon > t, \varepsilon\theta_x > t) \\ &\quad + \mathbf{1}(x^\varepsilon(t/\varepsilon) \in B, \varepsilon\zeta^\varepsilon > t, \varepsilon\theta_x \leq t),\end{aligned}$$

we get the following relation

$$\begin{aligned}\Phi_\varepsilon(t, x; B) &= \mathbb{P}(x^\varepsilon(t/\varepsilon) \in B, \varepsilon\theta_x > t \mid x^\varepsilon(0) = x) \\ &\quad + \mathbb{P}(x^\varepsilon(t/\varepsilon) \in B, \varepsilon\zeta^\varepsilon > t, \varepsilon\theta_x \leq t \mid x^\varepsilon(0) = x).\end{aligned}$$

The first term is calculated in an obvious way

$$\mathbb{P}(x^\varepsilon(t/\varepsilon) \in B, \varepsilon\theta_x > t \mid x^\varepsilon(0) = x) = \mathbf{1}_B(x) e^{-q(x)t/\varepsilon}.$$

The second term is calculated by using the total probabilities formula:

$$\mathbb{P}(x^\varepsilon(t/\varepsilon) \in B, \varepsilon\zeta^\varepsilon > t, \varepsilon\theta_x \leq t \mid x^\varepsilon(0) = x)$$

$$= q(x) \int_0^{t/\varepsilon} e^{-q(x)s} ds \int_E P^\varepsilon(x, dy) \Phi_\varepsilon(t - \varepsilon s, y; B)$$

$$= q(x) \int_0^{t/\varepsilon} e^{-q(x)(t/\varepsilon - s)} ds \int_E P^\varepsilon(x, dy) \Phi_\varepsilon(\varepsilon s, y; B).$$

Now differentiating with respect to t leads to Equation (8.4). □

Remark 8.1. We used the fact that the first sojourn time in state $x \in E$ is exponentially distributed with intensity $q(x)$:

$$\mathbb{P}(\theta_x > t) = e^{-q(x)t}.$$

Remark 8.2. Equation (8.4) has the following singular perturbing form

$$\frac{d}{dt}\Phi_\varepsilon = [\varepsilon^{-1}Q + Q_1]\Phi_\varepsilon, \tag{8.5}$$

where the perturbing operator Q_1 acts as follows

$$Q_1\varphi(x) = -q(x) \int_E P_1(x, dy)\varphi(y). \tag{8.6}$$

The generator

$$Q\varphi(x) = q(x) \int_E P(x, dy)[\varphi(y) - \varphi(x)],$$

defines the support Markov process $x(t)$, $t \geq 0$, on $(E, \mathcal{E})$, uniformly ergodic with stationary distribution

$$\pi(dx)q(x) = q\rho(dx), \quad q = \int \pi(dx)q(x).$$

The problem of singular perturbation considered in Chapter 5, can be applied for constructing a solution of Equation (8.5).

The formal scheme of the asymptotic solution of Equation (8.5) is the following. Let us consider the following asymptotic representation for a solution of Equation (8.5)

$$\Phi_\varepsilon(t, x) = \Phi(t) + \varepsilon\Phi_1(t, x) + \varepsilon\Theta_\varepsilon(t, x), \tag{8.7}$$

with negligible term $\Theta_\varepsilon(t,x)$ such that

$$|\Theta_\varepsilon(t,x)| \to 0, \quad \varepsilon \to 0.$$

Substituting (8.7) in (8.5) and comparing the terms with the same degrees in ε, yields the following equations:

$$Q\Phi(t) = 0, \qquad (8.8)$$

$$Q\Phi_1(t,x) = \frac{d}{dt}\Phi(t) - Q_1\Phi(t). \qquad (8.9)$$

The first equation (8.8) is satisfied evidently because, by assumption, $\Phi(t)$ does not depend on x.

The second equation (8.9) is valid if the following solvability condition is satisfied (see Section 1.6),

$$\Pi\left[\frac{d}{dt}\Phi(t) - Q_1\Phi(t)\right] = 0, \qquad (8.10)$$

or, in another form

$$\Pi\left[\frac{d}{dt}\Phi(t) - \widehat{Q}_1\Phi(t)\right] = 0, \qquad (8.11)$$

where the contracted operator $\widehat{Q}_1$ is determined by the relation (see Section 5.2)

$$\widehat{Q}_1\Pi = \Pi Q_1\Pi. \qquad (8.12)$$

The projector Π is defined by the relation

$$\Pi\varphi(x) = \int_E \pi(dx)\varphi(x) = \widehat{\varphi}\mathbf{1}(x) = 1, \ x \in E.$$

Calculation in (8.12) by using (8.6) gives the following

$$\widehat{Q}_1 = -\int_E \pi(dx)q(x)p(x) = -q\int_E \rho(dx)p(x) = -q\widehat{p},$$

where

$$\widehat{p} = \int_E \rho(dx)p(x)$$

is the *stationary averaged absorbing probability*.

8.1. ABSORPTION TIMES

Now, Equation (8.10) can be written

$$\frac{d}{dt}\Phi(t) + \Lambda\Phi(t) = 0, \quad \Lambda = q\widehat{p}. \tag{8.13}$$

Its initial value has to be chosen as follows

$$\Phi(0) = \Pi\Phi_\varepsilon(0,x) = \Pi 1_B(x) = \pi(B). \tag{8.14}$$

The solution of Equation (8.13) under initial condition (8.14) can be represented as follows

$$\Phi(t) = \pi(B)e^{-\Lambda t},$$

that is the limit representation of the transition probabilities (8.3)

$$\lim_{\varepsilon \to 0} \mathbb{P}(x^\varepsilon(t/\varepsilon) \in B, \varepsilon\zeta^\varepsilon > t \mid x^\varepsilon(0) = x) = \pi(B)e^{-\Lambda t}.$$

This scheme can be formally verified (see for instance [116]). In the next section this result will be obtained as a corollary in a more general scheme. The following result is obvious.

Corollary 8.1 *The following limits hold:*

$$\lim_{\varepsilon \to 0} \mathbb{P}(x^\varepsilon(t/\varepsilon) \in B \mid \zeta^\varepsilon > t/\varepsilon) = \pi(B),$$

$$\lim_{\varepsilon \to 0} \mathbb{P}(\varepsilon\zeta^\varepsilon > t) = e^{-\Lambda t}, \quad \Lambda = q\widehat{p}.$$

It is worth noticing that the limit conditional probability is the so-called *quasi-stationary distribution*.[148]

Heuristic principle. Corollary 8.1 can be interpreted as the heuristic principle of reliability for almost ergodic stochastic systems with stationary distribution $\pi(B), B \in \mathcal{E}$, on the set E of working states.

1. The stopping time ζ has exponential distribution

$$\mathbb{P}(\zeta > t) = e^{-\Lambda t}, \quad t \geq 0.$$

2. The intensity of the stopping time is

$$\Lambda = q\widehat{p},$$

where

$$q = \int_E \pi(dx)q(x),$$

is the stationary intensity of sojourn times in working states,

$$\widehat{p} = \int_E \pi(dx)p(x),$$

is the stationary probability of absorption, and $p(x)$ are the probabilities of absorption in state $x \in E$.

▷ **Example 8.1.** Consider a three state Markov process, $E^0 = \{0, 1, 2\}$, with generator matrix

$$Q^\varepsilon = \begin{pmatrix} 0 & 0 & 0 \\ \varepsilon\lambda & -(1+\varepsilon)\lambda & \lambda \\ \varepsilon\mu & \mu & -(1+\varepsilon)\mu \end{pmatrix} = \underbrace{\begin{pmatrix} 0 & 0 & 0 \\ 0 & -\lambda & \lambda \\ 0 & \mu & -\mu \end{pmatrix}}_{Q} + \varepsilon \underbrace{\begin{pmatrix} 0 & 0 & 0 \\ \lambda & -\lambda & 0 \\ \mu & 0 & -\mu \end{pmatrix}}_{Q_1}$$

The transition matrix of the embedded Markov chain is

$$P^\varepsilon = \begin{pmatrix} 1 & 0 & 0 \\ \varepsilon & 0 & 1-\varepsilon \\ \varepsilon & 1-\varepsilon & 0 \end{pmatrix} = \underbrace{\begin{pmatrix} 1 & 0 & 0 \\ 0 & 0 & 1 \\ 0 & 1 & 0 \end{pmatrix}}_{P} + \varepsilon \underbrace{\begin{pmatrix} 0 & 0 & 0 \\ 1 & 0 & -1 \\ 1 & -1 & 0 \end{pmatrix}}_{P_1}.$$

Now, for the ergodic process $x(t), t \geq 0$, taking values in $E = \{1, 2\}$, and generator Q, we have $\pi = (\frac{\mu}{\lambda+\mu}, \frac{\lambda}{\lambda+\mu})$. For the ergodic embedded Markov chain $x_n, n \geq 0$, we have $\rho = (1/2, 1/2)$.

Thus, since we have $p(1) = -P_1(1, E) = 1$ and $p(2) = -P_1(2, E) = 1$, the stoppage probability is $p = 1$.

On the other hand, we have $q(1) = \lambda$, and $q(2) = \mu$.
Hence

$$q = \pi_1 q(1) + \pi_2 q(2) = \frac{2\lambda\mu}{\lambda+\mu}$$

and

$$\Lambda = qp = \frac{2\lambda\mu}{\lambda+\mu}.$$

The limit of the distribution of the normalized absorption time is

$$\mathbb{P}(\zeta > t) = \exp(-\Lambda t).$$

8.2. STATIONARY PHASE MERGING

Remark 8.3. In order to obtain Equation (8.11) for the limit term $\Phi(t)$ in the asymptotic representation (8.7) we have to write Equation (8.9) for the next asymptotic term $\Phi_1(t,x)$ and use the solvability condition for this equation. This is the main scheme in the problem of singular perturbation approach (see Chapter 5).

8.2 Stationary Phase Merging

A surprising possibility of the phase merging scheme is that the simplification model can be constructed for the ergodic semi-Markov process $x(t), t \geq 0$, given by the semi-Markov kernel

$$Q(x, B, t) = P(x, B) F_x(t) = \mathbb{P}(x_{n+1} \in B \mid x_n = x) \mathbb{P}(\theta_{n+1} \leq t \mid x_n = x), \tag{8.15}$$

where $x \in E, B \in \mathcal{E}, t \geq 0$, and $x_n, \tau_n, n \geq 0$, is the Markov renewal process associated with the semi-Markov process $x(t), t \geq 0$, (see Section 1.3).

The main assumption is that the semi-Markov process $x(t), t \geq 0$, has a unique stationary distribution $\pi(B), B \in \mathcal{E}$. Let $\rho(B), B \in \mathcal{E}$, be the stationary distribution of the embedded Markov chain $x_n, n \geq 0$. As we have already point out, these two distributions are connected by the following relations:

$$\pi(dx) = \rho(dx)m(x)/m, \quad m(x) := \int_0^\infty \overline{F}_x(t)dt, \quad m := \int_E \rho(dx)m(x).$$

Let the split of phase space:

$$E = \bigcup_{k=1}^{N} E_k, \quad E_k \bigcap E_{k'} = \emptyset, \quad k \neq k' \tag{8.16}$$

be given such that the inequality holds

$$\rho(E_k) > 0, \quad 1 \leq k \leq N. \tag{8.17}$$

Introduce the merging function

$$v(x) = k, \quad x \in E_k, \quad 1 \leq k \leq N.$$

Definition 8.1 The *stationary phase merging process* is a Markov renewal process $\widehat{x}_n, \widehat{\theta}_n, n \geq 0$, on the merged phase space $\widehat{E} = \{1, ..., N\}$, defined by the merged semi-Markov matrix

$$\widehat{Q}(t) = [\widehat{Q}_{kr}(t); 1 \leq k, r \leq N],$$

given by the transition probabilities

$$\widehat{Q}_{kr}(t) = \mathbb{P}(\widehat{x}_{n+1} = r, \widehat{\theta}_{n+1} \leq t \mid \widehat{x}_n = k)$$
$$= \int_{E_k} \rho(dx) Q(x, E_r, t)/\rho(E_k), \quad 1 \leq k, r \leq N. \quad (8.18)$$

Particularly, the transition probabilities of the embedded Markov chain $\widehat{x}_n, n \geq 0$, are given by the relation

$$\widehat{p}_{kr} := \mathbb{P}(\widehat{x}_{n+1} = r \mid \widehat{x}_n = k)$$
$$= \int_{E_k} \rho(dx) P(x, E_r)/\rho(E_k), \quad 1 \leq k, r \leq N. \quad (8.19)$$

The distribution functions of the merged sojourn times are represented as follows

$$\widehat{F}_k(t) = \mathbb{P}(\widehat{\theta}_{n+1} \leq t \mid \widehat{x}_n = k)$$
$$= \int_{E_k} \rho(dx) F_x(t)/\rho(E_k), \quad 1 \leq k, r \leq N. \quad (8.20)$$

The problem is to determine how the Markov renewal process $\widehat{x}_n, \widehat{\tau}_n, n \geq 0$, is associated with the ergodic Markov renewal process $x_n, \tau_n, n \geq 0$. The solution of this problem is formulated as follows.

First define the restriction of the embedded Markov chain $x_n, n \geq 0$, to the set E_k, which is a Markov chain also denoted by $x_{\nu_m^{(k)}}, m \geq 0$, where

$$\nu_m^{(k)} := \inf\{n > \nu_{m-1}^{(k)} : x_n \in E_k\}, \quad \nu_0^{(k)} = 0, \quad m \geq 1,$$

are the observed jump times into E_k.

Theorem 8.1 *(Stationary phase merging [6]) Under the ergodicity condition for the embedded Markov chain $x_n, n \geq 0$, and the additional condition (8.17) the weak convergence holds*

$$\lim_{m \to \infty} \mathbb{P}(x_{\nu_m^{(k)}+1} \in E_r, \theta_{\nu_m^{(k)}+1} \leq t) = \widehat{Q}_{kr}(t), \quad 1 \leq k, r \leq N. \quad (8.21)$$

Especially,

$$\lim_{m \to \infty} \mathbb{P}(x_{\nu_m^{(k)}+1} \in E_r) = \widehat{p}_{kr}.$$

8.2. STATIONARY PHASE MERGING

Remark 8.4. Introduce the merged process

$$\tilde{x}(t) := v(x(t)), \quad t \geq 0, \tag{8.22}$$

on the merged phase space $\widehat{E} = \{1, ..., N\}$, and define its renewal moments

$$\tilde{\tau}_{n+1} := \inf\{t > \tilde{\tau}_n : \tilde{x}(t) \neq \tilde{x}(\tilde{\tau}_n)\}, \quad n \geq 0, \quad \tilde{\tau}_0 = 0.$$

Consider now the two-component process

$$\tilde{x}_n := \tilde{x}(\tilde{\tau}_n), \quad \tilde{\tau}_n, \quad n \geq 0.$$

The statement of Theorem 8.1 means that the weak convergence of

$$\tilde{x}_n, \quad \tilde{\theta}_n := \tilde{\tau}_n - \tilde{\tau}_{n-1}, \quad n \geq 1, \tag{8.23}$$

to the Markov renewal process $\widehat{x}_n, \widehat{\theta}_n, n \geq 0$, takes place as follows

$$(\tilde{x}_{n+m}, \tilde{\theta}_{n+m}) \Longrightarrow (\widehat{x}_n, \widehat{\theta}_n), \quad m \to \infty.$$

That is, in the steady-state regime, the merged process (8.22) can be considered as being associated with the stationary merged MRP $\widehat{x}_n, \widehat{\tau}_n, n \geq 0$.

PROOF OF THEOREM 8.1. The proof is based on considering the two component ergodic Markov chain

$$\zeta_n = (x_n, x_{n+1}), \quad n \geq 0, \tag{8.24}$$

on the phase space $E \times E$, with the stationary distribution

$$\tilde{\rho}(dx, dy) = \rho(dx) P(x, dy). \tag{8.25}$$

Set $A_k = E_k \times E$. The hitting times $\nu_m^{(k)}$ of the set E_k for the Markov chain $x_n, n \geq 0$, are the hitting times of the set A_k for the Markov chain $\zeta_n, n \geq 0$. The crucial property is that the thinned sequence $\zeta_{\nu_m^{(k)}}, m \geq 0$, is the ergodic Markov chain on the set A_k, with the stationary distribution

$$\rho_k(dx, dy) = \tilde{\rho}(dx, dy)/\tilde{\rho}(A_k). \tag{8.26}$$

From (8.25) and (8.26) this yields

$$\rho_k(dx, dy) = \rho(dx) P(x, dy)/\rho(E_k). \tag{8.27}$$

Now the ergodicity condition gives:

$$\lim_{m\to\infty} \mathbb{P}(x_{\nu_m^{(k)}+1} \in E_r) = \lim_{m\to\infty} \mathbb{P}(\zeta_{\nu_m^{(k)}} \in E_k \times E_r)$$
$$= \int_{E_k \times E_r} \rho_k(dx, dy)$$
$$= \int_{E_k} \rho(dx) P(x, E_r)/\rho(E_k)$$
$$= \widehat{p}_{kr}.$$

Taking into account the Markov property of renewal times $\nu_m^{(k)}, m \geq 1$, we calculate

$$\mathbb{P}(x_{\nu_m^{(k)}+1} \in E_r, \theta_{\nu_m^{(k)}+1} \leq t)$$
$$= \int_{E_k} \mathbb{P}(x_{\nu_m^{(k)}+1} \in E_r, \theta_{\nu_m^{(k)}+1} \leq t \mid x_{\nu_m^{(k)}} = x) P_m(dx),$$

with $P_m(dx) = \mathbb{P}(x_{\nu_m^{(k)}} \in dx)$.
Hence

$$\mathbb{P}(x_{\nu_m^{(k)}+1} \in E_r, \theta_{\nu_m^{(k)}+1} \leq t) = \int_{E_k} Q(x, E_r, t) P_m(dx), \qquad (8.28)$$

but the Markov chain $x_{\nu_m^{(k)}}, m \geq 0$, is ergodic on the set E_k with stationary distribution $\rho(dx)/\rho(E_k)$.

Passing to the limit in (8.28) we get (8.21). □

It is worth noticing that the merged embedded Markov chain $\widehat{x}_n, n \geq 0$, has, in general, a virtual jumps, which can be extracted in the well-known way. The transition probabilities of the merged embedded Markov chain $\widehat{x}_n^0, n \geq 0$, without virtual passages are represented as follows

$$\widehat{p}_{kr}^0 = \int_{E_k} \rho(dx) P(x, E_r) / \int_{E_k} \rho(dx) P(x, E \backslash E_k), \quad 1 \leq k, r \leq N.$$

The stationary phase merging principle introduced in Theorem 8.1 gives important hints for simplifying the analysis of semi-Markov systems.

The Markov renewal process with the ergodic embedded Markov chain can be merged for any arbitrary split phase space (8.16) which satisfies Condition (8.17). The merged process will again be a Markov renewal process given by the semi-Markov kernel calculated by (8.18).

As usual, the split phase space (8.16) is considered under some technical arguments, for example, states connected by some criterion (see Section 8.3).

8.3 Superposition of Two Renewal Processes

The semi-Markov model of stochastic system can be constructed by expanding the physical (technical) phase space up to the "semi-Markov" phase states. In other words, the technical states yield a natural split of the semi-Markov phase space (see Section 8.3).

8.3 Superposition of Two Renewal Processes

Let us consider two independent renewal processes, say τ^1 and τ^2, given by (see Fig. 8.2),

$$\tau_n^i = \sum_{k=0}^{n} \alpha_k^i, \quad n \geq 0, \quad i = 1, 2, \tag{8.29}$$

where $\alpha_k^i, k \geq 0, i = 1, 2$ are i.i.d. positive random variables, with distribution functions

$$F_i(t) := \mathbb{P}(\alpha_k^i \leq t), \quad F_i(0) = 0, \quad k \geq 0, i = 1, 2. \tag{8.30}$$

The counting renewal processes are defined by the relation

$$\nu_i(t) := \max\{n \geq 0 : \tau_n^i \leq t\}, \quad t \geq 0. \tag{8.31}$$

The superposition of the two independent renewal processes (8.29) is defined by

$$\nu(t) := \nu_1(t) + \nu_2(t), \quad t \geq 0. \tag{8.32}$$

Denote by $\tau_n, n \geq 0$, the renewal moments of the superposed process.

Certainly, the counting process (8.32) is a renewal process only in the case of exponential distributions (8.30). In that case, the three processes are homogeneous Poisson processes.

Nevertheless, the superposition (8.32) can be described by the Markov renewal process $x_n, \tau_n, n \geq 0$, on the phase space $E = E_1 \cup E_2$, $E_i = \{(i, x) : x \geq 0\}$, $i = 1, 2$, with the following formulas for sojourn times $\theta_{n+1} := \tau_{n+1} - \tau_n, n \geq 0$,

$$\theta_x^i = \alpha^i \wedge x. \tag{8.33}$$

The simplest way to explain this formula is to consider Fig. 8.2.

254   CHAPTER 8. APPLICATIONS I

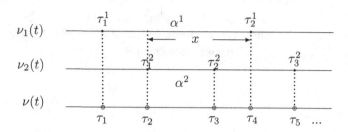

Fig. 8.2 Semi-Markov random walk

In order to fix the state of the embedded Markov chain $x_n, n \geq 0$, we have to introduce a residual time x from one renewal time τ_n^1 to the next nearly renewal time which can be τ_{n+1}^1 if $\alpha_{n+1}^1 \leq x$, or τ_m^2 if $\alpha_{n+1}^1 > x$.

So, renewal moments $\tau_n, n \geq 0$, can be described by the Markov chain $x_n, \zeta_n, n \geq 0$, with values in E, and which transition probabilities are given in the matrix

$$P(x, dy) = \begin{pmatrix} F_1(x - dy) & F_1(x + dy) \\ F_2(x + dy) & F_2(x - dy) \end{pmatrix}, \tag{8.34}$$

where:

$$F_1(x - dy) := \mathbb{P}(x_{n+1} = 1, \zeta_{n+1} \in dy \mid x_n = 1, \zeta_n = x),$$

$$F_1(x + dy) := \mathbb{P}(x_{n+1} = 2, \zeta_{n+1} \in dy \mid x_n = 1, \zeta_n = x),$$

$$F_2(x + dy) := \mathbb{P}(x_{n+1} = 1, \zeta_{n+1} \in dy \mid x_n = 2, \zeta_n = x),$$

$$F_2(x - dy) := \mathbb{P}(x_{n+1} = 2, \zeta_{n+1} \in dy \mid x_n = 2, \zeta_n = x).$$

Here the transition $((1, x), (1, dy))$ means that $\alpha^1 \in x - dy$, that is, $x - y < \alpha^1 \leq x - y + dy$. Similar interpretation holds for the other transitions.

The particularity of the Markov chain $x_n, \zeta_n, n \geq 0$, is that it has a stationary distribution determined by

$$\rho_1(dx) = a\overline{F}_2(x)dx, \quad \rho_2(dx) = a\overline{F}_1(x)dx. \tag{8.35}$$

where $a = 1/(a_1 + a_2)$, $a_i = \mathbb{E}\alpha^i, i = 1, 2.$

8.3. SUPERPOSITION OF TWO RENEWAL PROCESSES

It is worth noticing that the densities (8.35) can be defined by the stationary residual times α^{i*} according to the renewal theorem:

$$\rho_1(dx) = \rho_1 f_2^*(x)dx, \quad \rho_2(dx) = \rho_2 f_1^*(x)dx,$$

where:

$$\rho_i := \rho(E_i) = \frac{a_i}{a_1 + a_2} = a a_i,$$

$$f_i^*(x) = \overline{F}_i(x)/a_i.$$

The semi-Markov kernel of the Markov renewal process $x_n, \tau_n, n \geq 0$, can be calculated starting from (8.33) as follows. Set $Q_{ij}(x, dy, t)$ instead of $Q((i, x), (j, dy), t)$. We have:

$$Q_{12}(x, dy, t) = \mathbb{P}(\alpha^1 > x, \alpha^1 \in x + dy, \theta_x^1 \leq t)$$
$$= \mathbb{P}(\alpha^1 \in x + dy, x \leq t)$$
$$= F_1(x + dy)\mathbf{1}_{(x \leq t)},$$

and, similarly,

$$Q_{21}(x, dy, t) = F_2(x + dy)\mathbf{1}_{(x \leq t)}.$$

Next:

$$Q_{11}(x, dy, t) = \mathbb{P}(\alpha^1 \leq x, \alpha^1 \in x - dy, \theta_x^1 \leq t)$$
$$= \mathbb{P}(\alpha^1 \in x - dy, \alpha^1 \leq t)$$
$$= F_1(x - dy)\mathbf{1}_{(x \leq t+y)},$$

and, similarly,

$$Q_{22}(x, dy, t) = F_2(x - dy)\mathbf{1}_{(x \leq t+y)}.$$

In particular, we need to calculate in the stationary phase merging theorem

$$Q_{12}(x, E_2, t) = \overline{F}_1(x)\mathbf{1}_{(x \leq t)}$$
$$Q_{21}(x, E_1, t) = \overline{F}_2(x)\mathbf{1}_{(x \leq t)} \qquad (8.36)$$
$$Q_{ii}(x, E_i, t) = F_i(x \wedge t), \quad i = 1, 2.$$

Now, we can construct the stationary phase merging Markov renewal process describing the superposition of two independent renewal processes.

Proposition 8.1 (^{100}) *The stationary phase merging Markov renewal process $\widehat{x}_n, \widehat{\theta}_n, n \geq 0$, on the phase merged space $\widehat{E} = \{1, 2\}$ is defined by the equation for the sojourn times in states*

$$\widehat{\theta}_1 = \alpha^1 \wedge \alpha^{2*}, \quad \widehat{\theta}_2 = \alpha^{1*} \wedge \alpha^2, \tag{8.37}$$

where the upper index "*" means the stationary residual time, that is,

$$\mathbb{P}(\alpha^{i*} \in dx) = f_i^*(x)dx, \quad f_i^*(x) = \overline{F}_i(x)q_i, \quad q_i := 1/a_i, \quad i = 1, 2.$$

PROOF. The semi-Markov kernel for the stationary merged Markov renewal process is calculated by using representation (8.36) and the stationary distributions (8.35).

According to (8.18), we get:

$$\widehat{Q}_{12}(t) = q_2 \int_0^\infty f_2^*(x) \overline{F}_1(x) \mathbf{1}_{(x \leq t)} dx$$

$$= q_2 \int_0^t \overline{F}_2(x) \overline{F}_1(x) dx$$

$$=: q_2 F(t), \tag{8.38}$$

where, by definition,

$$F(t) := \int_0^t \overline{F}_2(x) \overline{F}_1(x) dx.$$

Similarly,

$$\widehat{Q}_{ii}(t) = \widehat{Q}_i(t) - \widehat{Q}_{ij}(t), \tag{8.39}$$

with:

$$\widehat{Q}_i(t) = 1 - \overline{F}_i(t) \overline{F}_j^*(t), \quad i = 1, 2, \quad j = 2, 1,$$

$$\overline{F}_j^*(t) := \mathbb{P}(\alpha^{j*} > t) = q_j \int_t^\infty \overline{F}_j(s) ds, \quad j = 2, 1.$$

Next:

$$\widehat{Q}_{11}(t) = q_2 \int_0^t \overline{F}_2(x) \overline{F}_1(x) dx$$

$$\widehat{Q}_{22}(t) = q_1 \int_0^t \overline{F}_2(x) \overline{F}_1(x) dx. \tag{8.40}$$

8.3. SUPERPOSITION OF TWO RENEWAL PROCESSES

Particularly, the transition probabilities of the embedded Markov chain $\widehat{x}_n, n \geq 0$, are represented as follows

$$\widehat{p}_{12} = \widehat{Q}_{12}(+\infty) = q_2\phi, \quad \widehat{p}_{21} = \widehat{Q}_{21}(+\infty) = q_1\phi,$$

$$\phi := \int_0^\infty \overline{F}_2(x)\overline{F}_1(x)dx = \mathbb{E}[\alpha^1 \wedge \alpha^2],$$

and, obviously,

$$\widehat{p}_{12} = 1 - \widehat{p}_{21}, \quad \widehat{p}_{22} = 1 - \widehat{p}_{21}.$$

Verification that the semi-Markov kernel $\widehat{Q}_{ij}(t), i,j = 1,2$, calculated in (8.37), of Proposition 8.1, coincides with that calculated in formulas (8.38)-(8.40), can be easily obtained in the backward way.

Let us calculate the semi-Markov kernel of the Markov renewal process given by the sojourn times (8.37):

$$\begin{aligned}
\widehat{Q}_{12}(t) &= \mathbb{P}(\widehat{\theta}_1 \leq t, \alpha^1 > \alpha^{2*}) \\
&= \mathbb{P}(\alpha^{2*} \leq t, \alpha^{2*} < \alpha^1) \\
&= q_2 \int_0^t \overline{F}_2(x)\overline{F}_1(x)dx \\
&=: q_2 F(t),
\end{aligned}$$

that is exactly (8.38).

Next:

$$\begin{aligned}
\widehat{Q}_{11}(t) &= \mathbb{P}(\widehat{\theta}_1 \leq t, \alpha^{2*} > \alpha^1) \\
&= \mathbb{P}(\alpha^1 \leq t, \alpha^1 < \alpha^{2*}) \\
&= q_2 \int_0^t \overline{F}_2(x)dF_1(x) \\
&= q_2 \int_0^t \overline{F}_2(x)\overline{F}_1(x)dx,
\end{aligned}$$

that is exactly (8.40). □

Comparing the stochastic representations (8.33) and (8.37) for the initial Markov renewal process $x_n, \theta_n, n \geq 0$, and the stationary merged Markov renewal process $\widehat{x}_n, \widehat{\theta}_n, n \geq 0$, we can conclude that going from (8.35) to (8.37) is very simple. The residual time in (8.35) is replaced by the stationarily distributed residual time α^*. That constitutes the *heuristic principle of stationary phase merging scheme*.

In the superposition of Markov renewal processes the residual times are transformed into the stationary residual times.

8.4 Semi-Markov Random Walks

The semi-Markov random walk (SMRW) is defined and the average, diffusion and Poisson approximation results are presented here.

The SMRWs, obtained by the superposition of two independent renewal processes constitute a special field in the theory of semi-Markov processes.

8.4.1 *Introduction*

Let $\nu_+(t), t \geq 0$ and $\nu_-(t), t \geq 0$, be the counting processes of the two renewal processes defined as follows (see Section 8.3):

$$\nu_\pm(t) = \max\{n : \sum_{k=1}^{n} \alpha_k^\pm \leq t\}, \quad t \geq 0. \tag{8.41}$$

The time intervals $\alpha_k^\pm, k \geq 1$, are jointly independent and are given by the distribution functions

$$P_\pm(t) = \mathbb{P}(\alpha_k^\pm \leq t), \quad t \geq 0.$$

The superposition of the two renewal processes is given by

$$\nu(t) = \nu^+(t) + \nu^-(t), \quad t \geq 0. \tag{8.42}$$

The SMRW is defined by the following sums

$$\zeta(t) = \xi_0 + \sum_{r=1}^{\nu^+(t)} \beta_r^+ - \sum_{r=1}^{\nu^-(t)} \beta_r^-, \quad t \geq 0. \tag{8.43}$$

The jumps $\beta_r^\pm, r \geq 1$, are jointly independent and are given by the distribution functions

$$G_\pm(u) = \mathbb{P}(\beta_r^\pm \leq u), \quad u \geq 0.$$

This kind of processes are interesting for various applied problems. They model the number of customers in the queue systems with given distribution functions of arrival and service times, and they can be interpreted

8.4. SEMI-MARKOV RANDOM WALKS

as storage processes with arbitrary distribution of time intervals between arrivals and departures of goods. The process (8.43) can also be considered as a mathematical model of risk with arbitrary distributions of intervals between times of payment of claims and the premium income.

The superposition of two renewal processes (8.41) can be described by the counting process $\nu(t) = \max\{n : \tau_n \leq t\}$, for the Markov renewal process $x_n, \tau_n, n \geq 0$, on the phase space $E = E_+ \cup E_-$, $E_\pm = \{(\pm, x) : x \geq 0\}$, with the following formula for sojourn times ($\theta_{n+1} := \tau_{n+1} - \tau_n, n \geq 0$).

$$\theta_x^\pm = \alpha^\pm \wedge x.$$

The transition probabilities of the embedded Markov chain $x_n, n \geq 0$, is defined by the matrix

$$P(x, dy) = \begin{bmatrix} P_+(x-dy) & P_+(x+dy) \\ P_-(x+dy) & P_-(x-dy) \end{bmatrix}. \tag{8.44}$$

The stationary distribution of embedded Markov chain has the density

$$\rho_\pm(t) = \overline{P}_\pm(t)/p, \quad p = p_+ + p_-, \quad p_\pm = \mathbb{E}\alpha^\pm, \tag{8.45}$$

where, as usual, $\overline{P}(t) := 1 - P(t)$.

The embedded SMRW $\zeta_n = \zeta(\tau_n), n \geq 0$, is defined by the relation

$$\zeta_{n+1} = \zeta_n + \mathbf{1}(x_{n+1} = +)\beta_{n+1}^+ - \mathbf{1}(x_{n+1} = -)\beta_{n+1}^-, \quad n \geq 0,$$

where $\mathbf{1}(A)$ is the indicator of a random event A. The SMRW (8.43) can be defined as follows: $\zeta(t) = \zeta_{\nu(t)}, \quad t \geq 0$.

8.4.2 The algorithms of approximation for SMRW

The algorithms of approximation for the SMRW (8.43) in the series scheme with the small series parameter $\varepsilon \to 0$ ($\varepsilon > 0$) are considered here. The average, diffusion and Poisson approximation schemes are investigated. The approximation algorithms are constructed by using the asymptotic expansion of the *compensating operator*, for the extended Markov renewal process $\zeta_n, x_n, \tau_n, n \geq 0$, and a solution of the singular perturbation problem for the generator of associated Markov process.

Introduce the notation for the first moments:

$$p_\pm = \mathbb{E}\alpha_k^\pm, \quad b_\pm = \mathbb{E}\beta_k^\pm.$$

The average drift per unit time for SMRW (8.43) is defined by the value

$$b = b_+/p_+ - b_-/p_-.$$

Note that we have by the limit theorem for renewal processes

$$\mathbb{E}\nu_\pm(t)/t \to 1/p_\pm, \quad t \to \infty.$$

Hence, the mean value of both sums on the right hand side of (8.43) have the following asymptotic evaluations

$$\mathbb{E} \sum_{r=1}^{\nu_\pm(t)} \beta_r^\pm / t \sim b_\pm/p_\pm, \quad t \to \infty.$$

Hence, the average drift per unit time for SMRW (8.43) is evaluated by

$$\mathbb{E}\xi(t) \sim u + bt, \quad t \to \infty.$$

The average algorithm. This algorithm is obtained for SMRWs in the following series scheme

$$\zeta_\varepsilon(t) = u + \varepsilon \left[\sum_{k=1}^{\nu^+(t/\varepsilon)} \beta_k^+ - \sum_{k=1}^{\nu^-(t/\varepsilon)} \beta_k^- \right], \quad t \geq 0. \qquad (8.46)$$

Theorem 8.2 *(Average) Under the condition $b \neq 0$ and the finiteness of the second moments of $\beta_k^\pm, k \geq 1$, the weak convergence*

$$\zeta_\varepsilon(t) \Longrightarrow \zeta_0(t) = u + bt, \quad \varepsilon \to 0,$$

takes place.

The algorithm of Poisson approximation. This algorithm is obtained for SMRWs in the following series scheme

$$\zeta^\varepsilon(t) = u + \left[\sum_{k=1}^{\nu^+(t/\varepsilon)} \beta_k^{\varepsilon,+} - \varepsilon \sum_{k=1}^{\nu^-(t/\varepsilon)} \beta_k^- \right], \quad t \geq 0.$$

The distribution function $G_+^\varepsilon(u)$ of the random variables $\beta_k^{\varepsilon,+}, k \geq 1$, satisfies the following Poisson approximation conditions:

PA1:

$$\int_0^\infty g(v) G_+^\varepsilon(dv) = \varepsilon \left[\int_0^\infty g(v) G_+(dv) + \theta_g^\varepsilon \right], \quad g(v) \in C_3(\mathbb{R}_+),$$

on the measure determining class $C_3(\mathbb{R}_+)$;
PA2:

$$\int_0^\infty v G_+^\varepsilon(dv) = \varepsilon[b_+ + \theta_b^\varepsilon], \quad \int_0^\infty v^2 G_+^\varepsilon(dv) = \varepsilon[c_+ + \theta_c^\varepsilon],$$

The negligible terms $\theta_g^\varepsilon, \theta_b^\varepsilon, \theta_c^\varepsilon$, satisfy the condition

$$|\theta_\bullet^\varepsilon| \to 0, \quad \varepsilon \to 0.$$

Theorem 8.3 *(Poisson Approximation) Under Conditions PA1 and PA2 and the finiteness of the third moments of $\beta_k^{\varepsilon,+}, k \geq 1$, the weak convergence*

$$\zeta^\varepsilon(t) \Longrightarrow \zeta^0(t) = u + b^0 t + \sum_{k=1}^{\nu^0(t)} \beta_k^0, t \geq 0,$$

takes place. The distribution function of the jumps $\beta_k^0, k \geq 1$, is determined by

$$G^0(u) = \mathbb{P}(\beta_k^0 \leq u) = G_+(u)/g_+, \quad g_+ = G_+(+\infty).$$

The intensity of the counting Poisson process $\nu^0(t), t \geq 0$, is defined by

$$\Lambda = g_+/p_+.$$

The deterministic drift b^0 of the compound Poisson process $\zeta^0(t), t \geq 0$, is defined by

$$b^0 = b - b_+^0/p_+, \quad b_+^0 = \int_0^\infty v G_+(dv). \tag{8.47}$$

Diffusion approximation scheme. This scheme is obtained for SMRWs in the following series scheme

$$\widetilde{\zeta}_\varepsilon(t) = u + \varepsilon \left[\sum_{k=1}^{\nu^+(t/\varepsilon^2)} \beta_k^+ - \sum_{k=1}^{\nu^-(t/\varepsilon^2)} \beta_k^- \right], \quad t \geq 0. \tag{8.48}$$

Theorem 8.4 *(Diffusion approximation) Under the balance condition $b = 0$, and the finiteness of the third moments of $\beta_k^\pm, k \geq 1$, the weak convergence*

$$\widetilde{\zeta}_\varepsilon(t) \Longrightarrow \widetilde{\zeta}_0(t) = u + \sigma w(t), \quad \varepsilon \to 0,$$

takes place. The process $w(t), t \geq 0$, is the standard Wiener process. The variance is determined by the formulas:

$$\sigma^2 = 2\sigma_0^2 + \sigma_2^2, \qquad \sigma_2^2 = c_+/p_+ + c_-/p_-,$$

$$c_\pm = E(\beta_k^\pm)^2, \qquad \sigma_0^2 = \int_0^\infty \rho(x)\sigma_0(x)dx,$$

$$\sigma_0(x) = \bar{b}_0(x)R_0\bar{b}_0(x).$$

The vectors $\bar{b}_0(x)$ are defined by

$$\bar{b}_0(x) = (b_+P_+(x) - b_-\overline{P}_+(x),\ b_+\overline{P}_-(x) - b_-P_-(x)).$$

The potential operator R_0 for the generator $Q = q(x)[P - I]$ of the embedded Markov chain $x_n, n \geq 0$, is defined by the relation

$$QR_0 = R_0Q = \Pi - I,$$

where the projector Π acts as follows on the real-valued vector-function $\varphi(x) = (\varphi_+(x), \varphi_-(x))$,

$$\Pi\varphi(x) = \int_0^\infty \rho_+(x)\varphi_+(x)dx + \int_0^\infty \rho_-(x)\varphi_-(x)dx.$$

8.4.3 Compensating Operators

The compensating operator on the real-valued vector test functions $\varphi(u, x) = (\varphi_+(u, x), \varphi_-(u, x))$ is given by the relation

$$\mathbb{L}\varphi(u, x) = q(x)[P\mathbb{G} - I]\varphi(u, x),$$

where the operator P is defined by the transition probabilities (8.44), and the integral operator $\mathbb{G}$ is defined by

$$\mathbb{G} = \begin{bmatrix} \mathbb{G}_+ & 0 \\ 0 & \mathbb{G}_- \end{bmatrix}, \qquad \mathbb{G}_\pm\varphi(u) = \int_0^\infty \mathbb{G}_\pm(dv)\varphi(u \pm v).$$

The product operator q is given by

$$q(x) = \begin{bmatrix} q_+(x) & 0 \\ 0 & q_-(x) \end{bmatrix}, \qquad q_\pm(x) = 1/p_\pm(x),$$

where

$$p_\pm(x) = \mathbb{E}(\alpha^\pm \wedge x) = \int_0^x \overline{P}_\pm(t)dt$$

8.4. SEMI-MARKOV RANDOM WALKS

Note that the following relation holds:

$$\mathbb{E}[\varphi_{\pm}(\zeta_{n+1}, x_{n+1})|\zeta_n = u, x_n = x = \pm] = P\mathbb{G}\varphi(u,x)$$

$$= \int_0^x P_{\pm}(dt) \int_0^\infty G_{\pm}(dv)\varphi_{\pm}(u \pm v, x - t)$$
$$+ \int_x^\infty P_{\pm}(dt) \int_0^\infty G_{\mp}(dv)\varphi_{\mp}(u \mp v, t - x).$$

So, the operator $P\mathbb{G} - I$ is the generator of the embedded Markov chain $\zeta_n, x_n, n \geq 0$.

The approximation algorithms, introduced in Theorems 8.2-8.4, are constructed by using the asymptotic expansion for the compensating operator in the corresponding series scheme.

The compensating operator in the average scheme (8.46) *is represented in the following form*

$$\mathbb{L}^\varepsilon \varphi = \varepsilon^{-1} q[P\mathbb{G}^\varepsilon - I]\varphi, \qquad (8.49)$$

where

$$\mathbb{G}^\varepsilon = \begin{bmatrix} \mathbb{G}^\varepsilon_+ & 0 \\ 0 & \mathbb{G}^\varepsilon_- \end{bmatrix}, \quad \mathbb{G}^\varepsilon_\pm \varphi(u) = \int_0^\infty \varphi(u \pm \varepsilon v) G_\pm(dv).$$

Lemma 8.2 *Under the condition of Theorem 8.2, the compensating operator* (8.49) *on the test function* $\varphi(u, \cdot) \in C^2(\mathbb{R}_+)$ *has the following asymptotic expansion*

$$\mathbb{L}^\varepsilon \varphi(u, x) = \varepsilon^{-1} Q\varphi(\cdot, x) + P\mathbb{B}\varphi'_u(u, x) + \varepsilon \theta^\varepsilon_a \varphi(u, x), \qquad (8.50)$$

where $\mathbb{B}$ *is the product operator*

$$\mathbb{B} = \begin{bmatrix} b_+ & 0 \\ 0 & -b_- \end{bmatrix}.$$

Note that the remaining operator θ^ε_a can be represented in an explicit form.

The compensating operator in the Poisson approximation scheme (8.47) *is represented in the following form*

$$\mathbb{L}^\varepsilon \varphi(u, x) = \varepsilon^{-1} q[P\mathbb{G}^\varepsilon - I]\varphi(u, x), \qquad (8.51)$$

where the integral operator $\mathbb{G}^\varepsilon$ is given by

$$\mathbb{G}^\varepsilon = \begin{bmatrix} \mathbb{G}^\varepsilon_+ & 0 \\ 0 & \mathbb{G}^\varepsilon_- \end{bmatrix},$$

where:

$$\mathbb{G}^\varepsilon_+ \varphi(u) = \int_0^\infty \varphi(u+v) G^\varepsilon_+(dv),$$

$$\mathbb{G}^\varepsilon_- \varphi(u) = \int_0^\infty \varphi(u-\varepsilon v) G_-(dv).$$

Lemma 8.3 *Under the conditions of Theorem 8.3, the compensating operator (8.51) on the test function $\varphi(u,\cdot) \in C^3(\mathbb{R}_+)$ has the following asymptotic representation*

$$\mathbb{L}^\varepsilon \varphi(u,x) = \varepsilon^{-1} Q\varphi(u,x) + P[\mathbb{B}\varphi'_u(u,x) + \mathbb{G}_+\varphi(u,x)] + \varepsilon \theta^\varepsilon_p \varphi(u,x), \tag{8.52}$$

where

$$\mathbb{G} = \begin{bmatrix} \mathbb{G}_+ & 0 \\ 0 & 0 \end{bmatrix}, \quad \mathbb{G}_+ \varphi(u) = \int_0^\infty [\varphi(u+v) - \varphi(u) - v\varphi'(u)] G_+(dv).$$

The remaining operator θ^ε_p can be represented in an explicit form.

The compensating operator in the diffusion approximation scheme (8.48) is represented in the following form

$$\mathbb{L}^\varepsilon \varphi = \varepsilon^{-2} q(x) [P\mathbb{G}^\varepsilon - I]\varphi, \tag{8.53}$$

where the matrix integral operator $\mathbb{G}^\varepsilon$ is the same as in the average scheme.

Lemma 8.4 *Under the conditions of Theorem 8.4, the compensating operator (8.53) on the test function $\varphi(u,\cdot) \in C^3(\mathbb{R})$ has the asymptotic representation*

$$\mathbb{L}^\varepsilon \varphi(u,x) = \varepsilon^{-2} Q\varphi(u,x) + \varepsilon^{-1} P\mathbb{B}\varphi'_u(u,x) + \frac{1}{2} P\mathbb{C}\varphi''(u,x) + \varepsilon \theta^\varepsilon_d \varphi(u,x), \tag{8.54}$$

with the product matrix operator $\mathbb{C} = \begin{bmatrix} c_+ & 0 \\ 0 & c_- \end{bmatrix}.$

The remaining operator θ^ε_d can be represented in an explicit form.

8.4.4 The singular perturbation problem

The average operator $\mathbb{L}$ for a SMRW in the average series scheme (8.50) is calculated by using a solution of the singular perturbation problem for the truncated operator

$$\mathbb{L}_0^\varepsilon = \varepsilon^{-1} Q + P\mathbb{B}.$$

According to Proposition 5.1, the average operator $\mathbb{L}$ is determined in the asymptotic representation of the compensating operator on the perturbed test function

$$\mathbb{L}_0^\varepsilon [\varphi(u) + \varepsilon \varphi_1(u,x)] = \mathbb{L}\varphi(u) + \theta_0^\varepsilon(x)\varphi(u),$$

where

$$\mathbb{L}\varphi(u) = \Pi P \mathbb{B} \Pi \varphi'(u).$$

After some computations we obtain

$$\mathbb{L}\varphi(u) = b\varphi'(u), \quad b = \Pi P \mathbb{B} \Pi.$$

The generator $\mathbb{L}$ defines the dynamical system

$$\begin{cases} \frac{d}{dt}\zeta_0(t) = b, & t \geq 0, \\ \zeta_0(0) = u. \end{cases} \quad (8.55)$$

Hence,

$$\zeta_0(t) = u + bt, \quad t \geq 0,$$

which is the limit process in the average scheme.

The Poisson limit operator in the series scheme (8.52) is calculated by using a solution of the singular perturbation problem for the truncated operator

$$\mathbb{L}_0^\varepsilon \varphi = \varepsilon^{-1} Q\varphi + P[\mathbb{B}\varphi'_u + \mathbb{G}\varphi].$$

According to Proposition 5.1 the limit operator is calculated by

$$\mathbb{L}\varphi(u) = \Pi P \mathbb{B} \Pi \varphi'(u) + \Pi \mathbb{G} \Pi \varphi(u).$$

Hence

$$\mathbb{L}\varphi(u) = b\varphi'(u) + p_+^{-1} \int_0^\infty [\varphi(u+v) - \varphi(u) - v\varphi'(u)]G_+(dv)$$

or, in another form, by using b^0 in (8.47)

$$\mathbb{L}\varphi(u) = b^0\varphi'(u) + \Lambda \int_0^\infty [\varphi(u+v) - \varphi(u)]G^0(dv).$$

This is the generator of the compound Poisson process with drift in Theorem 8.3.

The limit diffusion process in the approximation scheme (8.48) is defined by the generator calculated by using a solution of the perturbation problem for the truncated operator in Lemma 8.4

$$\mathbb{L}_0^\varepsilon \varphi = \varepsilon^{-2} Q\varphi + \varepsilon^{-1} P\mathbb{B}\varphi_u' + \frac{1}{2} P\mathbb{C}\varphi''.$$

According to Proposition 5.2 the limit operator $\mathbb{L}$ is determined in the asymptotic representation of the action of $\mathbb{L}_0^\varepsilon$ on the perturbed test function $\varphi(u) + \varepsilon\varphi_1(u,x) + \varepsilon^2\varphi_2(u,x)$

$$\mathbb{L}_0^\varepsilon[\varphi(u) + \varepsilon\varphi_1(u,x) + \varepsilon^2\varphi_2(u,x)] = \mathbb{L}\varphi(u) + \theta_0^\varepsilon(x)\varphi(u).$$

The operator $\mathbb{L}$ has the following representation

$$\mathbb{L}\varphi(u) = [\Pi P\mathbb{B} R_0 P\mathbb{B}\Pi + \frac{1}{2}\Pi P\mathbb{C}\Pi]\varphi''(u).$$

After some computation we get

$$\mathbb{L}\varphi(u) = \frac{1}{2}\sigma^2 \varphi''(u),$$

where the variance σ^2 is represented by the formulas in Theorem 8.4. This is the generator of the limit process $\tilde{\zeta}(t) = u + \sigma w(t)$ in Theorem 8.4.

Martingale characterization of the Markov renewal process. The proof of Theorem 8.2-8.4 is based on the following martingale characterization of the Markov renewal process $\zeta_n^\varepsilon, x_n^\varepsilon, n \geq 0$.

8.4. SEMI-MARKOV RANDOM WALKS

Lemma 8.5 *The Markov renewal process* $\zeta_n^\varepsilon, x_n^\varepsilon, n \geq 0$. *is characterized by the martingale*

$$\mu_{n+1}^\varepsilon = \varphi(\zeta_{n+1}^\varepsilon, x_{n+1}^\varepsilon) - \varphi(u,x) - \varepsilon \sum_{k=1}^n \theta_{k+1} \mathbb{L}^\varepsilon \varphi(\zeta_k^\varepsilon, x_k^\varepsilon)$$

See Section 1.3.5, Proposition 1.4.

The concluding step of the proof of Theorems 8.2-8.4 follows some familiar procedures adapted to the switching semi-Markov processes from Chapter 6. □

8.4.5 Stationary Phase Merging Scheme

The algorithm of the stationary phase merging scheme is realized by using the stationary distribution of the embedded Markov chain (8.45) and is based on the formulas given in Section 8.2. According to Section 8.2 the stationary phase merged superposition of two renewal processes is given in the merged phase space $\hat{E} = \{+, -\}$, by the formulas for sojourn times

$$\theta_\pm^* = \alpha^\pm \wedge \alpha^{\mp *}, \tag{8.56}$$

where $\alpha^{\pm *}$ are the stationary remaining times with densities

$$p_\pm^*(t) = \lambda_\pm \bar{P}_\pm(t), \quad \lambda = 1/p_\pm. \tag{8.57}$$

The stationary merged superposition of two renewal processes on the merged phase space $\hat{E}$ with given sojourn times (8.56) can be interpreted as being in the stationary regime.

The stationary merged IMC x_n^*, $n \geq 0$ is determined by the matrix of transition probabilities

$$P^* = \begin{bmatrix} 1 - p_+ & p_+ \\ p_- & 1 - p_- \end{bmatrix},$$

where

$$p_\pm = \mathbb{P}(\alpha^{\mp *} < \alpha^\pm) = \lambda_\mp p, \quad p = \int_0^\infty \bar{P}_+(t)\bar{P}_-(t)dt$$

The stationary merged embedded SMRW is defined by the relation

$$\zeta_{n+1}^* = \zeta_n^* + \mathbf{1}(x_{n+1}^* = +)\beta_{n+1}^+ - \mathbf{1}(x_{n+1}^* = -)\beta_{n+1}^-.$$

The stationary merged SMRW with continuous time can be defined as follows

$$\zeta^*(t) := \zeta^*_{\nu^*(t)}, \quad t \geq 0,$$

where $\nu^*(t)$, $t \geq 0$, is the counting process of renewal moments for the stationary merged superposition of two renewal processes. The algorithm of asymptotic approximation for the stationary merged SMRW can be formulated in a form analogous to the one presented in Section 8.4.2 by using the compensating operator of the extended Markov chain

$$\mathbb{L}^*\varphi(u) = q^*[P^*\mathbb{G} - I]\varphi(u),$$

where $\varphi(u) = (\varphi_+(u), \varphi_-(u))$. The product operator is

$$q^* = \begin{bmatrix} q^*_+ & 0 \\ 0 & q^*_- \end{bmatrix}, \quad q^*_\pm = 1/\mathbb{E}\theta^*_\pm.$$

The integral operator $\mathbb{G}$ is the same as in Section 8.4.3.

The asymptotic representations of the compensating operator in the approximation schemes are obtained in a form similar to Sections 8.4.3, but are actually simpler.

The asymptotic representations of the compensating operator can be realized in the same form as in Lemmas 8.1-8.4, with transition matrix P^* instead of P. The average and Poisson approximation algorithms give the same result as in Theorems 8.2-8.3. In the diffusion approximation scheme the variance σ^* of the limit diffusion in Theorem 8.4 is defined as follows

$$\sigma^{*2} = 2\sigma_1^{*2} + \sigma_2^2, \quad \sigma_1^{*2} = \overline{b}R_0\overline{b}^*.$$

The potential operator R_0^* is defined for the generator $Q^* = q^*[P^* - I]$, and

$$\overline{b} = b_+/p_+ - b_-/p_-, \quad \overline{b}_- = (\overline{b}^*_+, \overline{b}^*_-)$$
$$b^*_\pm = \pm(q_\pm b_\pm - p_\pm b_\mp).$$

Chapter 9

Applications II

In this chapter we continue the presentation of applied topics as in Chapter 8. In particular, we give a diffusion approximation for birth-and-death processes applied to storage and repairable systems theory, and a Lévy approximation result for impulsive processes.

9.1 Birth and Death Processes and Repairable Systems

9.1.1 *Introduction*

The Markov renewal system with finite identical devices working independently with intensities of working and repairing times depending on the switching ergodic Markov process is considered here. The diffusion approximation by the *Ornstein-Uhlenbeck diffusion* process when the number of devices tends to infinity is established.

The Markov Renewal System with finite devices working independently was considered in [47] and is called the *supplying energy system*. The functioning of each device is described by the alternative renewal process with exponentially distributed working and repairing times with respective intensities λ and μ.

In this section we consider the Markov renewal system with Markov switching in the diffusion approximation scheme when the number of devices tends to infinity.

Let $x(t)$, $t \geq 0$ be an ergodic jump Markov process in a standard state space $(E, \mathcal{E})$ with generator Q acting as follows

$$Q\varphi(x) = q(x) \int_E P(x, dy)[\varphi(y) - \varphi(x)], \qquad (9.1)$$

where $q(x)$ is the intensity of jumps and $P(x, dy)$ is a stochastic kernel defining the transition probabilities of jumps from the state x to the set of states dy. The stationary distribution $\pi(dx)$ of the process $x(t), t \geq 0$, defines the projector Π which acts as follows

$$\Pi\varphi(x) = \int_E \pi(dx)\varphi(x) = \widehat{\varphi}1(x), \qquad (9.2)$$

where $1(x) = 1$ for all $x \in E$.

The main assumption is that the intensities λ and μ depend on the state of the switching Markov process $x(t)$ as follows:

$$\lambda = \lambda(x(t/\varepsilon)), \quad \mu = \mu(x(t/\varepsilon)), \quad \varepsilon := 1/\sqrt{n}.$$

The states of the Markov renewal system are defined by the birth-and-death process $\nu^\varepsilon(t)$, $t \geq 0$, which describes the number of working devices at time t. The state space of the process $\nu^\varepsilon(t)$ is the finite set $E_n = \{0, 1, ..., n\}$. Let $\rho_n(u)$ be its stationary distribution.

9.1.2 Diffusion Approximation

The diffusion approximation scheme for the Markov renewal system with Markov switching can be obtained for the corresponding normalized process.

Let us introduce the following notation:

$$a(x) = \lambda(x) + \mu(x), \quad c(x) = \lambda(x) - \mu(x), \quad b(x) = \lambda(x)\mu(x)/a(x)$$

and

$$a = \int_E \pi(dx)a(x), \quad b = \int_E \pi(x)b(x), \quad v = b/a$$

$$p(x) = 1 - q(x) = \lambda(x)/a(x).$$

Theorem 9.1 *The normalized process*

$$\zeta^\varepsilon := \varepsilon\nu^\varepsilon(t) - \varepsilon^{-1}q(x(t/\varepsilon)) \qquad (9.3)$$

converges weakly, together with the stationary distribution, as $\varepsilon \to 0$, to the Ornstein-Uhlenbeck diffusion process $\zeta(t)$ with the generator

$$\mathbb{L}\varphi(u) = b\varphi''(u) - au\varphi'(u) \qquad (9.4)$$

9.1. BIRTH AND DEATH PROCESSES AND REPAIRABLE SYSTEMS

Note that the stationary distribution of the Ornstein-Uhlenbeck process with generator (9.4) has the density

$$\rho(x) = \frac{1}{\sqrt{2\pi v}} \exp\{-x^2/2v\} \tag{9.5}$$

Remark 9.1. The simple form of the stationary distribution of the limit process allows us to use this distribution as an approximation for the stationary distribution of the Markov Renewal System and calculate the functionals as follows

$$\int f(u)\rho_n(u)du \approx \int f(u)\rho(u)du.$$

This estimation formula gives us an exact enough result when the number of devices n is large enough.

The Markov renewal system can be considered with a finite number $m < n$ of repairing tools. It means that the number of repaired devices of $\nu_\varepsilon^-(t)$ satisfies a boundary condition such as

$$\nu_\varepsilon^-(t) = \begin{cases} n - \nu^\varepsilon(t), & \text{if } \nu^\varepsilon(t) \geq n - m \\ m, & \text{if } \nu^\varepsilon(t) < n - m \end{cases} \tag{9.6}$$

In the case where $\nu^\varepsilon(t) < n - m$, the $n - m - \nu^\varepsilon(t)$ devices are waiting to service. For such Markov renewal systems the diffusion approximation takes another form [99].

Theorem 9.2 *The normalized process (9.3) under the boundary condition (9.6) for $m = \varepsilon^{-2}[q(x) - \varepsilon c_0]$ converges weakly, together with the stationary distribution, as $\varepsilon \to 0$, to the diffusion process $\zeta_0(t)$ with the generator*

$$\mathbb{L}\varphi(u) = b\varphi''(u) - a_0(u)\varphi'(u),$$

where

$$a_0(u) = \begin{cases} au, & u > c_0 \\ \lambda_0 c_0 + \mu_0 u, & u \leq c_0 \end{cases}$$

with $\lambda_0 = \int_E \pi(dx)\lambda(x)$, $\mu_0 = \int_E \pi(dx)\mu(x)$, and c_0 is a constant such that $nc_0 \leq m$.

Remark 9.2. The density of stationary distribution of the limit process $\zeta_0(t)$ can be written

$$\rho_0(u) = \begin{cases} K\rho_1(u), & u > c_0 \\ K\rho_2(u), & u \leq c_0, \end{cases}$$

where $K = [\rho_1((-\infty, c_0]) + \rho_2((c_0, \infty))]^{-1}$, with $\rho_1(dx)$, $\rho_2(dx)$ the stationary distributions of the above diffusion process corresponding to the two branches of $a_0(u)$.

This density of stationary distribution ρ_n can be used in the optimization problem to choose the value of c_0 by

$$\min_{c_0} \int_{-\infty}^{\infty} f(u)\rho_0(u)du$$

with some (smooth) function $f(u)$.

▷ **Example 9.1.** Suppose that $x(t)$, $t \geq 0$, is a 2-state Markov process with generator matrix $Q = (q_{ij}; i, j = 1, 2)$ and $\mu(i) = \mu_i$, $\lambda(i) = \lambda_i$, $i = 1, 2$. Then $a = \pi_1(\lambda_1 + \mu_1) + \pi_2(\lambda_2 + \mu_2)$, $b = \frac{\pi_1\mu_1\lambda_1}{\mu_1+\lambda_1} + \frac{\pi_2\mu_2\lambda_2}{\mu_2+\lambda_2}$, $\lambda_0 = \pi_1\lambda_1 + \pi_2\lambda_2$ and $\mu_0 = \pi_1\mu_1 + \pi_2\mu_2$

As a numerical application for $-q_{11} = q_{12} = 0.1$, $q_{21} = -q_{22} = 0.2$, $2\mu_1 = \mu_2 = 0.2$ and $2\lambda_1 = \lambda_2 = 0.02$, we get $a = 0.0333$, $b = 0.0083$ and $v = 0.25$.

9.1.3 Proofs of the Theorems

PROOF OF THEOREM 9.1. The proof is divided into three steps.

As a first step we calculate the generator operator of the birth-and-death process $\zeta_x^\varepsilon(t)$ for a fixed value of x. The state space of the process $\zeta_x^\varepsilon(t)$ is the following

$$E_x = \{u_k = \varepsilon(k - \varepsilon^{-2}q(x)) : 0 \leq k \leq n\}.$$

It is worth noticing that the number of working devices is

$$k = n[q(x) + \varepsilon u_k], \quad 0 \leq k \leq n.$$

Hence, the number of repairing devices is

$$n - k = n[p(x) - \varepsilon u_k].$$

9.1. BIRTH AND DEATH PROCESSES AND REPAIRABLE SYSTEMS

The main step is the calculation of the jump intensities of $\zeta_x^\varepsilon(t)$. Note that the positive intensities have jumps only from state u_k to the states $u_k \pm \varepsilon$.

The intensity of jumps from u_k to $u_k + \varepsilon$ is determined by the intensity of the repairing time in the state u_k defined by the formula

$$\beta_x^{(k)} = \bigwedge_{j=1}^{n-k} \beta_x^j \tag{9.7}$$

where β_x^j, $j \geq 1$, are independent and identically exponentially distributed random variables with intensity $\mu(x)$.

Hence the intensity of the repairing time is defined by the formula

$$a_x^+(u_k) = n\mu(x)[p(x) - \varepsilon u_k].$$

The intensity of jumps from u_k to $u_k - \varepsilon$ is determined by the intensity of working time in state u_k defined as follows

$$\alpha_x^{(k)} = \bigwedge_{i=1}^{k} \alpha_x^i$$

where α_x^i, $i \geq 1$, are independent and identically exponentially distributed random variables with intensity $\lambda(x)$.

Hence the intensity of jumps from u_k to $u_k - \varepsilon$ is defined as

$$a_x^-(u_k) = n\mu(x)[q(x) + \varepsilon u_k].$$

Further we shall use the next notation:

$$a_x(u) = a_x^+(u) + a_x^-(u) = n[2b(x) + \varepsilon c(x)u]$$

and

$$a_x^+(u) - a_x^-(u) = -\varepsilon^{-1}c(x)u.$$

The next step is the calculation of the generator of the coupled Markov process $\zeta^\varepsilon(t)$, $x(t/\varepsilon)$, $t \geq 0$,

$$\mathbb{L}^\varepsilon \varphi(u, x) = [\varepsilon^{-1}Q + \mathbb{A}^\varepsilon(x)]\varphi(u, x)$$

where $\mathbb{A}^\varepsilon(x)$ is the generator operator of the process $\zeta_x^\varepsilon(t)$ which acts as follows

$$\mathbb{A}^\varepsilon(x)\varphi(u) = a_x^+(u)\varphi(u + \varepsilon) + a_x^-(u)\varphi(u - \varepsilon) - a_x(u)\varphi(u).$$

Now we calculate the asymptotic representation of the generator $\mathbb{A}^\varepsilon(x)$ using the Taylor's expansion for a twice differentiable function $\varphi(u)$

$$\mathbb{A}^\varepsilon(x)\varphi(u) = \mathbb{A}(x)\varphi(u) + \theta^\varepsilon(x)\varphi(u)$$

where

$$\mathbb{A}(x)\varphi(u) = b(x)\varphi''(u) - a(x)u\varphi'(u)$$

and the operator $\theta^\varepsilon(x)$ satisfies the following asymptotic condition

$$\|\theta^\varepsilon(x)\varphi\| \to 0, \quad \varepsilon \to 0,$$

on twice continuously differentiable functions $\varphi(u)$.

Now the martingale approach can be applied to complete the proof of theorem.

Let us introduce the martingale

$$\mu_t^\varepsilon = \varphi^\varepsilon(\zeta^\varepsilon(t), x(t/\varepsilon)) - \int_0^t \mathbb{L}^\varepsilon \varphi^\varepsilon(\zeta^\varepsilon(s), x(s/\varepsilon))ds \qquad (9.8)$$

with respect to the natural filtration $\mathcal{F}_t^\varepsilon$, $t \geq 0$, generated by the process $\zeta^\varepsilon(t), x(t/\varepsilon),\ t \geq 0$.

The main step is the construction of an asymptotic representation of the integral term in the martingale (9.8) by choosing the test functions φ^ε

$$\varphi^\varepsilon(u, x) = \varphi(u) + \varepsilon\varphi_1(u, x),$$

where $\varphi_1(u, x)$ is defined by the solution of

$$Q\varphi_1(u, x) = [\mathbb{L} - \mathbb{A}(x)]\varphi(u), \qquad (9.9)$$

in which the operator $\mathbb{A}$ is determined by the relation (see Proposition 5.1)

$$\mathbb{L}\Pi = \Pi\mathbb{A}(x)\Pi$$

This relation provides the solvability condition by using Equation (9.9). The function $\varphi(u)$ is chosen smooth enough. Here it is sufficient to consider a four times differentiable function.

Applying the generator $\mathbb{L}^\varepsilon$ to the test functions φ^ε and taking into account Equation (9.9), we obtain the following representation of the martingale (9.8)

$$\mu_t^\varepsilon = \varphi(\zeta^\varepsilon(t)) - \int_0^t \mathbb{L}\varphi(\zeta^\varepsilon(s))ds + \psi_t^\varepsilon \qquad (9.10)$$

9.1. BIRTH AND DEATH PROCESSES AND REPAIRABLE SYSTEMS

where the last term in the sum satisfies the following condition

$$\mathbb{E} \sup_{0 \leq t \leq T} |\psi_t^\varepsilon| \to 0, \quad \varepsilon \to 0.$$

Now we can use the standard arguments to establish the compactness of the family of the processes $\zeta^\varepsilon(t)$, $\varepsilon > 0$ (see Chapter 6).

Hence the weak limit

$$\zeta^\varepsilon(t) \Longrightarrow \zeta(t), \quad \varepsilon \to 0,$$

holds. The limit process $\zeta(t)$ is the solution of the following martingale problem

$$\varphi(\zeta(t)) - \int_0^t \mathbb{L}\varphi(\zeta(s))ds = \mu_t.$$

Thus, the process $\zeta(t)$, $t \geq 0$, is the Ornstein-Uhlenbeck process with generator $\mathbb{L}$ given in (9.4).

Now, in order to get the weak convergence of the stationary distributions, we establish the stochastic boundedness of the processes $\zeta^\varepsilon(t)$ [100].

For the Lyapounov function

$$V(u) = V_0 \int_0^u e^{b(z)} dz [V_1 - \int_0^z e^{-b(y)} dy],$$

with $V_0 > 0$ and $V_1 > \int_0^\infty e^{-b(y)} dy$, $b(z) = -\int_0^z a(u)du/\beta^2$, we have

$$\mathbb{L}V(u) = -\beta \{V_0 \frac{a(u)}{\beta^2} e^{b(u)} [V_1 - \int_0^u e^{-b(y)dy}] + V_0\}$$

$$-auV_0 e^{b(u)} [V_1 - \int_0^u e^{-b(y)dy}] \leq 0.$$

Hence we get $\pi^\varepsilon \Rightarrow \pi^0$. □

PROOF OF THEOREM 9.2. The proof of this theorem follows the same lines as that of Theorem 9.1. In this case, relation (9.11) becomes

$$\beta_x^{(k)} = \bigwedge_{j=1}^{(n-k)\wedge m} \beta_x^j. \qquad (9.11)$$

Thus we have:

$$a_x^+(u) - a_x^-(u) = \begin{cases} -\varepsilon^{-1} a(x)u, & u > c_0 \\ -\varepsilon^{-1}[c_0 \lambda(x) + \mu(x)u], & u \leq c_0 \end{cases}$$

and

$$a_x^+(u) + a_x^-(u) = \begin{cases} n[2b(x) - \varepsilon c(x)u], & u > c_0 \\ n[2b(x) - \varepsilon(c_0\lambda(x) + \mu(x)u)], & u \leq c_0 \end{cases}$$

where $c(x) = \lambda(x) - \mu(x)$.

From these, we can proceed as previously. $\square$

9.2 Lévy Approximation of Impulsive Processes

9.2.1 *Introduction*

The impulsive processes considered here are switched by Markov processes (see Sections 2.9.1, 7.2.1, 7.2.2 and 7.3.1).

Let us consider a family of random sequences $\alpha_k^\varepsilon(x), k = 1, 2, ..., x \in E$, where E is a non-empty set, indexed by the small parameter $\varepsilon > 0$, and a family of jump Markov processes $x^\varepsilon(t), t \geq 0$, with embedded Markov renewal process $x_k^\varepsilon, \tau_k^\varepsilon, k \geq 0$, and counting processes of jumps $\nu^\varepsilon(t), t \geq 0$. Thus, times $\tau_k^\varepsilon, k \geq 0$, are jump times, $x_k^\varepsilon := x^\varepsilon(\tau_k^\varepsilon)$, and $\nu^\varepsilon(t) := \max\{k \geq 0 : \tau_k^\varepsilon \leq t\}$.

Define now the impulsive process as partial sums in a series scheme, with series parameter $\varepsilon > 0$, by

$$\xi_\varepsilon(t) := \sum_{k=1}^{\nu^\varepsilon(t/\varepsilon)} \alpha_k^\varepsilon(x_k^\varepsilon).$$

The limit Lévy process, obtained here, has been used directly in[55] in order to model the time of ruin via defective renewal equation. So, results of the present section can be used directly in order to take into account a more general real situation, and results of [55] can be used in order to get ruin time probabilities for the limit Lévy process.

Since Lévy processes are now standard, Lévy approximation is quite useful for analyzing complex systems (see, e.g. [13,155]). Moreover they are involved in many applications, e.g., risk theory, finance, queueing, physics, etc. For a background on Lévy process see, e.g. [13,155,56].

Let $(E, \mathcal{E})$ be a standard state space. Let us consider an E-valued cadlag

9.2. LÉVY APPROXIMATION OF IMPULSIVE PROCESSES

Markov jump process $x(t), t \geq 0$, with generator Q, that is,

$$Q\varphi(x) = q(x) \int_E P(x, dy)[\varphi(y) - \varphi(x)],$$

and $x_n, \tau_n, n \geq 0$, the associated Markov renewal process to $x(t), t \geq 0$. The transition probability kernel of $x_n, n \geq 0$, is $P(x, B)$, $x \in E, B \in \mathcal{E}$. Let $\nu(t), t \geq 0$, be the counting process of jumps of $x(t), t \geq 0$, that is, $\nu(t) = \sup\{n \geq 0 : \tau_n \leq t\}$.

We suppose here that the process $x(t), t \geq 0$, is uniformly ergodic with stationary probability $\pi(B), B \in \mathcal{E}$. Thus the embedded Markov chain is uniformly ergodic too. Let $\rho(B), B \in \mathcal{E}$, denote the stationary probability measure of the embedded Markov chain $x_n, n \geq 0$. These two probability measures are related by the following relation

$$\pi(dx)q(x) = q\rho(dx), \quad q := \int_E \pi(dx)q(x).$$

Define the projector Π by

$$\Pi\varphi(x) := \int_E \pi(dy)\varphi(y)\mathbf{1}(x),$$

where $\mathbf{1}(x) = 1$ for all $x \in E$. Let us denote by R_0 the potential operator defined by (see Section 1.6)

$$R_0 Q = Q R_0 = \Pi - I. \tag{9.12}$$

Let $\varepsilon > 0$ be a small parameter and define the family of Markov processes $x^\varepsilon(t) := x(t/\varepsilon^2), t \geq 0$.

We formulate here a new result of approximation by a Lévy process of the following impulsive processes

$$\xi^\varepsilon(t) := \xi_0^\varepsilon + \sum_{k=1}^{\nu^\varepsilon(t/\varepsilon^2)} \alpha_k^\varepsilon(x_k^\varepsilon), \quad t \geq 0, \varepsilon > 0. \tag{9.13}$$

For any $\varepsilon > 0$, and any sequence $z_k, k \geq 1$, of elements of E, the random variables $\alpha_k^\varepsilon(z_k), k \geq 1$ are supposed to be independent. Let us denote by G_x^ε the distribution function of $\alpha_k^\varepsilon(x)$, that is,

$$G_x^\varepsilon(dv) := \mathbb{P}(\alpha_k^\varepsilon(x) \in dv), \quad k \geq 0, \varepsilon > 0, x \in E.$$

It is worth noticing that the coupled process $\xi^\varepsilon(t), x^\varepsilon(t), t \geq 0$, is a *Markov additive process* (see Section 2.5).

Let $\xi(t), t \geq 0$, be a Lévy process with *characteristic exponent* (cumulant) given by the *Lévy-Khintchine formula*

$$\psi(u) := \frac{1}{t}\mathbb{E}e^{iu\xi(t)} = ibu - \frac{1}{2}\sigma^2 u^2 + \int_{\mathbb{R}}[e^{iux} - 1 - iux\mathbf{1}_{\{|u|<1\}}]\mu(dx),$$

(9.14)

where $b \in \mathbb{R}$, $\sigma \geq 0$, and the positive measure μ, on $\mathbb{R} - \{0\}$, such that $\int (1 \wedge y^2)\mu(dy) < \infty$, is the *Lévy measure*. The law of the process $\xi(t)$ is completely determined by the function $\psi(u)$, that is, from the triplet (b, σ, μ), called the *characteristics* of $\xi(t)$. The infinitesimal generator $\mathbb{L}$ of the Lévy process $\xi(t)$, with exponent function $\psi(u)$ (9.14), is defined as follows ([56,13])

$$\mathbb{L}\varphi(x) = b\varphi'(x) + \frac{1}{2}\sigma^2\varphi''(x) + \int_{\mathbb{R}}[\varphi(x+y) - \varphi(x) - \varphi'(x)y\mathbf{1}_{\{|y|<1\}}]\mu(dy),$$

where φ is a twice continuously differentiable function which vanishes at infinity. The functions φ' and φ'' are first and second derivatives of φ, respectively.

9.2.2 Lévy Approximation Scheme

The results presented here concern the weak convergence of the $\mathbb{R}$-valued impulsive processes $\xi^\varepsilon(t), t \geq 0, \varepsilon > 0$, defined by (9.13). Generalization of these results to the $\mathbb{R}^d$ case $(d > 1)$ is straightforward.

We will need the following assumptions.

L1: We suppose that the Markov process $x(t), t \geq 0$, is uniformly ergodic.
L2: Initial value condition

$$\sup_{\varepsilon>0} \mathbb{E}|\xi_0^\varepsilon| \leq C < \infty.$$

L3: Approximation of the mean value

$$b^\varepsilon(x) := \int_{\mathbb{R}} v G_x^\varepsilon(dv) = \varepsilon b_1(x) + \varepsilon^2 b(x) + \varepsilon^2 \theta_b^\varepsilon(x),$$

where functions b_1 and b are bounded.
L4: Approximation of the second moment

$$c^\varepsilon(x) := \int_{\mathbb{R}} v^2 G_x^\varepsilon(dv) = \varepsilon^2[c(x) + \theta_c^\varepsilon(x)],$$

where the function c is bounded.

9.2. LÉVY APPROXIMATION OF IMPULSIVE PROCESSES

L5: Poisson approximation condition

$$\int_{\mathbb{R}} g(v) G_x^\varepsilon(dv) = \varepsilon^2 [\int_{\mathbb{R}} g(v) G_x(dv) + \theta_g^\varepsilon(x)],$$

where g belongs to the class of functions $C_3(\mathbb{R})$, (see Section 7.2)).

The above negligible terms, $\theta_b^\varepsilon(x), \theta_c^\varepsilon(x), \theta_g^\varepsilon(x)$, fulfill the following negligibility condition

$$\sup_{x \in E} |\theta^\varepsilon_\cdot(x)| \to 0, \quad \varepsilon \to 0.$$

L6: Balance condition

$$\int_E \rho(dx) b_1(x) = 0.$$

Remark 9.3. Assumptions L3, L5, and L6 together split jumps into three parts. The first terms in L3, together with L6, give the diffusion component, the second term in L3 gives the deterministic drift, and L5 gives the jumps of the limit process. The balance condition L6 means that the values of order $O(\varepsilon)$ are compensated in order to yield the additional diffusion part to the limit process.

The following theorem states the main result of this section.

Theorem 9.3 *Under Assumptions L1-L6, the following weak convergence holds*

$$\xi^\varepsilon(t) \Longrightarrow \xi^0(t), \quad \varepsilon \to 0,$$

provided that $\xi_0^\varepsilon \xrightarrow{P} \xi^0(0)$, *as* $\varepsilon \to 0$. *The limit process* $\xi^0(t), t \geq 0$, *is a Lévy process defined by the generator* $\mathbb{L}$ *as follows*

$$\mathbb{L}\varphi(u) = (b - b_0)\varphi'(u) + \frac{1}{2}\sigma^2 \varphi''(u) + \lambda \int_{\mathbb{R}} [\varphi(u+v) - \varphi(u)] G_0(dv), \quad (9.15)$$

with $\sigma^2 \geq 0$, *where:*

$$b = q \int_E \rho(dx) b(x), \quad b_0 = q \int_{\mathbb{R}} v G(dv),$$

$$\sigma^2 = 2q \int_E \rho(dx) [b_2(x) R_0 b_2(x) + \frac{1}{2} c(x)], \quad b_2(x) := \int_E P(x, dy) b_1(y),$$

$$G(dv) = \int_E \rho(dx)G_x(dv), \quad \lambda = qG(\mathbb{R}), \quad G_0(dv) = G(dv)/G(\mathbb{R}).$$

Remark 9.4. The jump part of the above limit process $\xi^0(t), t \geq 0$, is a compound Poisson process, that is,

$$\sum_{k=1}^{\nu^0(t)} \alpha_k^0, \quad t \geq 0,$$

where $\nu^0(t), t \geq 0$, is a Poisson process with intensity λ, and $\alpha_k^0, k \geq 1$, are i.i.d. random variables with common distribution function $G_0(dv)$.

Remark 9.5. The *characteristics* of the above limit Lévy process are the drift coefficient $b - b_0$, the diffusion coefficient σ^2, and the Lévy measure G defined on $\mathbb{R}_0$.

In the case of a finite set of values for the r.v.s $\alpha_k^\varepsilon(x)$, we get the following result.

▷ **Example 9.2.** Let:

$$\mathbb{P}(\alpha^\varepsilon(x) = \varepsilon a_1(x)) = p_0 - \varepsilon^2 p_1,$$

$$\mathbb{P}(\alpha^\varepsilon(x) = \varepsilon^2 a(x)) = q_0, \quad q_0 + p_0 = 1,$$

and

$$\mathbb{P}(\alpha^\varepsilon(x) = d(x)) = \varepsilon^2 p_1.$$

We suppose that the balance condition $\int_E \rho(dx) a_1(x) = 0$ is fulfilled. Calculation of the first two moments gives:

$$b^\varepsilon(x) := \mathbb{E}\alpha^\varepsilon(x) = \varepsilon^2 \left[\varepsilon^{-1} a_1(x) p_0 + [a(x)q_0 + d(x)p_1] \right] + o(\varepsilon^2),$$

$$c^\varepsilon(x) := \mathbb{E}[\alpha^\varepsilon(x)]^2 = \varepsilon^2 \left[a_1^2(x) p_0 + d^2(x) p_1 \right] + o(\varepsilon^2).$$

9.2. LÉVY APPROXIMATION OF IMPULSIVE PROCESSES

Hence:
$$b_1(x) = a_1(x)p_0, \quad b(x) = a(x)q_0 + d(x)p_1,$$
$$c(x) = a_1^2(x)p_0 + d^2(x)p_1.$$

Assumption L5 gives:
$$G_x^\varepsilon(g) := \int_{\mathbb{R}} g(v) G_x^\varepsilon(dv)$$
$$= [g(\varepsilon a_1(x))p_0 + g(\varepsilon^2 a(x))q_0 + g(c(x))p_1 \varepsilon^2]$$
$$= \varepsilon^2 g(c(x))p_1 + o(\varepsilon^2),$$

and hence
$$G_x(dv) = \delta_{c(x)}(dv)p_1.$$

Now, we get the characteristics of the limit operator in (9.15):
$$b = q[aq_0 + dp_1], \quad a := \int_E \rho(dx)a(x), \quad d := \int_E \rho(dx)d(x),$$
$$c := \int_E \rho(dx)c(x), \quad b_0 = qcp_1.$$

So, the deterministic drift is defined by
$$b - b_0 = q[aq_0 + (d - c)p_1].$$

The diffusion coefficient is
$$\sigma^2 = qp_0^2 \int_E \rho(dx) b_2(x) R_0 b_2(x), \quad b_2(x) = \int_E P(x, dy) b_1(y).$$

The Poisson part of limit process is determined by the generator
$$\mathbb{G}\varphi(u) = \lambda[\varphi(u + c) - \varphi(u)], \quad \lambda = qp_1.$$

The increasing jumps $\varepsilon a_1(x)$ are transformed into diffusion part with variance σ^2. The big jumps $c(x)$ with small probability $\varepsilon^2 p_1$ are transformed into Poisson process. The small jumps $\varepsilon^2 a(x)$ are transformed into deterministic drift.

Let us give now a result on the rate of convergence in the Lévy approximation scheme of Theorem 9.3.

Theorem 9.4 *Under assumptions of Theorem 9.3, the following inequality holds*

$$\sup_{0\leq t\leq T} |\mathbb{E}[\varphi(\xi^\varepsilon(t)) - \varphi(\xi^0(t))]| \leq \varepsilon C_T,$$

where $\varphi \in C_b^2(\mathbb{R})$, bounded twice continuously differentiable functions defined on $\mathbb{R}$, and C_T is a constant.

9.2.3   Proof of Theorems

Here we give the proofs of Theorems 9.3 and 9.4. The method of proof of Theorem 9.3 is as follows. We construct the compensating operator of the Markov additive process $\xi^\varepsilon(t), x^\varepsilon(t), t \geq 0$ (Lemmas 9.1-9.2), and obtain the generator of the limit process (Lemma 9.3) by a singular perturbation technique.

PROOF OF THEOREM 9.3. Let us start by the construction of the generator of the Markov additive process given in the following lemma.

Lemma 9.1 *The generator $\mathbb{L}^\varepsilon$ of the impulsive process $\xi^\varepsilon(t), x^\varepsilon(t), t \geq 0$, is given by*

$$\mathbb{L}^\varepsilon \varphi(u,x) = \varepsilon^{-2} q(x) \left[\int_E P(x,dy) \int_{\mathbb{R}} G_y^\varepsilon(dv) \varphi(u+v,y) - \varphi(u,x) \right]. \quad (9.16)$$

PROOF. By Definition 1.22 of the compensating operator and from a standard calculus we get the desired result. □

Lemma 9.2 *The main part in the asymptotic representation of the generator $\mathbb{L}^\varepsilon$ is as follows (with the same notation, $\mathbb{L}^\varepsilon$)*

$$\mathbb{L}^\varepsilon \varphi(u,x) = \varepsilon^{-2} Q\varphi(\cdot,x) + \varepsilon^{-1} Q_0 b_1(x) \varphi'_u(u,\cdot) + Q_0[b(x) - b_0(x)]\varphi'_u(u,\cdot)$$
$$+ \frac{1}{2} Q_0 c(x) \varphi''_{uu}(u,\cdot) + Q_0 \mathbb{\Gamma}_x \varphi(u,\cdot), \quad (9.17)$$

where:

$$Q_0\varphi(x) := q(x) \int_E P(x,dy)\varphi(y), \quad b_0(x) := \int_{\mathbb{R}} v G_x(dv),$$

$$\mathbb{\Gamma}_x\varphi(u) := \int_{\mathbb{R}} [\varphi(u+v) - \varphi(u)]G_x(dv).$$

9.2. LÉVY APPROXIMATION OF IMPULSIVE PROCESSES

PROOF. From (9.16), we can write

$$\mathbb{L}^\varepsilon \varphi(u,x) = \varepsilon^{-2} Q\varphi(\cdot,x) + \varepsilon^{-2} q(x) \int_E P(x,dy) \int_\mathbb{R} G^\varepsilon_y(dv)[\varphi(u+v,y) - \varphi(u,y)].$$

In order to apply Assumptions L3, L4 and L5, let us consider the operator

$$\mathbb{\Gamma}^\varepsilon_y \varphi(u) := \int_\mathbb{R} G^\varepsilon_y(dv)[\varphi(u+v) - \varphi(u)]$$

which is transformed as follows

$$\mathbb{\Gamma}^\varepsilon_y \varphi(u) = \int_\mathbb{R} G^\varepsilon_y(dv) g_\varphi(v) + b^\varepsilon(y)\varphi'(u) + \frac{1}{2} c^\varepsilon(y)\varphi''(u),$$

where the function

$$g_\varphi(v) := \varphi(u+v) - \varphi(u) - v\varphi'(u) - \frac{v^2}{2}\varphi''(u)$$

belongs to $C_3(\mathbb{R})$, and

$$b^\varepsilon(y) := \int_\mathbb{R} v G^\varepsilon_y(dv), \quad c^\varepsilon(y) := \int_\mathbb{R} v^2 G^\varepsilon_y(dv) = \varepsilon^2 [C(x) + \theta^\varepsilon_c(x)].$$

Then, using Assumptions L3-L5, we get

$$\mathbb{\Gamma}^\varepsilon_y \varphi(u) = \varepsilon^2 \Big\{ \int_\mathbb{R} G_y(dv) g_\varphi(v) + [\varepsilon^{-1} b_1(y) + b(y)]\varphi'(u) + \frac{1}{2} c(y)\varphi''(u) \Big\} + o(\varepsilon^2),$$

or, in another form

$$\mathbb{\Gamma}^\varepsilon_y \varphi(u) = \varepsilon \Big\{ \mathbb{\Gamma}_y \varphi(u) + \varepsilon^{-1} b_1(y)\varphi'(u) + [b(y) - b_0(y)]\varphi'(u) + \frac{1}{2} c(y)\varphi''(u) \Big\} + o(\varepsilon^2).$$

Putting this representation in (9.16), we get (9.17). □

We will now obtain the limit generator by solving the following singular perturbation problem for the reducible-invertible operator Q, according to Proposition 5.2,

$$\mathbb{L}^\varepsilon \varphi^\varepsilon(u,x) = \mathbb{L}\varphi(u) + \varepsilon \theta^\varepsilon(x), \qquad (9.18)$$

for a test function $\varphi^\varepsilon(u,x) = \varphi(u) + \varepsilon \varphi_1(u,x) + \varepsilon^2 \varphi_2(u,x)$.

Let us define the operators:

$$Q_1 := Q_0 \mathbb{B}_1(x) \quad \text{and} \quad Q_2 := Q_0[\mathbb{B}(x) + \mathbb{\Gamma}_x + \mathbb{C}(x)],$$

where $\mathbb{B}_1(x)\varphi(u) := b_1(x)\varphi'(u)$, and $\mathbb{B}(x)\varphi(u) = [b(x) - b_0(x)]\varphi'(u)$, and:

$$Q_3 := Q_2 + Q_1 R_0 Q_1, \quad \text{and} \quad \mathbb{C}(x)\varphi(u) := \frac{1}{2} c(x)\varphi''(u).$$

Note that under the balance condition we have

$$\Pi Q_1 \Pi = 0.$$

Lemma 9.3 *The asymptotic representation*

$$[\varepsilon^{-2}Q + \varepsilon^{-1}Q_1 + Q_2][\varphi + \varepsilon\varphi_1 + \varepsilon^2\varphi_2] = \mathbb{L}\varphi + \theta^\varepsilon(x),$$

is verified by

$$\varphi_1 = R_0 Q_1 \varphi,$$

$$\varphi_2 = R_0(\mathbb{L} - Q_3)\varphi,$$

with negligible term

$$\theta^\varepsilon(x) = [Q_1 + \varepsilon Q_2]\varphi_2 + Q_2\varphi_1.$$

The limit operator $\mathbb{L}$ *can be obtained by Proposition 5.2*

$$\mathbb{L}\varphi(u) = \Pi[Q_2 + Q_1 R_0 Q_1]\Pi\varphi(u). \tag{9.19}$$

Calculation of the limit operator. Taking into account $R_0 P = R_0 + \Pi - I$, and the balance condition L6, the limit operator (9.19) is represented by (9.15). Specifically, by a straightforward calculus, we obtain:

$$\Pi Q_0 \mathbb{B}(x)\varphi(u) = (b - b_0)\varphi'(u),$$

$$\Pi Q_0 \Gamma_x \varphi(u) = \lambda \int_\mathbb{R} [\varphi(u+v) - \varphi(u)]G_0(dv),$$

and

$$\Pi Q_0 b_1(x) R_0 P b_1(x) = \frac{1}{2}\sigma^2.$$

In order to prove the relative compactness of the family of the processes $\xi^\varepsilon(t), t \geq 0, \varepsilon > 0$, we can follow the lines of Chapter 6.

PROOF OF THEOREM 9.4. For the coupled Markov process $\xi^\varepsilon(t), x^\varepsilon(t), t \geq 0$, with generator $\mathbb{L}^\varepsilon$, we have the following equation

$$\varphi^\varepsilon(\xi^\varepsilon(t), x^\varepsilon(t)) = \varphi^\varepsilon(u, x) + \int_0^t \mathbb{L}^\varepsilon \varphi^\varepsilon(\xi^\varepsilon(s), x^\varepsilon(s))ds + y^\varepsilon(t), \tag{9.20}$$

9.2. LÉVY APPROXIMATION OF IMPULSIVE PROCESSES

where $y^\varepsilon(t)$ is an $\mathcal{F}_t^\varepsilon$-martingale, and the test functions φ^ε considered here are of the form

$$\varphi^\varepsilon(u,x) = \varphi(u) + \varepsilon\varphi_1(u,x). \tag{9.21}$$

For the limit process $\xi^0(t), t \geq 0$, in Theorem 9.4, we get

$$\varphi(\xi^0(t)) = \varphi(u) + \int_0^t \mathbb{L}\varphi(\xi^0(s))ds + y^0(t), \tag{9.22}$$

where $y^0(t), t \geq 0$, is an $\mathcal{F}_t^0 = \sigma(\xi^0(s), s \leq t)$-martingale.

From (9.20) and (9.21), we get

$$\varphi(\xi^\varepsilon(t)) = \varphi(u) + \int_0^t \mathbb{L}^\varepsilon\varphi^\varepsilon(\xi^\varepsilon(s), x^\varepsilon(s))ds + y^\varepsilon(t) + \varepsilon\theta_1^\varepsilon(t), \tag{9.23}$$

where

$$\theta_1^\varepsilon(t) := -\varphi_1(\xi^\varepsilon(t), x^\varepsilon(t)) + \varphi_1(u,x). \tag{9.24}$$

Now, from (9.22) and (9.23), we get

$$\varphi^\varepsilon(\xi^\varepsilon(t)) - \varphi(\xi^0(t)) = \int_0^t [\mathbb{L}^\varepsilon\varphi^\varepsilon(\xi^\varepsilon(s), x^\varepsilon(s)) - \mathbb{L}\varphi(\xi^0(t))]ds$$
$$+ y^\varepsilon(t) - y^0(t) + \varepsilon\theta_1^\varepsilon(t). \tag{9.25}$$

From Lemma 9.3, we have

$$\mathbb{L}^\varepsilon\varphi^\varepsilon(\xi^\varepsilon(s), x^\varepsilon(s)) = \mathbb{L}\varphi(\xi^0(t)) + \varepsilon\theta_2^\varepsilon(s). \tag{9.26}$$

From (9.25) and (9.26), we get

$$\varphi(\xi^\varepsilon(t)) - \varphi(\xi^0(t)) = y^\varepsilon(t) - y^0(t) + \varepsilon\theta_3^\varepsilon(t),$$

where $\theta_3^\varepsilon(t) := \theta_1^\varepsilon(t) + \int_0^t \theta_2^\varepsilon(s)ds$.

Since $\mathbb{E}[y^\varepsilon(t) - y^0(t)] = 0$, the result follows from the latter equality. $\square$

Problems to Solve

Here we give about fifty problems for the reader to solve. These problems propose results stated without proofs in the previous chapters, alternative proofs, and some extensions. They are classified following chapters.

Chapter 1

▷ **Problem 1.** Prove that the generators in examples in Section 1.2.2 are as stated there.

▷ **Problem 2.** Prove Proposition 1.3.

▷ **Problem 3.** Prove Proposition 1.4.

▷ **Problem 4.** Prove the Markov Renewal Equation (1.40).

▷ **Problem 5.** Prove identities of Proposition 1.6 for R_0 given by (1.66).

▷ **Problem 6.** Prove Proposition 1.7.

▷ **Problem 7.** Prove identity (1.59).

▷ **Problem 8.** Prove that the generator of the backward Markov process $x(t) = t - \tau(t), t \geq 0$, of a renewal process on $\mathbb{R}_+$, considered in Example 1.2, is a reducible-invertible operator.

▷ **Problem 9.** Let $x(t), t \geq 0$, be a regular semi-Markov process.

1) Prove that the process $x(t), \gamma(t), t \geq 0$, is a Markov process, (see Section 1.3.3).

2) Calculate its generator and prove that it is a reducible-invertible operator.

▷ **Problem 10.** Let $x(t), t \geq 0$, be a jump Markov process with a standard state space $(E, \mathcal{E})$. Let $q(x)$ be the intensity function, $m(x)$ the mean jump value at x, $m_2(x)$ the second moment of jump at x, and $Q(x, \cdot)$ the distribution of jumps at x. Prove that:

1) the compensator of the process $x(t), t \geq 0$, is

$$a(t) := \int_0^t q(x(s))m(x(s))ds, \quad t \geq 0,$$

2) the square characteristic of the local martingale $\mu(t) = x(t) - x(0) - a(t), t \geq 0$, is

$$\langle \mu \rangle_t = \int_0^t q(x(s))m_2(x(s))ds,$$

provided that $m_2(x)$ is finite for any $x \in E$.

Chapter 2

▷ **Problem 11.** Prove Lemma 2.2.

▷ **Problem 12.** Prove Lemma 2.3.

▷ **Problem 13.** Prove Proposition 2.1.

▷ **Problem 14.** Prove Corollary 2.1.

▷ **Problem 15.** Prove Corollary 2.3.

▷ **Problem 16.** ([174]) Let $x(t), t \geq 0$, be a semi-Markov process, let $x_n, \tau_n, n \geq 0$, be the corresponding Markov renewal process, and $\mathbb{L}$ be its compensating operator. Define the process $\xi(t), t \geq 0$, by

$$\xi(t) := \varphi(x_{\nu(t)+1}, \tau_{\nu(t)+1}) - \int_0^{\tau_{\nu(t)+1}} \mathbb{L}\varphi(x_{\nu(s)}, \tau_{\nu(s)}) ds, \quad n \geq 0,$$

and $\mathcal{F}_t := \sigma(\{\tau_n \leq t\} \cap \{(x_0, ..., x_n) \in \mathcal{B}_E^{n+1}\}; n \geq 0)$. The function φ is measurable, and φ and $\mathbb{L}\varphi$ are bounded.

Prove that:
1) for $t \geq 0$ and $s \geq 0$,

$$\mathbb{E}_x[\xi(t+s) - \xi(t) \mid \mathcal{F}_t] = 0, \quad a.s.$$

2) the process $\xi(t), t \geq 0$, is not measurable with respect to $\mathcal{F}_t$.

So, the process $\xi(t)$ is not an $\mathcal{F}_t$-martingale.

▷ **Problem 17.** Let $g : \mathbb{R}^d \to \mathbb{R}^d, d \geq 1$, be a globally Lipschitz continuous function, $f : \mathbb{R}^d \to \mathbb{R}$ be a function in $C^1(\mathbb{R}^d)$ and let $\mathbb{A}$ be a differential operator defined by

$$\mathbb{A}f(x) = \sum_{i=1}^d \frac{\partial f}{\partial x_i} g_i(x),$$

where $g(x) = (g_1(x), ..., g_d(x))$.

Consider the ordinary differential equations

$$\frac{d}{dt} x(t) = g(x(t)), \quad x(0) = x \in \mathbb{R}^d, \tag{9.27}$$

$$\frac{d}{dt} f(x(t)) = \mathbb{A}f(x(t)). \tag{9.28}$$

1) Prove that $x(t)$ is a solution of (9.27) if and only if it is a solution of (9.28).

2) Prove that the family $(\phi_t, t \in \mathbb{R})$, where $\phi_t : x \mapsto \phi(t, x) =: x(t)$, is a group, that is,

$$\phi_{t+s} = \phi_t \circ \phi_s,$$

that means $\phi(t+s, x) = \phi(t, \phi(s, x))$, $x \in \mathbb{R}^d$; and ϕ_t is a one-to-one mapping.

▷ **Problem 18.** Prove Corollary 2.4.

▷ **Problem 19.** Prove Proposition 2.11.

▷ **Problem 20.** Prove Proposition 2.12.

Chapter 3

▷ **Problem 21.** Prove Corollary 3.5.

▷ **Problem 22.** Let f be a probability density function on $\mathbb{R}_+$, whose first and second moments are finite and denoted by m and m_2 respectively. Show that $m_2 \geq 2m$ if f is completely monotone (see Theorem 3.3).
  Prove that in Theorem 3.3 we have $\sigma^2 > 0$.

▷ **Problem 23.** (*Liptser's formula*). Let $x(t), t \geq 0$, be an irreducible jump Markov process with finite state space $E = \{1, ..., d\}$ and generating matrix Q. Denote by $\pi_i, 1 \leq i \leq d$ its stationary distribution. Let a be a real-valued function defined on E. Let us define the following family of integral functionals

$$\alpha^\varepsilon(t) = \varepsilon^{-1} \int_0^t [a(x(s/\varepsilon^2)) - \mathbb{E}a(x(s/\varepsilon^2))]ds, \quad t \geq 0, \quad \varepsilon > 0.$$

Let $(v_1, ..., v_{d-1})$ be the solution of the following system of equations

$$\sum_{j=1}^{d-1} \widetilde{Q}_{ij} v_j = a(i) - a(d), \quad i = 1, ..., d-1,$$

where $\widetilde{Q}$ is a non-singular matrix defined as follows

$$\widetilde{Q}_{ij} = Q_{ij} - Q_{dj}, \quad i, j = 1, ..., d-1.$$

Deduce from Theorem 3.3 that

$$\alpha^\varepsilon(t) \Longrightarrow b\, w(t), \quad \varepsilon \to 0,$$

where $w(t), t \geq 0$, is a standard Wiener process, and

$$b^2 = \sum_{j=1}^{d-1} v_j^2 \sum_{i=1}^{d} |Q_{ij}|\pi_i - \sum_{\substack{i,j=1 \\ i \neq j}}^{d-1} v_i v_j (Q_{ij}\pi_i + Q_{ji}\pi_j).$$

▷ **Problem 24.** For the process in the previous problem, define the occupation time process of states

$$\Delta_i(t) = meas\{s : x(s) = i, 0 \leq s \leq t\}, \quad t \geq 0,$$

and the vector $\Delta^* = (\Delta_1^*, ..., \Delta_d^*)$, with

$$\Delta_i^* = \sqrt{t}\left(\frac{\Delta_i}{t} - \pi_i\right), \quad 1 \leq i \leq d.$$

Deduce from Liptser's formula that the vector Δ converges weakly to a multivariate normal distribution with mean zero and covariance matrix to be defined.

▷ **Problem 25.** Let us consider a family of contraction semigroups $\Gamma_t(x), t \geq 0, x \in E$, on a Banach space $\mathbf{B}$, and the random evolution operator

$$\phi(t) = \Gamma_{\tau_1}(x(0))\Gamma_{\tau_2-\tau_1}(x(\tau_1)) \cdots \Gamma_{t-\tau(t)}(x(\tau(t))).$$

1) Prove that $\phi(t)$ is a contraction operator on $\mathbf{B}$.
2) Prove that for any $f \in \mathbf{B}$ the mapping $t \mapsto \phi(t)f$ is continuous, and differentiable at points $t \neq \tau_1, \tau_2, ...$, and satisfies the following differential equation

$$\frac{d}{dt}\phi(t)f = \phi(t)\mathbb{\Gamma}(x(t))f,$$

where $\mathbb{\Gamma}(x)$ is the generator of $\Gamma_t(x)$, $x \in E$.
3) Define the expectation semigroup as follows

$$(\Pi(t)f)(x) := \mathbb{E}_x[\phi(t)f(x(t))].$$

Prove that $\Pi(t), t \geq 0$, is a contraction semigroup.

Chapter 4

▷ **Problem 26.** Construct the generator of the limit stochastic system $\widehat{U}(t), \widehat{x}(t), t \geq 0$, in Theorem 4.4.

▷ **Problem 27.** State and prove the result of Theorem 4.6, when the limit merged Markov process $\widehat{\widehat{x}}(t), t \geq 0$, has null generator $\widehat{\widehat{Q}}_2 = 0$.

▷ **Problem 28.** By ad hoc time scaling, give averaging results for system (4.35) in Theorem 4.6.

▷ **Problem 29.** State and prove the result of Theorem 4.10, when the limit merged Markov process $\widehat{\widetilde{x}}(t), t \geq 0$, is not conservative.

Chapter 5

▷ **Problem 30.** Prove Lemma 5.5 by using the following formula (Proposition 5.2).

$$\mathbb{L}\Pi = \Pi Q_2(v;x)\Pi + \Pi\widetilde{\mathbb{A}}(v;x)PR_0\widetilde{\mathbb{A}}(v;x)\Pi.$$

▷ **Problem 31.** Derive results of Proposition 5.17, for the following representation of the generator of the switching Markov process

$$Q^\varepsilon = Q + \varepsilon Q_1 + \varepsilon^2 Q_2 + \varepsilon^4 Q_3.$$

Give an interpretation of this scheme.

▷ **Problem 32.** Give a conclusion such as in Proposition 5.5 for the following generator

$$\mathbb{L}^\varepsilon = \varepsilon^{-k}Q + \varepsilon^{-k+1}Q_1 + ... + \varepsilon^{-2}Q_{k-2} + \varepsilon^{-1}Q_{k-1} + Q_k \quad (9.29)$$

for $k > 4$.

▷ **Problem 33.** Show that the generator of the random evolution in Proposition 4.1 is represented as follows

$$\mathbb{L}^\varepsilon = \varepsilon^{-3}Q + \varepsilon^{-2}Q_1 + \varepsilon^{-1}\widetilde{\mathbb{A}}(x),$$

where the operator

$$\widetilde{\mathbb{A}}(x)\varphi(u) = \widetilde{a}(x)\varphi'(u), \quad \widetilde{a}(x) := a(x) - \widehat{\widehat{a}},$$

satisfies the balance condition

$$\widehat{\Pi}\widetilde{\mathbb{A}}(x)\widehat{\Pi} = 0.$$

Prove that the limit generator is the following

$$\mathbb{L}\widehat{\Pi} = \widehat{\Pi}\widehat{\mathbb{A}}\widehat{R}_0\widehat{\mathbb{A}}\widehat{\Pi}.$$

Hint. Use Proposition 5.4.

▷ **Problem 34.** Prove that the generator of the random evolution in Proposition 4.2 is represented as follows

$$\mathbb{L}^\varepsilon = \varepsilon^{-4}Q + \varepsilon^{-3}Q_1 + \varepsilon^{-2}Q_2 + \varepsilon^{-1}\mathbb{A}(x),$$

where the generator

$$\mathbb{A}(x)\varphi(u) = a(x)\varphi'(u),$$

satisfies the balance condition

$$\widehat{\widehat{\widehat{\Pi}}}\Pi\mathbb{A}(x)\Pi\widehat{\widehat{\widehat{\Pi}}} = 0.$$

Prove that the limit generator is the following

$$\mathbb{L}\widehat{\widehat{\widehat{\Pi}}}\widehat{\widehat{\mathbb{A}}}\widehat{R}_0\widehat{\widehat{\mathbb{A}}}\widehat{\widehat{\widehat{\Pi}}}.$$

Hint. Use Proposition 5.5.

▷ **Problem 35.** Prove that the generator of the random evolution in Proposition 4.3 is represented as follows

$$\mathbb{L}^\varepsilon = \varepsilon^{-2}\widehat{\widehat{Q}} + \varepsilon^{-1}\widehat{\widehat{\mathbb{A}}},$$

where $\widehat{\widehat{Q}}$ is the generator of twice merged Markov process $\widehat{\widehat{x}}(t), t \geq 0$, and the operator

$$\widehat{\widehat{\mathbb{A}}} = \widehat{\Pi}\Pi\mathbb{A}(x)\Pi\widehat{\mathbb{A}},$$

satisfies the balance condition $\widehat{\widehat{\Pi}}\widehat{\widehat{\mathbb{A}}}\widehat{\widehat{\Pi}} = 0$.

The limit generator has the following form

$$\mathbb{L}\widehat{\widehat{\Pi}} = \widehat{\widehat{\Pi}}\widehat{\widehat{\mathbb{A}}}\widehat{R}_0\widehat{\widehat{\mathbb{A}}}\widehat{\widehat{\Pi}}.$$

Hint. Use Proposition 5.3.

▷ **Problem 36.** Prove that the generator of the random evolution in Proposition 4.4 is represented as follows

$$\mathbb{L}^\varepsilon = \varepsilon^{-4}Q + \varepsilon^{-3}Q_1 + \varepsilon^{-2}Q_2 + \varepsilon^{-1}\mathbb{A}(x),$$

where the operator

$$\widetilde{\mathbb{A}}(x)\varphi(u) = \widetilde{a}(x)\varphi'(u), \quad \widetilde{a}(x) := a(x) - \widehat{\widehat{a}},$$

satisfies the balance condition

$$\widehat{\widehat{\Pi}}\widehat{\Pi}\Pi\widetilde{\mathbb{A}}\Pi\widehat{\Pi}\widehat{\widehat{\Pi}} = 0.$$

Prove that the limit generator is the following

$$\mathbb{L}\widehat{\widehat{\Pi}} = \widehat{\widehat{\Pi}}\widetilde{\widehat{\mathbb{A}}}\widehat{\widehat{R}}_0\widetilde{\widehat{\mathbb{A}}}\widehat{\widehat{\Pi}}.$$

Hint. Use Proposition 5.5.

▷ **Problem 37.** Prove Theorem 4.6, by using the development in Section 5.6.1.

Chapter 6

▷ **Problem 38.** Let $z_n, n \geq 1$, be a sequence of i.i.d. centered random variables, and define the family of stochastic processes $x^\varepsilon(t), t \geq 0, \varepsilon > 0$, by

$$x^\varepsilon(t) := \varepsilon \sum_{k=1}^{[t/\varepsilon^2]} z_k, \quad t \geq 0.$$

Show that $x^\varepsilon(t) \Longrightarrow w(t)$, where $w(t), t \geq 0$, is a standard Wiener process.

▷ **Problem 39.** Prove the diffusion approximation result in Theorem 3.4, following calculus in Section 5.5.2 and Chapter 6.

▷ **Problem 40.** Let $x^\varepsilon(t/\varepsilon), t \geq 0, \varepsilon > 0$, be a family of semi-Markov processes with phase space E, split as follows

$$E = \cup_{k=1}^N E_k, \quad E_k \cap E_{k'} = \emptyset, \quad k \neq k'.$$

Let v be the merging function on E with values in $\{1, 2, ..., N\}$. Suppose that the following averaging principles are fulfilled:

$$v(x^\varepsilon(t/\varepsilon)) \Longrightarrow \widehat{x}(t),$$

$$\varepsilon \nu^\varepsilon(t/\varepsilon) \Longrightarrow \widehat{\nu}(t),$$

where $\widehat{x}(t), t \geq 0$ is a Markov process.

1) Show that the compensating operator of ν^ε is

$$\widetilde{\nu}^\varepsilon(t) = \int_0^t \lambda(x^\varepsilon(s), \gamma^\varepsilon(s)) ds,$$

and that of $\widehat{\nu}$ is

$$\widetilde{\nu}(t) = \int_0^t \widehat{q}(\widehat{x}(s)) ds.$$

2) Show that we have

$$\varepsilon \int_0^{t/\varepsilon} \lambda(x^\varepsilon(s), \gamma^\varepsilon(s)) ds \Longrightarrow \int_0^t \widehat{q}(\widehat{x}(s)) ds,$$

where

$$\widehat{q}(k) = \int_{E_k \times \mathbb{R}_+} \widetilde{\pi}(dx \times ds) \lambda(x, s),$$

and $\widetilde{\pi}$ is the stationary distribution of the Markov process $x^\varepsilon(t), t - \tau(t), t \geq 0$, on $E \times \mathbb{R}_+$.

Chapter 7

▷ **Problem 41.** Formulate the corresponding stochastic singular perturbation problem and prove Lemma 7.3.

▷ **Problem 42.** Show that the predictable characteristics $(B(t), \widehat{C}(t), \nu_t(g))$ of the semimartingale $\xi^0(t), t \geq 0$, in Theorem 7.2, relation (7.28), are given by:

$$B(t) = \int_0^t b(\widehat{x}(s)) ds, \quad b(v) = q_v \widehat{a}(v), \quad \widehat{a}(v) := \int_{E_v} \rho_v(dx) a(x).$$

The modified second characteristic is

$$\widehat{C}(t) = \int_0^t \widehat{C}(\widehat{x}(s))ds,$$

where

$$\widehat{C}(v) = q_v \int_{E_v} \rho_v(dx) C_0(x), \quad v \in V \quad \text{and} \quad C_0(x) = \int_{\mathbb{R}} u^2 \Phi_x(du).$$

The predictable measure is

$$\nu_t(g) = \int_0^t \widetilde{\Phi}_{\widehat{x}(s)}(g)ds, \quad \widetilde{\Phi}_v(g) = q_v \widehat{\Phi}_v(g),$$

where

$$\widehat{\Phi}_v(g) := \int_{E_v} \rho_v(dx) \Phi_x(g).$$

▷ **Problem 43.** Prove that the compensating operator formula is as stated in Lemma 7.5.

▷ **Problem 44.** Prove that under conditions of Theorem 7.1, the following convergence takes place

$$\mathcal{E}(\lambda \xi^\varepsilon)_t \Rightarrow \mathcal{E}(\lambda \xi^0)_t = \prod_{k=1}^{\nu^0(t)} [1 + \lambda \alpha_k^0] \exp(-tqa_0), \quad \varepsilon \to 0.$$

The limit stochastic exponential process $\mathcal{E}(\lambda \xi^0)_t$ is defined by the limit compound Poisson process (7.7).

Chapter 8

▷ **Problem 45.** Show that the stationary distribution of the MRP $x_n, \theta_n, n \geq 0$, with phase space $E = \{\pm, x \geq 0\}$, of the SMRW is given by

$$\rho_\pm(dx) = \overline{F}_\pm(x)dx/(a_+ + a_-),$$

where $a_\pm = \mathbb{E}\alpha_k^\pm$.

▷ **Problem 46.** Consider a centered SMRW defined as follows

$$\zeta^\varepsilon(t) = u + \varepsilon \left[\sum_{k=1}^{\nu^+(t/\varepsilon^2)} \beta_k^+ - \varepsilon \sum_{k=1}^{\nu^-(t/\varepsilon^2)} \beta_k^- \right]^+ - b\tau(t/\varepsilon^2),$$

where $b = b_+/p_+ - b_-/p_-$. Let $b \neq 0$ and the third moments $E[\beta_n^\pm]^3 < \infty$. For notation see Section 8.4.

Show that the weak convergence

$$\zeta^\varepsilon(t) \Rightarrow \zeta^0(t) = u + \sigma w(t), \quad \varepsilon \to 0 \tag{18}$$

takes place, and that the variance σ^2 is

$$\sigma^2 = \sigma_0^2 + \sigma_1^2,$$

where:

$$\sigma_0^2 = 2 \int_0^\infty [\overline{P}_-(x)\widetilde{b}_+^0(x) + \overline{P}_+(x)\widetilde{b}_-^0(x)]dx,$$

$$\widetilde{b}_\pm^0(x) := \widetilde{b}_\pm(x) R_0^\pm \widetilde{b}_\pm(x),$$

$$\sigma_1^2 = \int_0^\infty [\overline{P}_+(x) C_-(x) + \overline{P}_-(x) C_+(x)]dx,$$

$$C_\pm(x) := \mathbf{E}[\gamma_{n+1}^2 | x_n = x], \quad x \in E_\pm.$$

The potential operators $R_0^\pm$ are defined for the semi-Markov kernel $Q = q(x)[P - I]$. The process $w(t)$ is the standard Wiener process.

Chapter 9

▷ **Problem 47.** Let $x^\varepsilon(t), t \geq 0, \varepsilon > 0$, be a family of Markov processes, with embedded Markov chain $x_n^\varepsilon, n \geq 0$, with the standard state space $(E, \mathcal{E})$; let the process $\nu(t), t \geq 0$, be a Poisson process with intensity q.

Let the following autoregressive real-valued process $\alpha^\varepsilon(t), t \geq 0$, be defined by

$$\alpha^\varepsilon(t) = \alpha^\varepsilon(0) + \varepsilon \sum_{k=1}^{\nu(t/\varepsilon)} a(\alpha_k^\varepsilon; x_k^\varepsilon), \tag{9.30}$$

where a is a fixed real-valued function defined on $\mathbb{R} \times E$.

1) Prove that the generator of the coupled Markov process $\alpha^\varepsilon(t), x^\varepsilon(t), t \geq 0$, is

$$\mathbb{L}^\varepsilon = \varepsilon^{-1}[Q + Q_0(D^\varepsilon(x) - I)],$$

where $[D^\varepsilon(x) - I]\varphi(u) = \varphi(u + \varepsilon a(u; x)) - \varphi(u)$.

2) Formulate the singular perturbation problem and find out the limit generator.

3) Prove the following weak convergence result

$$\alpha^\varepsilon(t) \Rightarrow \alpha^0(t),$$

where the limit process $\alpha^0(t), t \geq 0$, (deterministic), is defined as a solution of the following evolutionary equation

$$\frac{d}{dt}\alpha^0(t) = \widehat{a}(\alpha^0(t)),$$

where $\widehat{a}(u) = q \int_E \rho(dx) a(u; x)$.

▷ **Problem 48.** Let the process $\alpha^\varepsilon(t), t \geq 0$, in the previous problem be scaled as follows

$$\alpha^\varepsilon(t) = \alpha^\varepsilon(0) + \varepsilon \sum_{k=1}^{\nu(t/\varepsilon^2)} a_\varepsilon(\alpha_k^\varepsilon; x_k^\varepsilon), \quad (9.31)$$

with

$$a_\varepsilon(u; x) = a(u; x) + \varepsilon a_1(u; x).$$

Prove that the following weak convergence takes place

$$\alpha^\varepsilon(t) \Rightarrow \alpha^0(t),$$

where the limit diffusion process $\alpha^0(t), t \geq 0$, is defined by the generator $\mathbb{L}$, defined as follows

$$\mathbb{L}\varphi(u) = b(u)\varphi'(u) + \frac{1}{2}B(u)\varphi''(u),$$

where, the drift coefficient is defined by

$$b(u) = q \int_E \rho(dx) a_1(u; x) + q \int_E \rho(dx) b_0(u; x),$$

with $b_0(u; x) = a(u; x) R_0 a'_u(u; x)$, and the diffusion coefficient is

$$B(u) = q \int_E \rho(dx) a_0(u; x),$$

with $a_0(u; x) = a(u; x) R_0 a(u; x)$.

General problems

▷ **Problem 49.** Let $\nu^\varepsilon(t), t \geq 0, \varepsilon > 0$, be a family of counting processes with intensities

$$\lambda(t) = \varepsilon^{-1} C(t; \varepsilon \nu^\varepsilon(t)), \quad t \geq 0, \quad \varepsilon > 0.$$

Prove that the following convergence holds

$$\varepsilon \nu^\varepsilon(t) \Longrightarrow x^0(t), \quad \varepsilon \to 0,$$

where $x^0(t), t \geq 0$ is the solution of

$$\frac{d}{dt} x(t) = C(t; x(t)), \quad x(0) = 0.$$

▷ **Problem 50.** Let us consider a birth and death process with state space $E_N = \{0, 1, ..., N\}$ and jumps intensities: $Q(i, +1) = (N - i)\lambda$, $0 \leq i < N$, and $Q(i, -1) = i\mu$, $0 < i \leq N$.
1) Put $\varepsilon = 1/N$ and define the normalized family of processes

$$\nu^\varepsilon(t) := \varepsilon \nu(t/\varepsilon), \quad t \geq 0, \quad \varepsilon > 0,$$

on the state spaces $E^\varepsilon = \{u = i\varepsilon : 0 \leq i \leq N\}$.
Prove that $\nu^\varepsilon(t) \Rightarrow \rho(t)$, as $\varepsilon \to 0$, where the limit process $\rho(t), t \geq 0$, is a deterministic function obtained as a solution of the following evolutional equation

$$\frac{d}{dt} \rho(t) = C(\rho(t)),$$

with $C(u) = \lambda(1 - u) - \mu u$.
2) Put $\varepsilon := N^{-1/2}$ and consider the normalized family of processes

$$\xi^\varepsilon(t) := \varepsilon \nu(t/\varepsilon^2) - \varepsilon^{-1} p,$$

where $p = \lambda/(\lambda + \mu)$.

Prove that $\xi^\varepsilon(t) \Rightarrow \xi^0(t)$, where $\xi^0(t), t \geq 0$ is the Ornstein-Uhlembeck diffusion process with generator defined as follows

$$\mathbb{L}^0\varphi(u) = -(\lambda+\mu)u\varphi'(u) + \frac{1}{2}B\varphi''(u),$$

where $B = 2\mu p$.

▷ **Problem 51.** Suppose that the process $x(t), t \geq 0$, satisfies the following mixing condition

$$\phi(t) := \sup_{s\geq 0} \sup_{\substack{A\in\mathcal{F}_t \\ B\in\mathcal{F}^{t+s}}} |\mathbb{P}(B \mid A) - \mathbb{P}(B)|,$$

where $\mathcal{F}_t := \sigma(x(s); s \leq t)$ and $\mathcal{F}^{t+s} := \sigma(x(u); u \geq t+s)$, and, for some $c > 0$,

$$\int_0^\infty e^{cu}\phi(u)du < +\infty.$$

Then the stochastic system $U^\varepsilon(t)$ defined by the solution of

$$\frac{d}{dt}U^\varepsilon(t) = C(x(t/\varepsilon))U^\varepsilon(t),$$

converges weakly, as $\varepsilon \to 0$, to the solution of the averaged system

$$\frac{d}{dt}U(t) = \widehat{C}U(t), \quad U(0) = u,$$

with $\widehat{C} := \mathbb{E}_0 C(x(t/\varepsilon))$.

Appendix A

Weak Convergence of Probability Measures

In this appendix we present some results on *weak convergence* of probability measures in *Polish spaces*, that is a complete and separable metric space. For example, the Euclidean space $\mathbb{R}^d$, the space $\mathbf{C}[0,\infty)$ with the topology of uniform convergence on bounded sets, the space $\mathbf{D}[0,\infty)$ with the Skorokhod metric are Polish spaces.

A.1 Weak Convergence

Let $C(E)$ be the set of real-valued bounded continuous functions on E, with the sup-norm.

Let us consider a Polish space E, with its Borel σ-algebra $\mathcal{E}$. So, the measurable space $(E,\mathcal{E})$ is a *standard space*. Let $M_1(E)$ be the space of all probability measures on $(E,\mathcal{E})$, endowed with the weak topology. In this topology the mappings $\mu \to \mu f$ are continuous for all $f \in C(E)$. The space $M_1(E)$, is a Polish space for the weak topology.

Definition A.1 Let $P_n, n \geq 1$, and P, be in $M_1(E)$. Then we say that (P_n) converges weakly to P, if for any real-valued bounded continuous function φ on E, we have

$$\int \varphi dP_n \longrightarrow \int \varphi dP, \quad n \to \infty.$$

It is denoted by

$$P_n \Rightarrow P, \quad n \to \infty.$$

Theorem A.1 *(Portmanteau) Let $P_n, n \geq 1$, and P in $M_1(E)$. Then the following assertions are equivalent.*

1) $P_n \Rightarrow P$, $n \to \infty$.
2) $\int \varphi dP_n \longrightarrow \int \varphi dP$, $n \to \infty$, *for any bounded and uniformly continuous function φ on E.*
3) $\int \varphi dP_n \longrightarrow \int \varphi dP$, $n \to \infty$, *for any bounded and measurable function φ on E.*
4) $\limsup_{n \to \infty} P_n(C) \leq P(C)$, *for any closed subset C of E.*
5) $\liminf_{n \to \infty} P_n(O) \geq P(O)$, *for any open subset O of E.*
6) $\lim_{n \to \infty} P_n(A) \leq P(A)$, *for any A of E, for which $P(\partial A) = 0$.*

Let S be a subset of $C(E)$.

Definition A.2 The set S is said to be *measure-determining* (class), if, whenever P and Q belong to $M_1(E)$,

$$\int \varphi dP = \int \varphi dQ, \quad \text{for all } \varphi \in S,$$

we have $P = Q$.

Definition A.3 The set S is said to be *convergence-determining* (class), if, whenever $P_n, n \geq 1$, and P belong to $M_1(E)$,

$$\int \varphi dP_n \longrightarrow \int \varphi dP, \quad \varphi \in S,$$

we have $P_n \Rightarrow P$.

A convergence-determining class is also a measure-determining class.

Definition A.4 For $(E, \mathcal{E})$-valued stochastic elements $x_n, n \geq 1$, and x, we say that x_n converges weakly to x, if $P_n \Rightarrow P$, where P_n is the probability distribution of x_n and P of x. We denote that by $x_n \Rightarrow x$.

It is not necessary for the above stochastic elements x_n and x to be defined on the same probability space. In case they are defined on the same probability space, say $(\Omega, \mathcal{F}, \mathbb{P})$, we have $P_n = \mathbb{P} \circ x_n^{-1}$ and $P = \mathbb{P} \circ x^{-1}$.

Corollary A.1 *Let $P, Q \in M_1(E)$.*
1) *If $\int \varphi dP = \int \varphi dQ$, for any bounded and uniformly continuous function φ on E, then $P = Q$.*
2) *If P and Q are limits of the same sequence in $M_1(E)$, then $P = Q$.*

A.2 Relative Compactness

Definition A.5 A subset M of $M_1(E)$ is said to be *relatively compact*, if every sequence in M has a convergent subsequence.

Theorem A.2 *Let $P_n, n \geq 1$, be a relatively compact sequence in $M_1(E)$, such that every convergent subsequence has the same limit P. Then $P_n \Rightarrow P$.*

Definition A.6
1) The probability measure $P \in M_1(E)$ is *tight* if for every $\varepsilon > 0$ there exists a compact subset K of E, such that $P(K^c) < \varepsilon$.

2) A subset M of $M_1(E)$, is *tight* if, for every $\varepsilon > 0$, there exists a compact subset K of E such that $P(K^c) < \varepsilon$, for every P in M.

Theorem A.3 *(Prohorov) A subset M of $M_1(E)$ is relatively compact (for the weak topology) if and only if it is tight.*

Theorem A.4 *Let $x_n(t), t \geq 0, n \geq 0$, a sequence of processes and let a process $x(t), t \geq 0$, be with simple paths in $\mathbf{D}[0, \infty)$.*
1) If $x_n(t) \Rightarrow x(t)$, then

$$(x_n(t_1), ..., x_n(t_k)) \Rightarrow (x_n(t_1), ..., x_n(t_k)), \quad n \to \infty, \quad (A.1)$$

for any finite set $\{t_1, ..., t_k\} \subset D_x := \{t \geq 0 : \mathbb{P}(x(t) = x(t-)) = 1\}$.

2) If the sequence $x_n(t)$ of processes is relatively compact and there exists a dense set $D \subset [0, \infty)$ such that (A.1) holds for every finite set $\{t_1, ..., t_k\} \subset D$, then

$$x_n(t) \Rightarrow x(t), \quad n \to \infty.$$

Theorem A.5 *(Skorokhod representation) Let $x_n, n \geq 1$, and x be E-valued stochastic elements, and suppose that $x_n \Rightarrow x$.*
Then there exist stochastic elements $\tilde{x}_n, n \geq 1$, and $\tilde{x}$, all defined on a common probability space, such that $\tilde{x}_n$ has the same distribution as x_n, and $\tilde{x}$ as x, and

$$x_n \xrightarrow{a.s.} x.$$

Remark A.1. The above definitions and more detailed results can be found, e.g., in [16,56,70,132].

Appendix B

Some Limit Theorems for Stochastic Processes

The present appendix gives three theorems used in proofs of theorems in Poisson approximation of Chapter 7, (Theorems B.1-B.2, from [70]) and in Lévy approximation of SMRW in Chapter 9 (Theorem B.3, from [169]).

B.1 Two Limit Theorems for Semimartingales

Let us consider the classes of functions $C_1(\mathbb{R}^d)$, $C_2(\mathbb{R}^d)$, and $C_3(\mathbb{R}^d)$ defined as follows (see [70], p. 354).

$C_2(\mathbb{R}^d)$ is the set of all real-valued continuous bounded functions defined on $\mathbb{R}^d$ which are zero around 0 and have a limit at infinity.

$C_1(\mathbb{R}^d)$ is the subclass of $C_2(\mathbb{R}^d)$ of all nonnegative functions $g_a(x) = (a|x|-1)^+ \wedge 1$ for all positive rationals a, and with the following property: let μ_n, μ be positive measures on $\mathbb{R}^d \setminus \{0\}$, finite on any complement of neighborhood of 0; then $\mu_n f \to \mu f$ for all $f \in C_1(\mathbb{R}^d)$ implies $\mu_n f \to \mu f$ for all $f \in C_2(\mathbb{R}^d)$. So, it is a *convergence-determining class*.

$C_3(\mathbb{R}^d)$ is the *measure-determining class* of functions φ, which are real-valued, bounded, and such that

$$\varphi(u)/|u|^2 \to 0, \quad |u| \to 0.$$

The above three classes satisfy the following inclusion relations:

$$C_1(\mathbb{R}^d) \subset C_2(\mathbb{R}^d) \subset C_3(\mathbb{R}^d).$$

Integral process ([70]). First we consider a random measure $\nu = \{\nu(\omega; dt, dx); \omega \in \Omega\}$ on $(\mathbb{R}_+ \times E, \mathcal{B}_+ \times \mathcal{E})$, such that $\nu(\{0\} \times E) \equiv 0$.

Let R be a measurable function on $(\Omega \times \mathbb{R}_+ \times E, \widetilde{O} \times \mathcal{E})$, where $\widetilde{O}$ is the optional σ-algebra on $\Omega \times \mathbb{R}_+$.

Define the integral process $R * \nu(\omega, t)$ by

$$R * \nu(\omega, t) = \int_{[0,t] \times E} R(\omega; s, x) \nu(\omega; ds, dx),$$

when $\int_{[0,t] \times E} |R(\omega; s, x)| \nu(\omega; ds, dx) < \infty$; and $R * \nu(\omega, t) = 0$ otherwise.

Theorem B.1 *(Theorem VIII.2.18, p. 423, in [70]) Let $x(t), t \geq 0$ be a semimartingale of an independent increment process continuous in probability and let $\nu^\varepsilon(t), t \geq 0$ and $\nu(t), t \geq 0$, be such that*

$$|x| * \nu^\varepsilon(t) < +\infty, \quad |x| * \nu(t) < +\infty \quad \text{for all } t \geq 0.$$

Define $B'^\varepsilon(t), t \geq 0$, and $B'(t), t \geq 0$, as follows

$$B'^\varepsilon(t) = B^\varepsilon(t) + (x - h(x)) * \nu^\varepsilon(t),$$

and $\widetilde{C}'^\varepsilon(t), t \geq 0, C'(t), t \geq 0$, as follows

$$\widetilde{C}'^{\varepsilon,jk}(t) = C^{\varepsilon,jk}(t) + (x^j x^k) * \nu^\varepsilon(t) - \sum_{s \leq t} \Delta B'^{\varepsilon,j}(s) \Delta B'^{\varepsilon,k}(s).$$

If

$$\sup_{s \leq t} |B'^\varepsilon(s) - B'(s)| \xrightarrow{P} 0, \quad \text{for all } t \geq 0,$$

and

$$\widetilde{C}'^\varepsilon(t) \xrightarrow{P} \widetilde{C}'(t), \quad \text{for all } t \geq 0,$$

$$g * \nu^\varepsilon(t) \xrightarrow{P} g * \nu(t), \quad \text{for all } t \geq 0, \quad g \in C_1(\mathbb{R}^d),$$

$$\lim_{a \uparrow \infty} \limsup_{\varepsilon > 0} \mathbb{P}^\varepsilon \left(|x|^2 \mathbf{1}_{\{|x| > a\}} * \nu^\varepsilon(t) > \epsilon \right), \quad \text{for all } \epsilon > 0, \quad t \in \mathbb{R}_+,$$

then

$$x^\varepsilon(t) \Longrightarrow x(t).$$

Moreover, in this case we have

$$g * \nu^\varepsilon(t) \xrightarrow{P} g * \nu(t), \quad \text{for all } t \geq 0, \quad g \in C_2(\mathbb{R}^d),$$

and
$$\sup_{s\leq t}\left|\widetilde{C}'^{\varepsilon}(s) - \widetilde{C}'(s)\right| \xrightarrow{P} 0, \quad \text{for all } t \geq 0.$$

Theorem B.2 *(Theorem IX.3.27, p. 507, in [70]) Let D be a dense set of $\mathbb{R}_+$, and assume that $C_1(\mathbb{R}^d)$ contains the positive and negative parts of the following functions*

$$g_a^i(x) = (x^i - h^i(x))(1 - g_a(x))$$
$$g_a^{ij}(x) = (x^i x^j - h^i(x)h^j(x))(1 - g_a(x)).$$

where $a \in \mathbb{Q}_+$ and $g_a(x) = (a|x| - 1)^+ \wedge 1)$.
Assume also that ν^ε and ν are such that

$$|x| * \nu^\varepsilon(t) < +\infty, \quad |x| * \nu(t) < +\infty \quad \text{for all } t \geq 0. \tag{B.1}$$

Let also the following conditions hold.

(1) *The strong majoration hypothesis: there is a continuous and deterministic increasing cadlag function F which strongly dominates the functions $\sum_{i \leq d} Var(B'^i(\alpha))$ and $\sum_{i \leq d} Var(\widetilde{C}'^{ii}(\alpha))$.*
(2) *The condition of big jumps:*

$$\limsup_{\substack{a\uparrow\infty \\ \alpha \in \Omega}} |x|^2 \mathbf{1}_{\{|x|>a\}} * \nu^\varepsilon(t;\alpha) = 0. \tag{B.2}$$

(3) *The uniqueness condition: there is a unique probability measure $\mathbb{P}$ on $(\Omega, \mathcal{F})$ such that the canonical process $x(t), t \geq 0$, is a semimartingale on $(\Omega, \mathcal{F}, \mathbf{F}, \mathbb{P})$ with characteristics (B, C, ν) and initial distribution η.*
(4) *The continuity condition: for any $t \in D$, $g \in C_1(\mathbb{R}^d)$ the functions $\omega \mapsto B'(t, \omega), \widetilde{C}'(t, \omega), g * \nu(t, \omega)$ are Skorokhod continuous on $\mathbf{D}(\mathbb{R}^d)$.*
(5) *$\mathcal{L}(x^\varepsilon(0)) \Rightarrow P_0$, where $\mathcal{L}(x^\varepsilon(0))$ is the law of the r.v. $x^\varepsilon(0)$, and P_0 is the initial distribution of the limit process.*
(6) *The following convergence in probability holds*

$$g * \nu^\varepsilon(t) - (g * \nu(t)) \circ x^\varepsilon \xrightarrow{P} 0 \quad \text{for all } t \in D,\ g \in C_1(\mathbb{R}^d), \tag{B.3}$$

and the three following conditions hold:

$$\sup_{s\leq t}|B'^\varepsilon(s) - B'(s) \circ x^\varepsilon| \xrightarrow{P} 0, \quad \text{for all } t \geq 0, \tag{B.4}$$

$$\widehat{C}'^{\varepsilon}(t) - \widehat{C}'(t) \circ x^{\varepsilon} \xrightarrow{P} 0, \quad \text{for all } t \in D, \tag{B.5}$$

$$\lim_{a \uparrow \infty} \limsup_{\varepsilon > 0} \mathbb{P}^{\varepsilon}\Big(|x|^2 \mathbf{1}_{\{|x|>a\}} * \nu^{\varepsilon}(t) > \epsilon\Big), \tag{B.6}$$

for all $\epsilon > 0$, $t \in \mathbb{R}_+$.

Then

$$\mathcal{L}(x^{\varepsilon}) \Rightarrow \mathbb{P}, \quad \varepsilon \to 0,$$

where $\mathcal{L}(x^{\varepsilon})$ is the law of the process $x^{\varepsilon}(t), t \geq 0$

B.2 A Limit Theorem for Composed Processes

Let $n^{\varepsilon}, \varepsilon > 0$, be a family of positive non random numbers, such that $n^{\varepsilon} \to \infty$, as $\varepsilon \to 0$; let $\alpha_k^{\varepsilon}, k = 1, 2, ..., \varepsilon > 0$, be a family of real-valued random variables and let a family of stochastic processes $\xi^{\varepsilon}(t), t \geq 0, \varepsilon > 0$ be defined as follows

$$\xi^{\varepsilon}(t) = \sum_{k=1}^{[tn^{\varepsilon}]} \alpha_k^{\varepsilon}, \quad t \geq 0, \varepsilon > 0. \tag{B.7}$$

Let further $\mu^{\varepsilon}, \varepsilon > 0$, be a family of non negative random variables, and set $\nu^{\varepsilon} := \mu^{\varepsilon}/n^{\varepsilon}$.

Define now the cadlag process

$$\zeta^{\varepsilon}(t) := \sum_{k \leq t\mu^{\varepsilon}} \alpha_k^{\varepsilon}, \quad \varepsilon > 0. \tag{B.8}$$

Then $\zeta^{\varepsilon}(t) = \xi^{\varepsilon}(t\nu^{\varepsilon})$ is a composition of the two stochastic processes: $\xi^{\varepsilon}(t)$ and $\nu^{\varepsilon}(t) = t\nu^{\varepsilon}$, and we have the following theorem [169].

Theorem B.3 *(Theorem 4.2.1, p. 241, in* [169]*) Let the following conditions hold:*

I. $(\nu^{\varepsilon}, \xi^{\varepsilon}(t), t \in U) \Longrightarrow (\nu^0, \xi^0(t), t \in U), \quad \varepsilon \to 0$, where:
   (1) ν^0 is a non-negative random variable.
   (2) $\xi^0(t), t \geq 0$, is a cadlag processes.
   (3) U is dense in $\mathbb{R}_+$ and contains the point 0.

II.
$$\lim_{c \to 0} \limsup_{\varepsilon \to 0} \mathbb{P}\Big(\Delta_J(\xi^\varepsilon, c, T) > \delta\Big) = 0, \quad \delta > 0, T > 0. \tag{B.9}$$

Then
$$(\zeta^\varepsilon(t), t \in W) \Longrightarrow (\zeta^0(t), t \in W), \quad \varepsilon \to 0,$$

where $W = \mathbb{R}_+ \setminus A$, A *is any set at most countable, and* $0 \in W$, *and* $\Delta_J(\xi^\varepsilon, c, T)$ *is the modulus of compactness, defined as follows:*

$$\Delta_J(\xi^\varepsilon, c, T) := \sup_{0 \vee (t-c) \leq t' \leq t \leq t'' \leq (t+c) \wedge T} \min\Big(|\xi^\varepsilon(t') - \xi^\varepsilon(t)|, |\xi^\varepsilon(t) - \xi^\varepsilon(t'')|\Big).$$

Remark B.1. Instead of relation (B.9), the process $\xi^\varepsilon(t), t \geq 0$, has to satisfy the *compact containment condition* (see Chapter 6).

Appendix C

Some Auxiliary Results

In this appendix we present some results useful in several theorems presented in this book. The first one (Lemma C.1) establishes the negligibility condition of the backward recurrence time of the jump times process of a semi-Markov family of processes. The second one (Lemma C.2) proves the positiveness of diffusion coefficients in limit theorems.

C.1 Backward Recurrence Time Negligibility

Lemma C.1 *Let the family of sojourn times $\theta_x, x \in E$, with distribution function F_x, satisfy the following conditions.*

A1: *Uniform integrability*

$$\sup_{x \in E} \int_T^\infty \bar{F}_x(t)dt \to 0, \quad T \to \infty.$$

A2: *For any $x \in E$, and $\varepsilon > 0$,*

$$\mathbb{E}e^{-\varepsilon \theta_x} \leq 1 - C\varepsilon.$$

Then the following estimation takes place

$$\mathbb{P}\left(\max_{0 \leq t \leq T} \gamma^\varepsilon(t) \geq \delta\right) \to 0, \quad \varepsilon \to 0,$$

for any $\delta > 0$, and any $T > 0$, where $\gamma^\varepsilon(t) := t - \tau^\varepsilon(t)$, (see Section 1.3.3).

PROOF. The regular property of the semi-Markov process provides the following estimation

$$\mathbb{P}(\tau^\varepsilon_{N/\varepsilon} \leq T) \to 0, \quad N \to \infty.$$

for $\varepsilon > 0$.

Indeed, let us calculate

$$\mathbb{P}(\tau^\varepsilon_{N/\varepsilon} \leq T) = \mathbb{P}(e^{-\tau^\varepsilon_{N/\varepsilon}} \geq e^{-T})$$
$$\leq \mathbb{E}e^{-\tau^\varepsilon_{N/\varepsilon}}e^T$$
$$= \mathbb{E}e^{-\varepsilon\tau_{N/\varepsilon}}e^T.$$

Using Condition A2, we estimate

$$\mathbb{E}e^{-\varepsilon\tau_{N/\varepsilon}} = \mathbb{E}e^{-\varepsilon\theta_{N/\varepsilon}}e^{-\varepsilon\tau_{N/\varepsilon-1}}$$
$$\leq (1 - C\varepsilon)\mathbb{E}e^{-\varepsilon\tau_{N/\varepsilon-1}}$$
$$\leq (1 - C\varepsilon)^{N/\varepsilon}$$
$$\leq e^{-CN} \to 0, \quad N \to \infty.$$

Under Condition A1 we estimate

$$\mathbb{P}(\max_{1\leq k \leq N/\varepsilon} \theta_k \geq \delta/\varepsilon) \leq \sum_{k=1}^{N/\varepsilon} \mathbb{P}(\theta_k \geq \delta/\varepsilon)$$
$$\leq \frac{N}{\varepsilon} \sup_{x \in E} \int_{\delta/\varepsilon}^{\infty} \bar{F}_x(t)dt$$
$$\leq \frac{N}{\varepsilon}\frac{\varepsilon}{\delta} \sup_{x \in E} \bar{F}_x(\delta/\varepsilon)$$
$$= \frac{N}{\delta} \sup_{x \in E} \bar{F}_x(\delta/\varepsilon) \to 0, \quad \varepsilon \to 0.$$

Now, we can estimate

$$\mathbb{P}(\max_{0\leq t \leq T} \gamma^\varepsilon(t) \geq \delta) \leq \mathbb{P}(\max_{0 \leq t \leq T} \theta_{\nu^\varepsilon_+(t)} \geq \delta/\varepsilon)$$
$$\leq \mathbb{P}(\max_{1\leq k \leq N/\varepsilon} \theta_{k+1} \geq \delta/\varepsilon, \tau^\varepsilon_{N/\varepsilon} > T) + \mathbb{P}(\tau^\varepsilon_{N/\varepsilon} \leq T)$$
$$\leq \mathbb{P}(\max_{1\leq k \leq N/\varepsilon} \theta_{k+1} \geq \delta/\varepsilon) + \mathbb{P}(\tau^\varepsilon_{N/\varepsilon} \leq T) \to 0,$$

as $\varepsilon \to 0$, $N \to \infty$. $\square$

C.2 Positiveness of Diffusion Coefficients

For an ergodic Markov process $x(t), t \geq 0$, with state space $(E, \mathcal{E})$, generator Q, stationary distribution π, and potential operator R_0, defined in Section 1.6, let us prove the following result.

Lemma C.2 *Let us consider two functions, say φ and ψ, such that $Q\varphi = \psi$. Then we have*

$$b := \int_E \pi(dx)\psi(x)R_0\psi(x) \leq 0.$$

Remark C.1. This result is important since it proves that the variance, expressed as $\sigma = -b$, in the diffusion approximation scheme is nonnegative.

PROOF. From the martingale characterization of Markov processes, we have that

$$M_t := \varphi(x(t)) - \int_0^t Q\varphi(x(s))ds,$$

is a martingale with square characteristic (see Theorem 1.2),

$$\langle M \rangle_t = \int_0^t [Q\varphi^2(x(s)) - 2\varphi(x(s))Q\varphi(x(s))]ds, \quad t \geq 0.$$

Let us calculate

$$\begin{aligned}\Pi\langle M\rangle_t &:= \int_E \pi(dx)\langle M\rangle_t \\ &= \int_0^t [\Pi Q\varphi^2(x) - 2\Pi\varphi(x)Q\varphi(x)]ds, \quad (\Pi Q = 0) \\ &= -2t\Pi\varphi(x)Q\varphi(x) \\ &= -2t\int_E \Pi(dx)\varphi(x)Q\varphi(x) \geq 0.\end{aligned}$$

Hence

$$\int_E \Pi(dx)\varphi(x)Q\varphi(x) \leq 0,$$

and, using the fact that $\varphi Q\varphi = -\psi R_0\psi$, we get

$$\int_E \Pi(dx)\psi(x)R_0\psi(x) \geq 0.$$

□

Bibliography

1. Anantharam V., Konstantopoulos T. (1995). A functional central limit theorem for the jump counts of Markov processes with an application to Jackson networks, *Adv. Appl. Prob.*, 27, 476–509.
2. Anisimov V.V. (1977). Switching processes, *Cybernetics*, 13 (4), 590–595.
3. Anisimov V.V. (1995). Switching processes: averaging principle, diffusion approximation and applications, *Acta Aplicandae Mathematica*, **40**, 95–141.
4. Anisimov V.V. (2000). J-convergence for switching processes with rare perturbations to diffusion processes with Poisson type jumps. In *Skorokhod's Ideas in Probability Theory*, V.S. Korolyuk, N. Portenko, H. Syta (Eds), Institute of Mathematics, Kiev, pp 81–98.
5. Anisimov V.V. (2004). Averaging in Markov models with fast semi-Markov switches and applications, *Commun. Statist.- Theory Meth.*, 33(3), 517–531.
6. Arjas E., Korolyuk V.S. (1980). Stationary phase merging of Markov renewal processes, (in Russian), *DAN of Ukraine*, **8**, 3–5.
7. Asmussen S. (1987). *Applied Probability and Queues*, Wiley, Chichester.
8. Asmussen S. (2000). *Ruin probabilities*, World Scientific.
9. Assing S., Schmidt W.M. (1998). *Continuous Strong Markov Processes in Dimension One*, Lecure Notes in Mathematics, 1688, Springer, Berlin.
10. Barbour A.D., Chryssaphinou O. (2001). Compound Poisson approximation: a user's guide, *Ann. Appl. Probab.*, vol. 11, no. 3, 964–1002.
11. Barbour A.D., Holst L., Janson S. (1992). *Poisson Approximation*, Oxford University Press, Oxford.
12. Bensoussan A., Lions J.-L., Papanicolaou G.C. (1978). *Asymptotic Analysis of Periodic Structures*, North-Holland, Amsterdam.
13. Bertoin J. (1996). *Lévy processes*. Cambridge Tracts in Mathematics, 121. Cambridge University Press, Cambridge.
14. Bhattacharya R.N. (1982). On the functional central limit theorem and the law of the iterated logarithm for Markov processes, *Z. Wahrsch. verw. Gebiete*, 60, 185–201.
15. Bhattacharya R.N., Waymire E.C. (1990). *Stochastic Processes with Applications*, Wiley, N.Y.
16. Billingsley P. (1968). *Convergence of Probability Measures*, Wiley, New

York.
17. Blankenship G.L., Papanicolaou G.C. (1978). Stability and control of stochastic systems with wide band noise disturbances, I, *SIAM J. Appl. Math.*, 34, 437–476.
18. Boccara N. (2004). *Modeling Complex Systems*, Springer, Berlin.
19. Borovkov A.A. (1998). *Ergodicity and Stability of Stochastic Processes*, Wiley, Chichester.
20. Borovkov K., Novikov A. (2001). On a piece-wise deterministic Markov processes model, *Statist. Probab. Lett.*, 53, 421–428.
21. Borovshkikh Y. V., Korolyuk V. S. (1997). *Martingale Approximation*, VSP, Utrecht, The Netherlands.
22. Bouleau N., Lépingle D. (1994). *Numerical methods for stochastic processes*, Wiley, N.Y.
23. Breiman L. (1968). *Probability*, Addison-Wesley, Mass.
24. Brémaud P. (1981). *Point Processes and Queues*, Springer, Berlin.
25. Chetouani H., Korolyuk V.S. (2000). Stationary distribution for repairable systems, *Appl. Stoch. Models Bus. Ind.*, 16, 179–196.
26. Chung K.L. (1982). *Lectures from Markov Processes to Brownian Motion*, Springer, N.Y.
27. Çinlar E., Jacod J., Protter P. and Sharpe M.J. (1980). Semimartingale and Markov processes, *Z. Wahrschein. verw. Gebiete*, 54, 161–219.
28. Cogburn E. (1984). The ergodic theory of Markov chains in random environnement, *Z. Wahrsch. Verw. Gebiete*, 66, 109–128.
29. Cogburn R., Hersh R. (1973). The limit theorems for random differential equations, *Indiana Univ. Math. J.*, 22, 1067–1089.
30. Comets F., Pardoux E. (Eds.) (2001). *Milieux Aléatoires*, Société Mathématique de France, No 12.
31. Crauel H., Gundlach M. (Eds) (1999). *Stochastic Dynamics*, Springer, N.Y.
32. Dautray R., Cessenat M., Ledanois G., Lions P.-L., Pardoux E., Sentis R. (1989). *Méthodes Probabilistes pour les Equations de la Physique*, Eyrolles, Paris.
33. Davis M.H.A. (1984). Piecewise-deterministic Markov processes: A general class of non-diffusion stochastic processes, *J. Roy. Statist. Soc.*, B46, 353–388.
34. Davis M.H.A. (1993), *Markov Models and Optimization*, Chapman & Hall.
35. Dellacherie C., Meyer P. A. (1982). *Probabilities and Potential*, B. Noth-Holland.
36. Devooght J. (1997). Dynamic reliability, *Advances in Nuclear Science and Technology*, 25, 215–278.
37. Doob J.L. (1954). *Stochastic Processes*, Wiley, N.Y.
38. Ducan P. (1994). *Mixing: Properties and Examples*, LNS no 85, Springer, N.Y.
39. Dudley R.M. (1989). *Real Analysis and Probability*, Wadsworth.
40. Dunford N., Schwartz J. (1958). *Linear Operators. General Theory*, Interscience.
41. Dynkin E.B. (1965). *Markov Processes*, Springer-Verlag.

42. Elliott R.J. (1982). *Stochastic Calculus and applications*, Springer, Berlin.
43. Ellis R., Rosenkrantz W. (1977). Diffusion approximation for transport processes with boundary conditions, Preprint, University of Massachussetts.
44. Embrechts P., Klüppelberg C., Mikosc T. (1999). *Modelling Extremal Events for Insurance and Finance*, Springer, Berlin.
45. Ethier S.N., Kurtz T.G. (1986). *Markov Processes: Characterization and convergence*, J. Wiley & Sons, New York.
46. Ezhov I.I., Skorokhod A.V. (1969). Markov processes homogeneous with respect to the second component I, *Theory Probab. Appl.*, 14, 3–14; II, ibid, 679–692.
47. Feller W. (1966). *An introduction to probability theory and its applications*, vol. 1 and 2, J. Wiley, NY.
48. Fleming W.H., Mete Soner H. (1993). *Controlled Markov Processes and Viscosity Solutions*, Springer-Verlag, N.Y.
49. Freidlin, M.I. (Ed.) (1994). *The Dynkin Festschrift: Markov Processes and their Applications*, Birkäuser, Boston.
50. Freidlin M.I. (1996). *Markov Processes and Differntial Equations: Asymptotic Problems*, Lectures in Mathematics, ETH Zürich, Birkhäuser.
51. Freidlin, M.I. and Wentzell, A.D. (1998). *Random Perturbations of Dynamical Systems*, 2nd Edition, Springer, N.Y.
52. Fristedt B., Gray L. (1997). *A Modern Approach to Probability Theory*, Birkhäuser, Boston.
53. Furrer H., Michna Z., Weron A. (1997). Stable Lévy motion approximation in collective risk theory, *Insurance: Mathem. & Econ.*, 20, 97–114.
54. Garnier L. (1997). Multi-scaled diffusion-approximation. Applications to wave propagation in random media, *ENSAIM: Probab. Statist.*, 1, 183–206.
55. Gerber H.U. (1970). An extension of the renewal equation and its applications in the collective theory of risk, *Skandinavisk Aktuarietidskrift*, 205–210.
56. Gihman, I.I., Skorohod, A.V. (1974). *Theory of stochastic processes*, vol. 1,2, & 3, Springer, Berlin.
57. Glynn, P.W. (1990). Diffusion approximation, In *Hanbook in Operations Research and Management Science*, Vol. 2, Stochastic Models, D.P. Heyman, M.J. Sobel (Eds), Noth Holland, Amsterdam, pp 145-198.
58. Glynn, P.W., Haas P.J. (2004). On functional central limit theorems for semi-Markov and related processes, *Commun. Statist.- Theory Meth.*, 33(3), 487–506.
59. Griego, R., Hersh, R. (1969). Random evolutions, Markov chains, and Systems of partial differential equations, *Proc. Nat. Acad. Sci. U.S.A.*, 62, 305–308.
60. Gut A. (1988). *Stopped Random Walks*, Springer-Verlag, N.Y..
61. Hall P., Heyde C. (1980). *Martingale Limit Theorems and its Applications*, Academic Press, N.Y.
62. He S.W., Wang J.G. and Yan J.A. (1992). *Semimartingale theory and stochastic calculus*, Science Press and CRC Press, Hong Kong.
63. Hersh R. (1974). Random evolutions: a Survey of results and problems, *Rocky Mountain J. Math.*, 4, 443–477.

64. Hersh R. (2003). The birth of random evolutions, *Mathematical Intelligencer*, 25(1), 53–60.
65. Hersh R., Papanicolaou G. (1972). Non-commuting random evolutions, and an operator-valued Feynman-Kac formula, *Comm. Pure Appl. Math.*, 30, 337–367.
66. Hersh R., Pinsky M. (1972). Random evolutions are asymptotically Gaussian, *Comm. Pure Appl. Math.*, 25, 33–44.
67. Iglehart D.L. (1969). Diffusion approximation in collective risk theory, *J. Appl. Probab.*, 6, 285–292.
68. Iscoe I., McDonald D. (1994). Asymptotics of exit times for Markov jump processes I, *Ann. Prob.*, 22(1), 372-397.
69. Ito K., McKean Jr M.P. (1996). *Diffusion Processes and their Sample Paths*, Springer, Berlin.
70. Jacod J., Shiryaev A.N. (1987). *Limit Theorems for Stochastic Processes*, Springer-Verlang, Berlin.
71. Jacod J., Skorokhod A.V. (1996). Jumping Markov processes, *Ann. Inst. H. Poincaré*, **32**, (1), pp 11–67.
72. Jefferies B. (1996). *Evolution Processes and the Feynman-Kac Formula*, Kluwer, Dordrecht.
73. Kabanov Y., Pergamenshchiskov S. (2003). *Two-Scale Stochastic Systems: Asymptotic Analysis and Control*, Springer, Berlin.
74. Kallenberg (1975). *Random Measures*, Akademie, Verlag, Berlin.
75. Kaniovski Yu., Pflug G. (1997). Limit theorems for stationary distributions of birth-and-death processes, Interim Report IR-97-041/July, IIASA, Laxenburg, Austria.
76. Kaniovski Yu. (1998). On misapplications of diffusion approximations in birth and death processes of noisy evolution, Interim Report IR-98-050/August, IIASA, Laxenburg, Austria.
77. Kannan D., Lakshmikantham V. (Eds.) (2002). *Handbook of Stochastic Analysis and Applications*, M. Dekker, N.Y.
78. Karatzas I., Shreve S.E. (1988). *Brownian Motion and Stochastic Calculus*, Springer-Verlag.
79. Karlin S., Taylor H.M. (1981). *A Second Course in Stochastic Processes*, Academic Press, San Diego.
80. Kartashov N.V. (1996). *Strong Stable Markov Chains*, VSP Utrecht, TBiMC Kiev.
81. Karr A. F. (1991). *Point Processes and their Statistical Inference*, 2nd Edition, Marcel Dekker, N.Y.
82. Kato T. (1980). *Perturbation Theory for Linear Operators*, Springer, Berlin.
83. Kazamaki N. (1994). *Continuous Exponential Martingales and BMO*, LNM no 1579, Springer, Berlin.
84. Keilson J. (1979). *Markov Chains Models - Rarity and Exponentiality*, Springer-Verlag, N.Y.
85. Keepler M. (1998). Random evolutions processes induced by discrete time Markov chains, Portugaliae Mathematica, 55(4), 391–400.
86. Khasminskii R. (1969). *Stability of the solutions of systems of differential*

equations under random disturbance of their parameters, Nauka, Moscow.
87. Khoshnevisan D. (2002). *Multiparameter Processes*, Springer, N.Y.
88. Kifer Y. (1988). *Random Perturbations of Dynamical Systems*, Birkhäuser, Boston.
89. Kimura M. (1964). Diffusion models in population genetics, *J. Appl. Prob.*, 1, 177–232.
90. Kingman J.F.C. (1993). *Poisson Processes*, Clarendon Press, Oxford.
91. Kipnis C., Varadhan S.R.S. (1986). Central limit theorem for additive functionals of reversible Markov processes and applications to simple excursions, *Commun. Math. Phys.*, 104(1), 1–19.
92. Kleinrock L. (1975). *Qeueing Systems. Vol. 1: Theory*, Wiley, N.Y.
93. Klebaner F. C. (1998). *Introduction to Stochastic Calculus with Applications*, Imperial College Press, London.
94. Kluppelberg C., Mikosch T. (1995). Explosive Poisson shot processes with applications to risk reserves, *Bernoulli*, 1, (1& 2), pp 125–147.
95. Kokotović P., Khalil H.K., O'Reilly J. (1999). *Singular Perturbation Methods in Control: Analysis and Design*, SIAM, Classics in Applied Mathematics, Philadelphia.
96. Korolyuk V.S. (1990). Central limit theorem for semi-Markov random evolutions, *Comp. Math. Appl.*, 83–88.
97. Korolyuk V.S. (1998). Stability of stochastic systems in diffusion approximation scheme, *Ukrainian Math. J.*, 50, N 1, 36–47.
98. Korolyuk V. S. (1999). Semi-Markov random walk, In *Semi-Markov Models and Applications*, J. Janssen, N. Limnios (Eds.), pp 61– 75, Kluwer, Dordrecht.
99. Korolyuk V. S., Derzko N.A., Korolyuk V. V. (1999). Markovian repairman problems. Classification and approximation, In *Statistical and Probabilistic Models in Reliability*, D.C. Ionescu, N. Limnios (Eds), pp 143–151, Birkhäuser, Boston.
100. Korolyuk V. S., Korolyuk V. V. (1999). *Stochastic Models of Systems*, Kluwer, Dordrecht.
101. Korolyuk V.S., Limnios N. (1999). A singular perturbation approach for Liptser's functional limit theorem and some extensions, *Theory Probab. and Math. Statist.*, 58, pp 83–87.
102. Korolyuk V.S., Limnios N. (1999). Diffusion approximation of integral functionals in merging and averaging scheme, *Theory Probab. and Math. Statist.*, 59, pp 91-98.
103. Korolyuk V.S., Limnios N. (2000). Diffusion approximation of integral functionals in double merging and averaging scheme, *Theory Probab. and Math. Statist.* 60, pp 87–94.
104. Korolyuk V.S., Limnios N. (2000). Evolutionary systems in an asymptotic split state space, in *Recent Advances in Reliability Theory: Methodology, Practice and Inference*, N. Limnios & M. Nikulin (Eds), pp 145–161, Birkhauser, Boston.
105. Korolyuk V.S., Limnios N. (2002). Poisson Approximation of Homogeneous Stochastic Additive Functionals with Semi-Markov Switching, *Theory of*

Probability and Mathematical Statistics, 64, pp 75–84.
106. Korolyuk V.S., Limnios N. (2003). Increment Processes and its Stochastic Exponential with Markov Switching in Poisson Approximation Scheme, *Computers and Mathematics Applications*, 46 (7), 1073-1080.
107. Korolyuk V.S., Limnios N. (2004). Average and diffusion approximation for evolutionary systems in an asymptotic split phase space, *Annals Appl. Probab.*, 14(1), pp 489–516.
108. Korolyuk V.S., Limnios N. (2002). Markov additive processes in a phase merging scheme, *Theory Stochastic Processes*, vol. 8, no 24, pp 213–226.
109. Korolyuk V.S., Limnios N. (2004). Semi-Markov random walk in Poisson approximation scheme, *Communication in Statistics - Theory and Methods*, 33(3), pp 507–516.
110. Korolyuk V.S., Limnios N. (2004). Lévy approximation of increment processes with Markov switching, *Stochastics and Stochastic Reports*, 76(5), 383–374.
111. Korolyuk V.S., Limnios N. (2004). Poisson Approximation of Increment Processes with Markov Switching, *Theory Probab. Appl.*, 49(4), 1–18.
112. Korolyuk V.S., Limnios N. (2005). Diffusion approximation for evolutionary systems with equilibrium in asymptotic split phase space, *Theory of Probability and Mathematical Statistics*, 70.
113. Korolyuk V.S., Limnios N. (2005). Centered semi-Markov random walk in diffusion approximation scheme, Proc. *Intern. Symposium on Applied Stoch. Proc. Data Anal.*, ASMDA-2005, Brest.
114. Korolyuk V.S, Momonova A. (2003). *DAN of Ukraine.*
115. Korolyuk V.S., Portenko N., Syta H. (Eds.) (2000). *Skorokhod's Ideas in Probability Theory*, Institute of Mathematics, Kiev.
116. Korolyuk V.S., Turbin A.F. (1993). *Mathematical foundations of the state lumping of large systems*, Kluwer Academic Publ., Dordtrecht.
117. Korolyuk V.S., Swishchuk A. (1995), *Semi-Markov random evolution*, Kluwer Academic Publ., Dordrecht.
118. Korolyuk V.S., Swishchuk A. (1995), *Evolution of System in Random Media*, CRC Press.
119. Kovalenko I.N., Kuznetsov N.Yu., Pegg P. A. (1997). *Mathematical Theory of Reliability of Time Dependent Systems with Practical Applications*, Wiley, Chichester.
120. Kurtz T.G. (1972). A random trotter formula, *Proceeding of the American Mathematical Society*, 35(1), 147–154.
121. Kushner H.J. (1990). *Weak Convergence Methods and Singular Perturbed Stochastic Control and Filtering Problems*, Birkhäuser, Boston.
122. Kushner H.J., Clark D.S. (1978). *Stochastic Approximation Methods for Constrained and Unconstrained Systems*, Springer-Verlag, N.Y., 1978.
123. Kushner H.J., Dupuis P.G. (1992). *Numerical Methods for Stochastic Control Problems in Continuous Time*, Springer-Verlag, N.Y.
124. Iglehart D.L. (1969). Diffusion approximations in collective risk theory, *J. Appl. Prob.*, 6, 285-292.
125. Lamperti J. (1977). *Stochastic Processes*, Springer-Verlag, N.Y.

126. Lapeyre B., Pardoux E., Sentis R. (1998). *Méthodes de Monte-Carlo pour les Equations de Transport et de Diffusion*, Springer, Paris.
127. Limnios N., Oprişan G. (2001). *Semi-Markov Processes and Reliability*, Birkhäuser, Boston.
128. Lindvall T. (1973). Weak convergence of probability measures and random function space $D[0, +\infty)$, *J. Appl. Prob.*, 10, 109–121.
129. Lindvall T. (1992). *Lectures on the Coupling Method*, Wiley, N.Y.
130. Liptser R. Sh. (1984), On a functional limit theorem for finite state space Markov processes, in *Steklov Seminar on Statistics and Control of Stochastic Processes*, pp 305-316, Optimization Software, Inc., N.Y.
131. Liptser R. Sh. (1994). The Bogolubov averaging principle for semimartingales, *Proceedings of the Steklov Institute of Mathematics*, Moscow, No 4, 12 pages.
132. Liptser R. Sh., Shiryayev A. N. (1989). *Theory of Martingales*, Kluwer Academic Publishers, Dordrecht, The Netherlands.
133. Liptser R. Sh., Shiryayev A. N. (1991). Martingale and limit problems for theorems for stochastic processes, in *Encyclopaedia of Mathematical Sciences. Probability Theory III*, Yu. Prokhorov and A.N. Shiryaev (Eds), Springer, pp. 158–247.
134. Maglaras C., Zeevi A. (2004). Diffusion approximation for a multiclass markovian service system with "guaranteed" and "best-effort" service level, *Math. Oper. Res.*, 29(4), 786–813.
135. Maxwell M., Woodroofe M. (2000). Central limit theorems for additive functionals of Markov chains, *Ann. Probab.*, 28(2), 713–724.
136. Métivier M. (1982). *Semimartingales. A course on Stochastic Processes*, Walter de Gruyter, Berlin.
137. Meyn S.P., Tweedie R.L. (1993). *Markov Chains and Stochastic Stability*, Springer, N.Y.
138. Murdock J.A. (1999). *Perturbations: Theory and Methods*, SIAM, Classics in Applied Mathematics, Philadelphia.
139. Nummelin E. (1984). *General Irreducible Markov Chains and Non-negative Operators*, Cambridge University Press, Cambridge.
140. Oksendal B. (1998). *Stochastic Differential Equations*, Fifth Edition, Springer, Berlin.
141. Orey S. (1971). *Lecture Notes on Limit Theorems for Markov Chain Transition Probabilities*, Van Nostrand Reinhold, London.
142. Papanicolaou G. (1987). *Random media*, Springer-Verlag, Berlin.
143. Papanicolaou G., Kohler W., White B. (1991). *Random media*, Lectures in Applied Math., 27, SIAM, Philadelphia.
144. Parzen E. (1999). *Stochastic Processes*, SIAM Classics, Philadelphia.
145. Petrov V.V (1995). *Limit Theorems of Probability Theory*, Clarendon, Oxford.
146. Pinsky M. (1991), *Lectures on Random Evolutions*, World Scientific, Singapore.
147. Pollard D. (1984). *Convergence of Stochastic Processes*, Springer, N.Y.
148. Pollett P.K. (1996). Quasistationary distributions bibliography.

149. Prabhu N.U. (1980). *Stochastic Storage Processes*, Springer-Verlag, Berlin.
150. Revuz D. (1975). Markov Chains, North-Holland, Amsterdam.
151. Revuz D., Yor M. (1999). *Continuous Martingales and Brownian Motion*, Springer, 3rd Edition, Berlin.
152. Royden H.L. (1988). *Real Analysis*, 3rd Ed., McMilan, N.Y.
153. Rogers L.C.G., Williams D. (1994). Diffusions, Markov Processes, and Martingales, vol. 1 & 2, J. Wiley & Sons, Chichester, U.K.
154. Rudin W. (1991). Functional Analysis, McGraw-Hill, N.Y.
155. Sato K.-I. (1999). *Lévy processes and infinitely divisible distributions*. Cambridge Studies in Advanced Mathematics, 68. Cambridge University Press, Cambridge.
156. Shanbhag D.N., Rao C.R. (Eds.) (2001). *Stochastic Processes: Theory and Methods*, Handbook of Statistics, Vol. 19, Elsevier.
157. Shiryaev A.N. (1999). *Essentials of Stochastic Finance: Facts, Models, Theory*, World Scientific, Singapore.
158. Shurenkov V.M. (1984). On the theory of Markov renewal, *Theor. Probab. Appl.*, 19(2), 247–265.
159. Shurenkov V.M. (1989). *The Ergodic Theory of Markov Processes*, (in Russian), Nayka, Moscow.
160. Shurenkov V.M. (1998). *Ergodic Theorems and Related Problems*, VSP, Utrecht.
161. Skorokhod A.V. (1984). *Random Linear Operators*, R. Reidel-Kluwer, Dordrecht.
162. Skorokhod A.V. (1988). *Stochastic Equations for Complex Systems*, R. Reidel-Kluwer, Dordrecht.
163. Skorokhod A.V. (1989). *Asymptotic Methods in the Theory of Stochastic Differential Equations*, AMS, vol. 78, Providence.
164. Skorokhod A.V. (1991). *Random Processes with Independent Increments*, Kluwer, Dordrecht.
165. Skorokhod A.V. (1991). *Lectures on the Theory of Stochastic Processes*, VSP, Utrecht.
166. Skorokhod A.V., Borovskikh Yu. V. (Eds.) (1995). *Exploring Stochastic Laws*, VSP.
167. Skorokhod A.V., Hoppensteadt F.C., Salehi H. (2002). *Random Perturbation Methods with Applications in Science and Engineering*, Springer-Verlang, N.Y.
168. Silvestrov D. S. (2001). The perturbed renewal equation and diffusion type approximation for risk processes, *Theory of Probability and Mathematical Statistics*, 62, 145–156.
169. Silvestrov D. S. (2004). *Limit Theorems for Randomly Stopped Stochastic Processes*. Series: Probability and its Applications, Springer.
170. Sobsczyk K. (1991). *Stochastic Differential Equations*, Kluwer, Dordrecht.
171. Stone C. (1963). Weak convergence of stochastic processes defined on semi-infinite time intervals, *Proc. Amer. Math. Soc.*, 14, 694–696.
172. Stroock D.W. (1993). *Probability Theory: An Analytic View*, Cambridge University Press, Cambridge.

173. Stroock D.W., Varadhan S.R.S. (1979). *Multidimensional Diffusion Processes*, Springer-Verlag, Berlin.
174. Sviridenko M.N. (1998). Martingale characterization of limit distributions in the space of functions without discontinuities of second kind, *Math. Notes*, **43**, No 5, pp 398–402.
175. Sviridenko M.N. (1986). Martingale approach to limit theorems for semi-Markov processes, *Theor. Probab. Appl.*, pp 540–545.
176. Swishchuk A. (1997). *Random evolutions and their applications*, Kluwer, Dordrecht.
177. Van Pul M.C.J. (1991). *Statistical analysis of software reliability models*, CWI Tract, Amsterdam.
178. Vishyk M.I., Lyusternik L.A. (1960). On solutions of some problems related to perturbations in the case of matrices and selfadjoint or non-selfadjoint differential equations, Uspekhi Mat. Nauk, 15(3), 3–80, (in Russian).
179. Watanabe H. (1984). Diffusion approximation of some stochastic difference equations, *Adv. Probab. Appl.*, 7, 439–546.
180. Watkins J. (1984). A central limit theorem in random evolution, *Ann. Probab.*, 12, 480–514.
181. Whitt W. (2002). *Stochastic-Process Limits: An Introduction to Stochastic-Process Limits and their Applications to Queues*, Springer, N.Y.
182. Ye J.J. (1997). Dynamic programming and the maximum principle for control of piecewise deterministic Markov processes, In *Mathematics of Stochastic Manufacturing Systems*, Yin G.G., Zhang Q. (Eds), AMS, Lectures in Applied Mathematics, vol. 33, Providence.
183. Yin G.G. (2001). On limit results for a class of singularly perturbed switching diffusions, *J. Theor. Probab.*, 14(3), 673–697.
184. Yin G.G., Zhang Q. (1998). *Continuous-Time Markov Chains and Applications. A singular perturbation approach*, Springer, N.Y.
185. Yin G.G., Zhang Q. (2005). *Discrete-Time Markov Chains. Two-Time-Scale Methods and Applications*, Springer, N.Y.
186. Yosida K. (1980). *Functional Analysis*, Springer, Berlin.
187. Yu P.S. (1977). On accuracy improvement and applicability conditions of diffusion approximation with applications to modelling of computer systems, Technical Report No. 129, Digital Systems Lab., Stanford University.
188. Zhenting H., Qingfeng G. (1988). *Homogeneous Denumerable Markov Processes*, Springer, Berlin and Science Press, Beijing.

Notation

$\mathbb{N}$	the set of natural numbers
$\mathbb{Z}$	the set of relative integers
$\mathbb{Q}$	the set of rational numbers
$\mathbb{R}$	the set of real numbers
$\mathbb{R}_0$	$\mathbb{R} \setminus \{0\}$
$\overline{\mathbb{R}}$	the set $[-\infty, +\infty]$
$\mathbb{R}_+$	the set of real positive numbers $[0, +\infty)$
F_x	distribution function of the sojourn time in state $x \in E$
	$\overline{F}_x = 1 - F_x$
$\varphi_t(\lambda)$	characteristic function of increments
$\psi(\lambda)$	cumulant function
v^*	the transpose of vector $v \in \mathbb{R}^d$
δ_a	Dirac measure in $a \in \mathbb{R}^d$
$(\Omega, \mathcal{F}, \mathbb{P})$	probability space
$\mathbf{1}_A, \mathbf{1}(A)$	indicator function of set A
$\mathbb{E}X$	mathematical expectation of X
$Var X$	variance of X
Φ_t	random evolution
$\mathbf{F} = (\mathcal{F}_t, t \geq 0)$	filtration
$\Im$	stochastic basis $\Im = (\Omega, \mathcal{F}, \mathbf{F}, \mathbb{P})$
$(E, \mathcal{E})$	state space, a measurable space
$P(x, B)$	transition kernel of Markov chain
$P_t(x, B)$	transition function of a Markov process
P	transition operator associated to $P(x, B)$
P_t	transition operator associated to $P_t(x, B)$

Q	generator of a Markov process, semi-Markov kernel
R_0	potential operator of the generator Q
π	stationary distribution of (semi-) Markov processes
ρ	stationary distribution of the Markov chain $x_n, n \geq 0$
Π	projection operator on the null space of generator Q
E_k	a class of split state space $E = \cup_{k=1}^{N} E_k$
π_k	stationary distribution of Markov process on E_k
$m(x)$	mean sojourn time in state x
m	mean times, ρ-mean: $m = \int_E \rho(dx) m(x)$
$x_n, n \geq 0$	embedded Markov chain of a semi-Markov process
$\tau_n, n \geq 0$	jump times of a semi-Markov process, $\tau_0 = 0$
$\theta_n, n \geq 1$	inter-jump times of a semi-Markov process
θ_x	sojourn time in state $x \in E$ of a semi-Markov process
$\nu(t), t \geq 0$	counting process of jumps
$\tau(t)$	the time of the last jump before t, $\tau(t) = \tau_{\nu(t)}$
$\nu_+(t)$	the number of the first jump after t, $\nu_+(t) = \nu(t) + 1$
$\tau_+(t)$	the time of the first jump after t, $\tau_+(t) = \tau_{\nu_+(t)}$
$\langle M \rangle$	square characteristic of the martingale M
T	$T = (B(t), C(t), \Gamma_g(t))$ the predictable characteristics of semimartingales
B	the Banach space of real measurable bounded functions defined on E
$C(E)$	the space of real-valued continuous bounded functions defined on E
$C_0(E)$	the space of bounded continuous functions on E vanishing at infinity
$C_0^k(E)$	the space of bounded continuous functions on E having continuous derivatives of order up to and including k vanishing at infinity
$C_0^\infty(E)$	the space of bounded continuous functions on E having continuous derivatives of all orders vanishing at infinity
$C_\kappa^k(\mathbb{R}^d)$	the space of k-th times continuously differentiable functions on E with compact support
$C^{2,0}(\mathbb{R}^d \times E)$	the space of twice differentiable function in the first argument and continuous in the second
$C_3(\mathbb{R})$	the class of real-valued bounded continuous functions g with $g(u)/u^2 \to 0$ as $|u| \to 0$
$\mathbf{C}_E[0, \infty)$	the space of E-valued continuous functions defined on $\mathbb{R}_+$

INDEX

$\mathbf{D}_E[0,\infty)$	the space of E-valued right continuous functions having left limits (cadlag), defined on $\mathbb{R}_+$
$\mathcal{D}_Q, \mathcal{D}(Q)$	the domain of the operator Q
$a \vee b$	maximum of a and b
$a \wedge b$	minimum of a and b
$[x]$	the integer part of the real number x
Δx_n	increment of sequence (x_n), $\Delta x_n = x_n - x_{n-1}$
$\Delta x(t)$	jump of the process $x(t)$ at t, $\Delta x(t) = x(t) - x(t-)$
ϑ	the shift operator on the sequences: $\vartheta(x_0, x_1, \ldots) = (x_1, x_2, \ldots)$
$o(x)$	small o of x in $x_0 \in \mathbb{R}$, $\lim_{x \to x_0} o(x)/x = 0$
$O(x)$	big O of x in $x_0 \in \mathbb{R}$: $\lim_{x \to x_0} O(x)/x = c \in \mathbb{R}$
φ	a test function; $\varphi'(u), \varphi''(u), \ldots$ are derivatives of function φ, $\varphi'_u(u,v), \varphi''_{uu}(u,v), \ldots$ derivatives with respect to the first variable
a.e.	almost everywhere
a.s.	almost surely
i.i.d.	independent and identically distributed
r.v.(s)	random variable(s)
PII	process with independent increment
PSII	process with stationary independent increment
PLII	process with locally independent increments
MAP	Markov additive process
SMP	Semi-Markov process
MRP	Markov renewal process
$\xrightarrow{a.s.}$	almost sure convergence
$\xrightarrow{P}$	convergence in probability
$\xrightarrow{d}$	convergence in distribution
$\xrightarrow{L^p}$	convergence in norm L^p
$\Longrightarrow$	weak convergence in the sense of Skorokhod topology in the space $\mathbf{D}[0,\infty)$ or in Sup norm in $\mathbf{C}[0,\infty)$.

Index

absorption condition, 110
absorption time, 112, 121, 243
additive functional, 15, 36, 47, 61, 67, 74–77, 84, 86, 90, 94, 104, 122, 123, 126, 128, 130, 132, 150, 152, 154, 175, 228, 230, 237, 240
  with equilibrium, 93
arithmetic distribution, 22
auxiliary processes, 23
average approximation, 67–69, 71–76, 79, 80, 82, 84, 85, 90, 91, 93, 97, 103, 116, 117, 119–121, 139, 150–152, 154–160, 182, 186, 201, 203, 205, 216, 217, 219, 220, 258, 259, 264, 265, 268

Banach space, 1
birth and death process, 269, 270
Brownian motion, 10, 12

Chapman-Kolmogorov equation, 3, 6
characteristic exponent, 278
characteristic function, 12
compact containment condition, 200, 209, 210, 240, 309
compensating measure of jumps, 63
compensating operator, 24, 25, 36, 39, 40, 42, 56–60, 67–70, 72, 73, 76, 80, 100–102, 153, 155, 157, 163, 166, 176–178, 180, 182, 184–186, 188, 196, 201, 212, 213, 220, 237, 262–265, 268, 282, 289, 295

extended -, 59
compound Poisson process, 12, 104
convergence-determining class, 302, 305
counting process, 8, 20, 28
Cramér's condition, 229
cumulant, 13

diffusion
  coefficient, 10
  operator, 10
  time-homogeneous -, 10
diffusion approximation, 67, 70, 71, 73, 81, 82, 84, 88, 89, 95, 97, 112, 122, 128, 139, 160, 161, 165, 166, 168, 170, 172, 173, 175, 176, 180, 188, 189, 200, 202, 203, 206, 217, 219, 228, 261, 264, 268, 270
  with equilibrium, 90, 173
diffusion process, 10
direct Riemann integrable, 22
domain, 7
domain of generator, 7
drift coefficient, 10
dynamical system, 104
Dynkin formula, 16

equilibrium process, 97
ergodic merging, 104
evolutionary equation, 13, 36
evolutionary system, 43
exponential process, 224

extended Markov renewal process, 25

Factorization theorem, 64
filtration, 1

generalized diffusion, 27
generator, 7, 10

impulsive process, 61, 221–223, 225, 276
increment process, 40
integral functional, 36, 104, 134
invariant probability, 4

Lévy approximation, 276
Lévy process, 12
  characteristics -, 280
Lévy-Khintchine formula, 12, 13, 278
limit distribution, 23
Liptser formula, 135, 290
local martingale, 16
Lyapounov function, 275

Markov additive process, 46, 277
Markov additive semimartingale, 61, 63, 65
Markov chain, 3
  ψ-irreducible -, 4
  aperiodic -, 4
  embedded -, 7
  ergodic -, 4
  Harris positive -, 4
  periodic -, 4
  positive -, 4
  strong -, 3
  uniformly irreducible -, 4
Markov kernel, 2, 3
Markov process, 7
  egodic -, 6
  jump -, 7
  martingale characterization, 17
  pure jump -, 9
  regular -, 7, 21
  time-homogeneous -, 6
  weak differentiable -, 14
  with locally independent increments, 14
Markov property, 3, 5, 6
  strong -, 3, 6
Markov renewal equation, 21, 23
Markov renewal process, 20
  extended -, 59
  martingale characterization - , 25
Markov renewal theorem, 21–23
Markov semigroup, 7
Markov transition function, 2
martingale, 16
  Doob-Mayer decomposition, 16
  problem, 17
  square integrable, 16
measure-determining class, 305
measure-determining class, 219, 302
merging
  condition, 123
  double -, 112, 121, 137, 185
  function, 106, 114, 226, 231, 249
  state space, 103
  with absorption, 110
modulated process, 35

operator
  closed -, 32
  densely defined -, 32
  inverse -, 32
  normally solvable -, 32
  potential -, 33
  reducible-invertible -, 32
Ornstein-Uhlenbeck process, 97, 269, 270, 275

phase merging principle, 75, 139
phase space, 2
piecewise–deterministic Markov processes, 14
Poisson
  equation, 33
Poisson approximation, 193, 219, 220, 230, 231, 241, 258, 260, 261, 263, 268
  condition, 222, 229, 260, 279
Poisson process, 12

INDEX

Polish space, 1
potential operator, 5
predictable characteristics, 26, 27, 63
predictable process, 25
process
    adapted, 2
    impulsive, 61
    independent increments -, 11
    LII, 14
    locally independent increment -, 47
    stationary and independent
        increments -, 11
projector, 32

random evolution, 38, 50
    continuous -, 50
    coupled -, 165
    coupled -, 52, 71
    jump -, 50, 54
    Markov jump -, 55
    semi-Markov -, 56
reducible-invertible operator, 31, 139
relatively compact, 303
reliability, 243
renewal process, 258
renewal processes
    superposition, 253
repairable system, 269

second modified characteristic, 62
semi-Markov
    kernel, 19
    random walk, 258
semi-Markov process, 19, 20
    regular -, 21
semigroup
    uniformly continuous -, 7
semimartingale, 25, 64, 65
    special -, 25
shift operator, 41
signed kernel, 3, 106
singular perturbation, 139
solvability condition, 33
split
    double -, 121
split and merging

    ergodic -, 123
split condition, 123
split state space, 103
square-integrability condition, 222
standard state space, 1, 301
state space, 2
stationary distribution, 4, 6, 8
stationary phase merging, 249
stationary projector, 5, 6
stochastic basis, 1
stochastic integral functional, 30
stopping time, 2
sub-Markov kernel, 2
switched
    process, 35
switching
    Markov -, 42, 47, 49, 52, 64, 65, 71,
        73, 77, 83, 93, 94, 99, 118,
        120, 121, 123, 125–130, 132,
        134, 151, 154, 159–161, 172,
        173, 175, 188, 193, 196, 209,
        219–221, 225, 227, 228, 231,
        232, 269, 270
    process, 35
    semi-Markov -, 36, 38, 41–44, 50,
        52, 54, 59, 61, 67, 68,
        73–75, 79, 81, 82, 98, 104,
        116, 117, 119, 120, 153, 154,
        156–158, 165, 166, 168–170,
        173, 176, 180, 182, 186, 188,
        189, 193, 201, 212, 216, 217,
        219, 220, 228, 230, 237, 267

tight, 303
truncation function, 26

uniform square-integrability, 229
usual conditions, 1

weak convergence, 301
weak topology, 301
Wiener process, 10

Printed in the United States
By Bookmasters